Botulinum Toxins: New Uses in the Treatment of Diseases

Botulinum Toxins: New Uses in the Treatment of Diseases

Editor

Siro Luvisetto

MDPI

Basel • Beijing • Wuhan • Barcelona • Belgrade • Novi Sad • Cluj • Manchester

Editor
Siro Luvisetto
National Research Council
(CNR) of Italy
Monterotondo Scalo, Italy

Editorial Office
MDPI
St. Alban-Anlage 66
4052 Basel, Switzerland

This is a reprint of articles from the Special Issue published online in the open access journal *Toxins* (ISSN 2072-6651) (available at: https://www.mdpi.com/journal/toxins/special_issues/Botulinum_Uses).

For citation purposes, cite each article independently as indicated on the article page online and as indicated below:

Lastname, A.A.; Lastname, B.B. Article Title. *Journal Name* **Year**, *Volume Number*, Page Range.

ISBN 978-3-0365-8828-5 (Hbk)
ISBN 978-3-0365-8829-2 (PDF)
doi.org/10.3390/books978-3-0365-8829-2

Cover image courtesy of Sara Marinelli

Contents

About the Editor

Siro Luvisetto

Siro Luvisetto (senior researcher at National Research Council of Italy, CNR) obtained his degree in chemistry at the University of Padova (Italy). After graduating in 1984, he started his scientific career at the Institute of General Pathology (University of Padova, Italy), where he studied mitochondrial physiology with an emphasis on energy conversion during mitochondrial oxidative phosphorylation. In 1987, he was a visiting scientist at the Department of Membrane Research of the Weizmann Institute of Science (Rehovot, Israel). In 1988, he gained a permanent position as a researcher at the National Research Council of Italy (CNR) at the Center for the Study of Mitochondrial Physiology (Padua, Italy). In 1997, he joined the CNR Centre for Biomembranes Study, where he shifted his research interests toward the biophysical characterization of high-voltage-activated neuronal calcium channels responsible for the familial hemiplegic migraine. In 2001, he moved to Rome, where he worked at the CNR Institute of Neuroscience, before working at CNR Institute of Cell Biology; presently, he works at the CNR Institute of Biochemistry and Cell Biology where he studies the pharmacology of pain modulation and peripheral nerve regeneration after peripheral nerve injury in animal models. Specifically, his main interests are the effects of botulinum neurotoxins on peripheral regeneration after peripheral nerve injury. Dr. Luvisetto has authored and co-authored more than 90 peer-reviewed journal articles, including reviews and book chapters. He has been an invited speaker at several international and national conferences and meetings. He is an active member of some national and international professional associations. He has served as a reviewer for many scientific journals. He has worked as a Guest Editor for three Special Issues published by Toxins. He is a Section Board Member (bacterial toxins section) of the Editorial Board of Toxins and a Reviewer Board Member at *Toxins*.

Preface

This book collates 13 scientific articles, covering both animal studies and clinical reports, derived from laboratories involved in studies in which the use of botulinum neurotoxins as a therapeutic drug is extensively investigated. It is well known that the therapeutic efficacy of botulinum neurotoxins in many pathologies lies in their biological activity as inhibitors of cholinergic transmission neuromuscular junction level. These insights have allowed extraordinary developments in the clinical use of BoNTs in a variety of pathologies, which are characterized by excessive muscle contractions and hypercholinergic dysfunctions, as well as being poorly treated by conventional drugs. Nowadays, the list of human pathologies in which treatment using botulinum neurotoxins produces beneficial effects is constantly growing. This book, entitled "Botulinum Neurotoxins: New Uses in the Treatment of Diseases", is the printed version of a Special Issue (the third in a series of three Special Issues) published by Toxins regarding the new therapeutic applications of botulinum neurotoxins. This Special Issue is an updated version of the previous two editions: "Botulinum Neurotoxins in the Nervous System: Future Challenges for Novel Indications" and "Effects of Botulinum Toxin on Functional Recovery after Injuries of Nervous System". This book is targeted at an audience that may not be familiar with the action of botulinum neurotoxins; therefore, readers will gain new insights into the multiple applications of this neurotoxin. On the other hand, for professional purposes, this book provides an updated picture of the state of the art regarding the possible development of new therapeutic treatments involving botulinum neurotoxins.

Siro Luvisetto
Editor

toxins

MDPI

Editorial

Introduction to the *Toxins* Special Issue on Botulinum Toxins: New Uses in the Treatment of Diseases

Siro Luvisetto

National Research Council of Italy—CNR, Institute of Biochemistry and Cell Biology (IBBC), Via Ercole Ramarini 32, Monterotondo Scalo, 00015 Roma, Italy; siro.luvisetto@cnr.it

Studies on animals and humans have amply demonstrated the therapeutic efficacy of botulinum neurotoxins (BoNTs) in many pathologies. BoNTs act as blockers of cholinergic transmission at the neuromuscular junction (NMJ). This has led to the development of their clinical use in a variety of hypercholinergic disorders, such as dystonia, torticollis, blepharospasm and many others characterized by an abnormal release of acetylcholine at the NMJ. In the last decade, many studies have provided evidence of the efficacy of BoNTs in the treatment of pathologies where the expected therapeutic action is not limited only to NMJ, but is also attributable to the interaction of BoNTs with other structures or with neurotransmitters other than acetylcholine. Nowadays, the list of human pathologies in which treatment with BoNT produces beneficial effects is constantly increasing.

This Special Issue (SI), entitled: "Botulinum neurotoxins: new uses in the treatment of diseases", is the third SI published by *Toxins* of which I am honored to be the Guest Editor. The previous SIs were dedicated to "Botulinum neurotoxins in the nervous system: future challenges for new indications" [1] and the "Effects of botulinum toxin on functional recovery after nervous system injuries", whereby this SI, containing a collection of research on new therapy with BoNTs (mainly of the serotype A, BoNT/A), represents an update. This Editorial intends to introduce the 13 articles collected in this SI, which I strongly recommend reading in their original version.

The first contribution is a research article from Mueller et al. [2] reporting a human study on the effects of Xeomin® (IncobotulinumtoxinA; a BoNT/A formulation specifically approved for the treatment of sialorrhea), concomitantly used with the radioligand actinium-225-PSMA in the therapy of prostate cancer metastases [2]. Cancer therapy with this radioligand compound is highly effective, as is its counterpart sialotoxic at the level of the parotid and submandibular glands, which regulate salivary secretions. Starting from the consideration that the injection into the salivary glands of Xeomin® represents the elective therapy for chronic drooling, the authors had the excellent idea of using it to prevent excessive salivation during cancer therapy. The presented data demonstrated that an injection of a high dose, up to a total dose of 250 units of Xeomin®, into the salivary glands is well tolerated without causing severe systemic side effects. These findings paved the road to future trials that have included BoNT/A as a component for salivary gland protection in all of the radioligand cancer therapies, inducing permanent salivary gland dysfunction.

The next article, by Munoz-Lora et al. [3], presents new findings on the effect of BoNT/A in pain therapy, another booming BoNT/A therapeutic treatment. The authors analyzed the mechanism through which BoNT/A exerts antinociceptive effects on a rat model of temporomandibular joint (TMJ) rheumatoid arthritis, a chronic inflammatory pain model induced by repeated intra-articular injections of methylated bovine serum albumin in TMJ. The efficacy of two commercial pharmaceutical formulations of BoNT/A, namely Dysport® (AbobotulinumtoxinA) and Botox® (OnabotulinumtoxinA), was tested. Both formulations were able to reduce the pain-related behavior and the mechanical allodynia of the hypernociceptive rats. The antinociceptive effects of BoNT/A were correlated with

Citation: Luvisetto, S. Introduction to the *Toxins* Special Issue on Botulinum Toxins: New Uses in the Treatment of Diseases. *Toxins* **2023**, *15*, 540. https://doi.org/10.3390/toxins15090540

Received: 27 August 2023
Accepted: 30 August 2023
Published: 1 September 2023

the appearance of cleaved-SNAP-25 in the trigeminal nucleus caudalis, demonstrating an axonal transport of toxins to the central nociceptive sensory areas, together with reductions in c-Fos and GFAP upregulation, demonstrating a reduction in neuronal and glial activation.

Two other studies, by Yi et al. [4,5], provide anatomical information on the intramuscular neural arborization of the serratus anterior muscle, which is commonly injected with BoNTs for the treatment of myofascial pain syndrome, and various muscles surrounding the nasal region, which are frequently injected with BoNTs to increase effectiveness in removing wrinkles in the nose region. These two articles do not suggest a new BoNT treatment. However, they help to identify the exact anatomical distribution of specific muscles, such as the serratus anterior and the muscles of the nose region, which aids clinicians in electromyographic guidance of BoNT injections with the specific purpose of carefully targeting the correct muscles, avoiding adverse effects due to the probable spread of the toxin, especially after repeated treatments.

Another contribution of the present SI on the effects of BoNTs in the treatment of pain comes from a review article by Lippi et al. [6]. The authors reported a comparative analysis of 12 human studies published between 2006 and 2020, showing a significant effect of administering BoNT/A, various commercial preparations, or BoNT/B in patients suffering from neuropathic pain due to post-herpetic neuralgia, spinal cord injury, peripheral nerves, diabetic neuropathy, post-traumatic/post-operative neuropathies and carpal tunnel syndrome.

An innovative contribution regards the use of Botox® (OnabotulinumtoxinA) in the field of dental occlusal treatment, specifically in a clinical application of the toxin to reconstruct the physiological homeostasis of the masticatory complex in patients affected by short-faced syndrome, presented by Li et al. [7]. The authors showed that by means of targeted injections of Botox® into masseter muscle, to relax excessive powerful and thick muscle in short-faced patients, it was possible to restore the physiological homeostasis of the masticatory complex avoiding dental complications, such as temporomandibular disorders, bruxism, periodontitis, etc., often observed with other treatments.

A very exhaustive review, presented by Gazerani [8], is devoted to the effects of BoNTs in counteracting pruritus in a variety of conditions in which chronic itch is the main symptom. In the review, evidence of the beneficial effects of BoNTs is systematically analyzed using studies of both animal models or with healthy volunteers subjected to various forms of induced pruritus, or using off-label studies with humans subjected to chronic itchy conditions originating from different pathologies, whether dermatological or not. A mechanism of action of BoNTs is proposed, based on the blockade of the peripheral components of pruritus, i.e., the release of pruritic mediators and the activation of immune cells, blocking the vasomotor and autonomic components of pruritus.

A review article from Luvisetto [9] collected data on the interaction of BoNT/A with glial cells, both at the central (astrocytes, microglia and oligodendrocytes) and peripheral level (Schwann cells and satellite glial cells). An analysis of the data confirmed that BoNT/A can block the release of neuroactive substances not only from neuronal cells but also from glial cells. This effect is not limited to microglia and astrocytes but is also demonstrated in Schwan cells and oligodendrocytes, non-neuronal cells involved in the reconstruction of the myelin sheath damaged by traumatic injury. These results pave the way for an extraordinary application of BoNT/A in the treatment of spinal cord injuries, which will be the subject of further experimental research presented in this SI.

In another contribution, Wenninger et al. [10] analyzed the frequency of hemorrhagic side effects after BoNT/A injections, of different commercial preparations, for the treatment of benign essential blepharospasm and hemifacial spasm in patients taking antithrombotic drugs. The authors observed that, during the treatment of blepharospasm and hemifacial spasm with BoNT/A in patients undergoing concomitant treatment with antithrombotic drugs, the unwanted side effects, such as hematoma frequency, were not significantly different between patients treated with antithrombotic drugs or not. The main conclusion of this report was that the two treatments, both antithrombotic drugs and BoNT/A injection, can be safely performed together.

The next article is concerned with human studies combining the effect of BoNTs and the aid of electronic devices. Bertoncelli et al. [11] explained the use of a statistical machine learning algorithm, named BTX-PredictMed. The algorithm is based on a model that identifies clinical phenotypes for the prognosis of cerebral palsy in children that can benefit from an injection of BoNTs. The clinical features that the algorithm analyzes comprise neuromuscular scoliosis, equine foots, upper limbs and trunk muscle tone disorders, and the presence of other forms of spasticity, dystonia and dysplasia. The importance of this algorithm resides in the fact that it could be very helpful for professionals needing to make informed medical decisions.

In another study, Schulte-Baukloh et al. [12] reported an analysis of the exact onset of the action of Botox® in patients affected by an overactive bladder. By means of a Diary Pod app, automated bladder diary equipment essentially constituting a measuring device connected through a mobile app to the remote dashboard of the clinical portal, the voiding frequency was accurately documented. A trend of reduction was observed 4 days after treatment with the toxin, with a significant change from the 5th day after toxin injection.

The extraordinary role of BoNT/A as an inducer of nerve regeneration in a mouse model of spinal cord injury was presented in the research article of Mastrorilli et al. [13]. In this study, a single dose of Xeomin®, injected directly into the spinal cord during the acute phase of traumatic injury, was able to induce complete regeneration and functional recovery, with the restoration of walking capacity, in paraplegic mice.

Finally, Beret et al. [14] show the effect of an intra-articular injection of Xeomin® in a rat model of TMJ osteoarthritis induced via an intra-articular injection of monosodium iodoacetate. Behavioral, histological and radiographic analyses, with the latter performed using positron emission tomography imaging, showed reduced pain and decreased joint inflammation in the BoNT/A-treated animals.

In conclusion, the research and review articles included in this SI of *Toxins* contribute to advancing the state of the art on the novel therapeutic uses of BoNTs. Furthermore, many of the published studies focus on emerging or less-investigated applications of BoNTs in uncommon pathologies, thus providing the scientific community with new data supporting better knowledge of the contribution that can be made by BoNTs in improving human health.

Acknowledgments: As Guest Editor, I wish to thank all of the authors and colleagues who contributed to the success of this SI of *Toxins*, and all expert reviewers who performed careful and rigorous evaluations of the article submissions. The valuable editorial support from all of the Editorial staff and MDPI management team is also acknowledged.

Conflicts of Interest: The author declares no conflict of interest.

References

1. Luvisetto, S. Introduction to the Toxins Special Issue on Botulinum Neurotoxins in the Nervous System: Future Challenges for Novel Indications. *Toxins* **2020**, *12*, 601. [CrossRef]
2. Mueller, J.; Langbein, T.; Mishra, A.; Baum, R.P. Safety of High-Dose Botulinum Toxin Injections for Parotid and Submandibular Gland Radioprotection. *Toxins* **2022**, *14*, 64. [CrossRef]
3. Muñoz-Lora, V.R.M.; Dugonjić Okroša, A.; Matak, I.; Del Bel Cury, A.A.; Kalinichev, M.; Lacković, Z. Antinociceptive Actions of Botulinum Toxin A1 on Immunogenic Hypersensitivity in Temporomandibular Joint of Rats. *Toxins* **2022**, *14*, 161. [CrossRef]
4. Yi, K.-H.; Lee, J.-H.; Kim, H.-J. Intramuscular Neural Distribution of the Serratus Anterior Muscle: Regarding Botulinum Neurotoxin Injection for Treating Myofascial Pain Syndrome. *Toxins* **2022**, *14*, 271. [CrossRef] [PubMed]
5. Yi, K.-H.; Lee, J.-H.; Hu, H.-W.; Kim, H.-J. Novel Anatomical Guidelines on Botulinum Neurotoxin Injection for Wrinkles in the Nose Region. *Toxins* **2022**, *14*, 342. [CrossRef] [PubMed]
6. Lippi, L.; de Sire, A.; Folli, A.; D'Abrosca, F.; Grana, E.; Baricich, A.; Carda, S.; Invernizzi, M. Multidimensional Effectiveness of Botulinum Toxin in Neuropathic Pain: A Systematic Review of Randomized Clinical Trials. *Toxins* **2022**, *14*, 308. [CrossRef] [PubMed]

7. Li, X.; Feng, X.; Li, J.; Bao, X.; Xu, J.; Lin, J. Can Botulinum Toxin-A Contribute to Reconstructing the Physiological Homeostasis of the Masticatory Complex in Short-Faced Patients during Occlusal Therapy? A Prospective Pilot Study. *Toxins* **2022**, *14*, 374. [CrossRef] [PubMed]
8. Gazerani, P. How Does Botulinum Toxin Inhibit Itch? *Toxins* **2022**, *14*, 701. [CrossRef]
9. Luvisetto, S. Botulinum Neurotoxins beyond Neurons: Interplay with Glial Cells. *Toxins* **2022**, *14*, 704. [CrossRef] [PubMed]
10. Wenninger, F.C.; Wabbels, B. Frequency of Hemorrhagic Side Effects of Botulinum Neurotoxin Treatment in Patients with Blepharospasm and Hemifacial Spasm on Antithrombotic Medication. *Toxins* **2022**, *14*, 769. [PubMed]
11. Bertoncelli, C.M.; Latalski, M.; Bertoncelli, D.; Bagui, S.; Bagui, S.C.; Gautier, D.; Solla, F. Prediction Model for Identifying Computational Phenotypes of Children with Cerebral Palsy Needing Neurotoxin Treatments. *Toxins* **2023**, *15*, 20. [CrossRef] [PubMed]
12. Schulte-Baukloh, H.; Weiß, C.; Weinberger, S.; Hubatsch, M.; Schlomm, T.; Ralla, B. Real-Time Documentation of the Effect of Onabotulinumtoxin A Detrusor Injection in OAB Patients—Preliminary Results. *Toxins* **2023**, *15*, 30. [CrossRef] [PubMed]
13. Mastrorilli, V.; De Angelis, F.; Vacca, V.; Pavone, F.; Luvisetto, S.; Marinelli, S. Xeomin®, a Commercial Formulation of Botulinum Neurotoxin Type A, Promotes Regeneration in a Preclinical Model of Spinal Cord Injury. *Toxins* **2023**, *15*, 248. [CrossRef] [PubMed]
14. Béret, M.; Barry, F.; Garcia-Fernandez, M.-J.; Chijcheapaza-Flores, H.; Blanchemain, N.; Chai, F.; Nicot, R. Efficacy of Intra-Articular Injection of Botulinum Toxin Type A (IncobotulinumtoxinA) in Temporomandibular Joint Osteoarthritis: A Three-Arm Controlled Trial in Rats. *Toxins* **2023**, *15*, 261. [CrossRef] [PubMed]

toxins

MDPI

Article

Safety of High-Dose Botulinum Toxin Injections for Parotid and Submandibular Gland Radioprotection

Joerg Mueller [1,*], Thomas Langbein [2], Aditi Mishra [3] and Richard P. Baum [3]

1 Department of Neurology, Vivantes Klinikum Spandau, Neue Bergstrasse 6, D-13585 Berlin, Germany
2 Department of Nuclear Medicine, Technical University of Munich, Klinikum rechts der Isar, Ismaninger Strasse 22, D-81675 München, Germany; thomas.langbein@tum.de
3 Department of Nuclear Medicine, Curanosticum Wiesbaden, Deutsche Klinik für Diagnostik, Aukammallee 33, D-65191 Wiesbaden, Germany; mishra@curanosticum.de (A.M.); baum@curanosticum.de (R.P.B.)
* Correspondence: j.mueller@berlin.de

Abstract: Botulinum Toxin injections into salivary glands (SG) up to a total dose of 100 units IncobotulinumtoxinA (IncoA) represent the treatment of choice for sialorrhea. However, BTX might also protect SG against sialotoxic radioligand cancer therapies. The radioligand Actinium-225-PSMA effectively targets Prostate Cancer (PCa) metastases but inevitably destroys SG due to unintended gland uptake. A preliminary case series with regular-dose IncoA failed to reduce SG PSMA-radioligand uptake. We therefore increased IncoA dosage in combination with transdermal scopolamine until a clinically relevant SG PSMA-radioligand uptake reduction was achieved. Ten consecutive men with metastasized PCa refractory to all other cancer therapies received gradually increasing IncoA dosages as part of a compassionate use PSMA-radioligand-therapy trial. The parotid gland received six and the submandibular gland three injection points under ultrasound control, up to a maximum of 30 units IncoA per injection point. A maximum total dose of 250 units IncoA was applied with up to 170 units per parotid and 80 units per submandibular gland. Treatment was well tolerated and all side-effects were non-serious. The most frequent side-effect was dry mouth of mild severity. No dysphagia, facial weakness, chewing difficulties or systemic side-effects were observed. SG injections with IncoA up to a total dose of 250 units are safe when distributed among several injection-points under ultrasound control by an experienced physician. These preliminary findings lay the basis for future trials including BTX as major component for SG protection in established as well as newly emerging radioligand cancer therapies.

Keywords: high-dose botulinum toxin; salivary glands; radioprotection

Key Contribution: Safety of high-dose incobotulinum toxin for salivary gland radioprotection.

Citation: Mueller, J.; Langbein, T.; Mishra, A.; Baum, R.P. Safety of High-Dose Botulinum Toxin Injections for Parotid and Submandibular Gland Radioprotection. *Toxins* **2022**, *14*, 64. https://doi.org/10.3390/toxins14010064

Received: 26 November 2021
Accepted: 10 January 2022
Published: 17 January 2022

1. Background

In healthy adults, the parotid and submandibular glands account for 95% of the total salivary secretion. Botulinum Toxin (BTX) injections into these major salivary glands are considered the treatment of choice for chronic neurogenic sialorrhea [1]. Among the type-A botulinum toxin preparations, only IncobotulinumtoxinA (IncoA) is approved for the treatment of sialorrhea in the US and EU. The recommended total dose is 75 to 100 units IncoA, with up to 30 units given to each parotid and up to 20 units to each submandibular gland. Apart from neurological indications, the use of BTX as a salivary gland protective agent in cancer patients with normal salivary gland function might emerge as a promising new BTX-indication. Radioligand therapies against prostate or thyroid cancer frequently induce gland destruction due to unintended radioligand-uptake into the healthy salivary glands. Severe sialotoxicity represents the treatment-limiting side-effect of the highly effective prostate cancer radioligand therapy with Actinium-225-PSMA (Ac-PSMA) [2,3].

All previous attempts with various drugs including monotherapy with anticholinergics failed to overcome this therapy-limiting side-effect [3,4]. Patients with differentiated thyroid cancer treated with radioactive iodine (^{131}I) are also at risk of permanent xerostomia. Selvakumar et al. reported that almost 50% of patients from a nationwide study in the Netherlands suffered from permanent salivary gland dysfunction with moderate to severe xerostomia in 35% at median 11 years following radioiodine therapy with ^{131}I [5]. Salivary gland destruction is associated with severe xerostomia inducing a plethora of complications resulting in a significant reduction in quality of life.

A recent publication proved the relationship between salivary gland activity and salivary gland uptake of Prostate-Specific Membrane Antibody (PSMA) radioligands [6].

BTX reduces salivary gland activity in a dose-dependent manner with subsequent and fully reversible ductal and acinar apoptosis [7,8]. However, in a preliminary case series with regular-dose IncoA distributed across the submandibular and parotid glands we observed no reduction of Lutetium-144-PSMA (Lu-PSMA) radioligand uptake in any of the injected glands. Therefore, the present study intended to assess safety and tolerability of gradually increasing SG BTX-A doses required to achieve a clinically relevant PSMA-uptake reduction. Toxic Ac-PSMA salivary gland uptake correlates with the amount of gland destruction and clinical xerostomia [9] occurs when salivary gland function is reduced by about 50% [10]. Here, we report and discuss the preliminary safety data from 10 advanced prostate cancer (PCa) patients who received increasing IncoA dosages as part of a compassionate use PSMA-radioligand therapy trial. The oncological and Positron emission tomography (PET) outcome data will be reported elsewhere (manuscript in preparation).

2. Patients and Methods

10 consecutive men with a mean age of 68 years [range 53–80] were treated with increasing dosages of IncoA by means of a compassionate use trial. All men suffered from advanced metastasized PCa for whom all other treatment options had failed. All patients were refractory to previous surgery, external radiation, hormone- and various chemotherapies as well as to previous radioligand monotherapies with Lu-PSMA and thus scheduled for a compassionate use trial with a combined Lu-PSMA plus Ac-PSMA "tandem" radioligand therapy. Since Ac-PSMA therapy is known to induce immediate complete and irreversible salivary gland destruction as side-effect and no established gland protective therapy has been available, patients were offered BTX injections as part of the compassionate trial. All patients gave written informed consent following detailed explanation of the procedure and possible side-effects—in particular dysphagia. All procedures were conducted in accordance with the Declaration of Helsinki.

Measures of saliva production were performed prior to IncoA and after radioligand therapies. Unprovoked salivary flow was measured with four dental rolls placed in the orifice of the mouth and then retained there for five minutes. Provoked salivary flow was measured by chewing on a folded compress placed on the tongue for two minutes [11]. The difference in weight between the dry and wet rolls was calculated in grams (g) and both measures were added up.

Each 100 unit IncoA (Xeomin®, Merz Pharmaceuticals, Frankfurt, Germany) vial was reconstituted with 1 mL normal 0.9% saline using a 1 mL syringe. BTX was injected percutaneously (26 G, 25 mm needle length) into the salivary glands under ultrasound control using a GE 7.5 MHz Linear Array Ultrasound Probe (GE Health Care, Chicago, IL, USA). The parotid gland received six injection points (between-point distance around 10 mm, each injection point with the same dose). The submandibular gland received three injection points, each with the same dose. A maximum dose of 30 units IncoA per injection point was not exceeded.

All patients received IncoA-injections into two of the four main salivary glands (right parotid plus contralateral left submandibular gland). Injections were performed median 14 days [range 5–33 days] before tandem-PSMA radioligand therapy by a neurologist with more than 25 years of experience with BTX.

Patients were closely monitored by phone interviews and repeat clinical follow-up visits. Predefined adverse events of special interest were dry mouth, dysphagia, facial weakness, eyelid-drop, conjunctivitis, stomatitis, chewing and breathing difficulties as well as generalized weakness. Side-effects were recorded using a 0 to 3 scale (none/mild/moderate/severe). The concomitant permanent medication was recorded at the time of BTX injection.

3. Results

Table 1 summarizes BTX treatment details and side-effects for each patient. A maximum total dose of 250 units IncoA was applied with up to 170 units into a single parotid gland and 80 units IncoA into a single submandibular gland. Treatment was well tolerated and all side-effects were non-serious.

Table 1. Botulinum Toxin (IncoA) treatment details and adverse events.

Patient	PG Right IncoA (Units)	SMG Left IncoA (Units)	Total Dose IncoA (Units)	Adverse Event	AE Severity
1	100	50	150	Painful swallowing	Mild
2	100	50	150	-	-
3	120	70	190	Dry mouth	Mild
4	120	70	190	-	-
5	130	70	200	Dry mouth	Mild
6	170	80	250	Dry mouth *	Moderate *
7	170	80	250	Dry mouth	Mild
8	170	80	250	Dry mouth *	Moderate *
9	170	80	250	Dry mouth	Mild
10	170	80	250	Dry mouth	Mild

PG = Parotid gland, SMG = Submandibular gland, * moderate pre-existing xerostomia.

Mild to moderate injection pain was experienced by all patients and resolved immediately. Injection pain was more pronounced in the parotid glands. One injection-related local hematoma occurred in a patient with thrombocytopenia.

The most frequent BTX-related side-effect was dry mouth of mild severity. Patient 1 who received a total dose of 150 units IncoA reported a mild flu-like painful swallowing episode which lasted for one week and resolved without intervention. Patients 6 + 8 had pre-existing moderate xerostomia after previous chemo- and radiation therapies that remained unchanged (see Table 1).

No dysphagia, facial weakness, eyelid-drop, conjunctivitis, stomatitis, chewing or breathing difficulties were observed. No patient showed distant or systemic side-effects.

Follow-up: at the time of manuscript submission, four of the ten patients had completed two Ac-PSMA cycles and were scheduled for their third and fourth Actinium-225-Prostate Specific Membrane Antigen (Ac-PSMA) radiotherapy (RT). Combined unprovoked and provoked saliva production 16–20 weeks after the first IncoA injection resulted in a mean 29% [range 25–31%] loss (or mean 71% preservation) of saliva production as compared to around 60–70% salivary gland destruction following two cycles of unprotected Ac-PSMA [3].

Concomitant medication: the concomitant medication is summarized in Table 2. All patients received endocrine therapies and all patients used at least one pain medication constantly. Seven patients received additional transdermal scopolamine during a period of 72 h prior PSMA-radioligand therapy.

Table 2. Concomitant permanent medication.

Any Drug Treatment	*N* = 10
Antiandrogenes	*N* = 5
GnRH-analogues	*N* = 5
Steroids	*N* = 4
RANK-ligand-Inhibitors	*N* = 5
Bisphosphonates	*N* = 1

Table 2. *Cont.*

Any Drug Treatment	$N = 10$
Opioids	$N = 6$
NSAIDs/COX-2 inhibitors	$N = 9$
Gabapentinoids	$N = 3$
Antihypertensives	$N = 5$
Statins	$N = 1$
Erythropoietin	$N = 1$
Antacids	$N = 1$

4. Discussion

To the best of our knowledge, these are the first data to report the safety and tolerability of high-dose BTX-A injections into the salivary glands up to a total dose of 250 units IncoA. The individual glands received four to five times the dose regularly used to treat sialorrhea. Notably, the injected dosage of 170 units IncoA (Xeomin®) into a single parotid gland was tolerated without any swallowing or chewing problems or systemic side-effects. Likewise, injected dosages of 80 units IncoA into a single submandibular gland did not result in swallowing problems. The latter result is of particular interest considering the glands proximity to the supra-hyoidal muscles critical for swallowing function. Our results also indicate that high-dose salivary gland treatment with IncoA under ultrasound guidance using a maximum of 30 units per injection point is not associated with clinically relevant local or systemic toxin spread. A mandatory requirement for the application of high-dose BTX into salivary glands is ultrasound-guidance in the hand of an experienced neurologist or Otolaryngologist. Previous studies demonstrated the superiority of ultrasound guided injections over anatomical landmark-oriented injections. A post-mortem study showed an almost doubled hit-rate of ultrasound-guided injections in the submandibular glands compared to free-hand injections [12]. The high error-rate even of experienced injectors might be attributable to the variance of submandibular gland size as well as to different inter- and intra-individual distances between gland and mandible [1,13].

It remains a matter of debate whether more than one injection point per gland improves outcome and/or reduces the risk of side-effects from local toxin spread. However, histological analyses show that BTX affects mainly those salivary gland acinar cells located around the site of injection [7]. BTX mediated salivary gland protection from radioligand toxicity is intended to cover the gland tissue as completely as possible. Therefore, we deliberately chose six injection points for each parotid and three injection points for each submandibular gland and advocate this injection technique for salivary gland treatment, especially with the use of higher BTX-dosages. The long-term safety of local BTX injections has been proven for many years with full recovery of the target glands within around 16 weeks following BTX injection. The prevalence of neutralizing antibodies was not tested in this study but there is clear evidence for the particularly low antigenicity of IncoA even with higher dosages used in patients with spasticity and dystonia as well as in non-neurological indications [14–17].

Patients included in the present study were heavily pretreated and showed rapid PCa progression after exhausting all established cancer therapies and thus safety as well as tolerability of the gland protective BTX therapy was of primary importance. A recent series of 26 patients treated with Ac-PSMA after failure of Lu-PSMA reported xerostomia in 100% of patients and 23% stopped Ac-PSMA-RT because of intolerable xerostomia following two RT cycles [18]. These results are in line with a previous report of around 66% loss of parotid and submandibular gland function following two unprotected Ac-PSMA RT cycles [3]. By contrast, our preliminary data suggest that high-dose IncoA in combination with transdermal scopolamine provides a clinically relevant SG protection accompanied by good tolerability.

However, the modalities of SG-protective BTX therapies may vary depending on cancer type and prognosis. In differentiated thyroid cancer for example, more than 70% of patients require only up to two ^{131}I radioligand treatment sessions to achieve long-term remission and survival [5]. Whether such a patient group with excellent oncological long-term prognosis or PCa patients in earlier lines of PSMA-therapy might benefit from modified BTX treatment regimen in combination with systemic anticholinergics with the objective to protect all four major SG from radioligand toxicity remains to be examined in future studies.

Interestingly, the strong PSMA radioligand accumulation in salivary glands does not correspond to high PSMA expression levels in the glands, indicating unspecific PSMA radioligand uptake into SG [19]. BTX-A induces reversible cellular changes in salivary glands with decreased mucosal and serous acini diameter and temporary glandular atrophy [8,20]. These effects are probably due to fully reversible glandular denervation resulting in reduced radiation sensitivity [20]. In addition, combined blockage of parasympathetic gland innervation by BTX-A and anticholinergics reduces SG blood flow, resulting in lower accumulation of toxic PSMA and iodine radioligands inside the glands.

In conclusion, salivary gland injections with IncobotulinumtoxinA up to a total dose of 250 units are safe when injected under ultrasound control by an experienced physician, even in seriously ill patients in a fragile medical condition. These novel findings of our preliminary study lay the basis for future trials including botulinum toxin as major component for salivary gland protection in established as well as newly emerging radioligand cancer therapies.

Author Contributions: Conceptualization, J.M. and R.P.B.; methodology, J.M., T.L. and R.P.B.; formal analysis, J.M., T.L. and A.M.; investigation, J.M., A.M. and R.P.B.; writing—original draft preparation, J.M.; writing—review and editing, J.M. and R.P.B. All authors have read and agreed to the published version of the manuscript.

Funding: Non sponsor funded research.

Institutional Review Board Statement: The study was performed in accordance with the regulations of the German Federal Agency for Radiation Protection and ethical approval was obtained from the State Medical Association Thuringia, registration number 23333/2018/106, date 12 September 2018.

Informed Consent Statement: Written informed consent was obtained from all patients.

Data Availability Statement: Data available on request due to restrictions e.g. privacy or ethical. The data presented in this study are available on request from the corresponding author.

Conflicts of Interest: J.M. has received speaker honoraria from Merz Pharmaceuticals®. R.P.B., A.M. and T.L. have nothing to declare. All authors do not have any conflict of interest with the present study.

References

1. Jost, W.H.; Bäumer, T.; Laskawi, R.; Slawek, J. Therapy of sialorrhea with Botulinum Neurotoxin. *Neurol. Ther.* **2019**, *8*, 273–288. [CrossRef] [PubMed]
2. Kratchowil, C.; Bruchertseifer, F.; Rathke, H.; Hohenfellner, M.; Giesel, F.L.; Haberkorn, U.; Morgenstern, A. Targeted Alpha Therapy of mCRPC with ^{225}Actinium-PSMA-617: Swimmer-Plot analysis suggests efficacy regarding duration of tumor control. *J. Nucl. Med.* **2018**, *59*, 795–802. [CrossRef] [PubMed]
3. Langbein, T.; Chaussé, G.; Baum, R.P. Salivary Gland Toxicity of PSMA Radioligand Therapy: Relevance and Preventive Strategies. *J. Nucl. Med.* **2018**, *59*, 1172–1173. [CrossRef] [PubMed]
4. Mohan, V.; Bruin, N.M.; Tesselaar, M.E.T.; de Boer, J.P.; Vegt, E.; Hendrikx, J.J.M.A.; Al-Mamgani, A.; van de Kamer, J.B.; Sonke, J.J.; Vogel, W.V. Muscarinic inhibition of salivary glands with glycopyrronium bromide does not reduce the uptake of PSMA-ligands or radioiodine. *EJNMMI Res.* **2021**, *11*, 25. [CrossRef] [PubMed]
5. Selvakumar, T.; Nies, M.; Klein Hesselink, M.S.; Brouwers, A.H.; van der Horst-Schrivers, A.N.A.; Tissing, W.J.E.; Vissink, A.; Links, T.P. Long-Term Effects of Radioiodine Treatment on Salivary Gland Function in Adult Survivors of Pediatric Differentiated Thyroid Carcinoma. *J. Nucl. Med.* **2019**, *60*, 122–177. [CrossRef] [PubMed]
6. Huang, K.; Baur, A.D.; Hupperts, H.; Feldhaus, F.W.; Brenner, W.; Furth, C.; Mueller, J.; Amthauer, H. PSMA-Positivität der Speicheldrüsen in der Ga-68-PSMA-PET: Spielt Nüchternheit eine Rolle? *Nuklearmedizin* **2020**, *59*, 103.
7. Shan, X.F.; Xu, H.; Cai, Z.G.; Wu, L.L.; Yu, G.Y. Botulinum Toxin A inhibits salivary secretion of rabbit submandibular gland. *Int. J. Oral Sci.* **2013**, *5*, 217–223. [CrossRef] [PubMed]

8. Regueira, L.S.; Baratella-Evencio, L.; de Oliveira, J.B.; da Silva, P.S.O.C.; Valenca, G.A.L.; Conceicao, J.C.P.; da Cruz Perez, D.E. Effects of chronic treatment with botulinum toxin type A in salivary glands of rats: Histological and immunohistochemical analyses. *J. Oral Pathol. Med.* **2019**, *48*, 728–734. [CrossRef]
9. Feuerecker, B.; Gafita, A.; Tauber, R.; Alessandria, C.D.; Seidl, C.; Bruchertseifer, F.; Retz, M.; Weber, W.; Morgenstern, A.; Elber, M. Effekte eines Zyklus Actinium-225-PSMA-617 (AcPSMA) auf die Speicheldrüsen—Vorläufige Ergebnisse. *Nuklearmedizin* **2020**, *59*, 161.
10. Löfgren, C.D.; Wickström, C.; Christersson, C. A systemic review of methods to diagnose oral dryness and salivary gland function. *BMC Oral Health* **2012**, *12*, 29. [CrossRef] [PubMed]
11. Kohler, P.F.; Winter, M.E. A quantitative test for xerostomia. *The Saxon test, an oral equivalent of the Schirmer test. Arthritis Rheum.* **1985**, *28*, 1128–1132. [PubMed]
12. So, J.I.; Song, D.H.; Park, J.H.; Choi, E.; Yoon, J.Y.; Yoo, Y.; Chung, M.E. Accuracy of ultrasound-guided and non-ultrasound-guided botulinum toxin injection into cadaver salivary glands. *Ann. Rehab. Med.* **2017**, *41*, 51–57. [CrossRef] [PubMed]
13. Loens, S.; Brüggemann, N.; Steffen, A.; Bäumer, T. Localization of Salivary Glands for Botulinum Toxin Treatment: Ultrasound Versus Landmark Guidance. *Mov. Disord. Clin. Pract.* **2020**, *7*, 194–198. [CrossRef] [PubMed]
14. Hefter, H.; Brauns, R.; Uerer, B.; Rosenthal, D.; Albrecht, P. Effective long-term treatment with incobotulinum (Xeomin®) without neutralizing antibody induction: A monocentric, cross-sectional study. *J. Neurol.* **2020**, *267*, 1340–1347. [CrossRef] [PubMed]
15. Samadzadeh, S.; Uerer, B.; Brauns, R.; Rosenthal, D.; Lee Ji Albrecht, P.; Hefter, H. Clinical implications of difference in antigenicity of different Botulinum Neurotoxin Type A preparations: Clinical Take-Home Messages from our Research Pool and Literature. *Toxins* **2020**, *12*, 499. [CrossRef] [PubMed]
16. Ramirez-Castaneda, J.; Jankovic, J. Long-term efficacy and safety of botulinum toxin injections in dystonia. *Toxins* **2013**, *5*, 249–266. [CrossRef] [PubMed]
17. Kaufmann-Janette, J.; Cox, S.E.; Dayan, S.; Joseph, J. Botulinum Toxin Type A for Glabellar Frown Lines: What impact of Higher Doses on Outcomes? *Toxins* **2021**, *13*, 494. [CrossRef] [PubMed]
18. Feuerecker, B.; Tauber, R.; Knorr, K.; Heck, M.; Beheshti, A.; Seidl, C.; Bruchertseifer, F.; Pickhard, A.; Gafita, A.; Kratchowil, C.; et al. Activity and Adverse Events of Actinium-225-PSMA-617 in Advanced Metastatic Castration-resistant Prostate Cancer After Failure of Lutetium-177-PSMA. *Eur. Urol.* **2021**, *3*, 343–350. [CrossRef] [PubMed]
19. Rupp, N.J.; Umbricht, C.A.; Pizzuto, D.A.; Lenggenhager, D.; Töpfer, A.; Müller, J.; Muehlematter, U.J.; Ferraro, D.A.; Messerli, M.; Morand, G.B.; et al. First Clinicopathologic Evidence of a Non–PSMA-Related Uptake Mechanism for ^{68}Ga-PSMA-11 in Salivary Glands. *J. Nucl. Med.* **2019**, *60*, 1270–1276. [CrossRef]
20. Teymoortash, A.; Sommer, F.; Mandic, R.; Schulz, S.; Bette, M.; Aumüller, G.; Werner, J.A. Intraglandular application of botulinum toxin leads to structural and functional changes in rat acinar cells. *Br. J. Pharmacol.* **2007**, *152*, 161–167. [CrossRef]

toxins MDPI

Article

Antinociceptive Actions of Botulinum Toxin A1 on Immunogenic Hypersensitivity in Temporomandibular Joint of Rats

Victor Ricardo Manuel Muñoz-Lora [1,2,3], Ana Dugonjić Okroša [4], Ivica Matak [1], Altair Antoninha Del Bel Cury [2], Mikhail Kalinichev [5] and Zdravko Lacković [1,*]

1 Laboratory of Molecular Neuropharmacology, Department of Pharmacology, School of Medicine, University of Zagreb, 10000 Zagreb, Croatia; victor_9874@hotmail.com (V.R.M.M.-L.); ivica.matak@mef.hr (I.M.)
2 Department of Prosthodontics and Periodontology, Piracicaba Dental School, University of Campinas (UNICAMP), Piracicaba 13414-903, Brazil; altair@unicamp.br
3 Dental Research Division, School of Dentistry, Ibirapuera University, São Paulo 04661-100, Brazil
4 Department of Pharmacology, Faculty of Pharmacy and Biochemistry, University of Zagreb, 10000 Zagreb, Croatia; anadugonjic2008@gmail.com
5 Ipsen Innovation, 91940 Les Ulis, France; mikhail.kalinichev@icloud.com
* Correspondence: lac@mef.hr; Tel.: +385-1-4566-843

Abstract: Botulinum neurotoxin type A1 (BoNT-A) reduces the peripheral peptide and cytokine upregulation in rats with antigen-evoked persistent immunogenic hypersensitivity (PIH) of the temporomandibular joint (TMJ). Herein, we examined the effects of two preparations of BoNT-A, abobotulinumtoxinA (aboBoNT-A; Dysport) and onabotulinumtoxinA (onaBoNT-A; Botox), on spontaneous and evoked nociceptive behaviors, as well as on central neuronal and astroglial activation. The antigen-evoked PIH was induced in rats via repeated systemic and unilateral intra-articular (i.a.) injections of methylated bovine serum albumin (mBSA). Rats were subsequently injected with unilateral i.a. aboBoNT-A (14 U/kg), onaBoNT-A (7 U/kg), or the vehicle (saline). After i.a. treatments, spontaneous and mechanically evoked nocifensive behaviors were assessed before and after the low-dose i.a. formalin (0.5%) challenge. The central effects of BoNT-A were assessed by an immunohistochemical analysis of cleaved synaptosomal-associated protein 25 (cSNAP-25) presence, c-Fos, GFAP, and CGRP expression in the trigeminal nucleus caudalis (TNC). Both BoNT-A preparations similarly reduced the formalin-induced spontaneous pain-related behaviors and mechanical allodynia of the hypernociceptive rats. Likewise, their effects were associated with the central occurrence of cSNAP-25 and reduction of c-Fos and GFAP upregulation in the TNC. BoNT-A antinociceptive activity on the PIH is associated with the toxin axonal transport to trigeminal sensory areas and reduction of neuronal and glial activation in central nociceptive regions.

Keywords: botulinum neurotoxin type A1; nociception; microglia; astrocytes; central nervous system; analgesia

Key Contribution: Herein, it was shown that botulinum toxin type A1-based pharmaceutical preparations alleviate the experimental joint pain associated with immunogenic hypersensitivity. Reduction of neuronal and astrocyte activation suggest the toxin's central actions.

Citation: Muñoz-Lora, V.R.M.; Dugonjić Okroša, A.; Matak, I.; Del Bel Cury, A.A.; Kalinichev, M.; Lacković, Z. Antinociceptive Actions of Botulinum Toxin A1 on Immunogenic Hypersensitivity in Temporomandibular Joint of Rats. *Toxins* **2022**, *14*, 161. https://doi.org/10.3390/toxins14030161

Received: 26 January 2022
Accepted: 18 February 2022
Published: 23 February 2022

1. Introduction

Temporomandibular joint (TMJ) rheumatoid arthritis (RA) is a chronic inflammatory condition involving a range of local and systemic proinflammatory mediators. TMJ-RA produces structural and functional deterioration of the affected TMJ and associated tissues [1]. TMJ-related symptoms are present in almost 65% of RA patients and are typically bilateral [2,3]. Along with the impairment of normal functions of the jaw joint, the TMJ-RA is often associated with periods of alternating nonpainful and painful phases [1]. Active painful phases are usually provoked by an otherwise nonpainful stimulus (e.g., mandibular functional movements: speaking, eating, etc.), with a negative impact on the patient's quality of life [4,5]. Notwithstanding, treatments with nonsteroidal anti-inflammatory drugs (NSAIDs) and local anesthetic into the TMJ compartment have proven to relieve RA pain just temporarily [6].

Animal models of RA-related pain, including TMJ-RA, have been proposed with the goal to improve the understanding of its etiology and develop new treatment options [7–9]. When used as an antigen, systemic and unilateral intra-TMJ administrations of methylated bovine serum albumin (mBSA) in rats result in the development of persistent immunogenic hypersensitivity (PIH) and inflammatory changes of the exposed TMJ (Figure 1) [10,11]. Interestingly, even though this model resembles human TMJ-RA, it has a monoarthritic (single joint) presentation, since it is induced locally inside the mBSA-stimulated joint [8,10,11]. The hypersensitivity appears as an exaggerated pain-related response (facial grooming and head flinching) to a normally nonpainful chemical stimulation of the TMJ, such as a low-dose (0.5%) formalin challenge [8,12]. This arthritic model produces a delayed-type hypersensitivity (viz., type IV hypersensitivity) and is characterized by central sensitization due to immune and neuronal cells activation, leading to the local release of a variety pronociceptive factors [10,11,13,14]. This fact increases the validity of the model to study the progression of pain on chronic inflammatory conditions and to test promising treatment options.

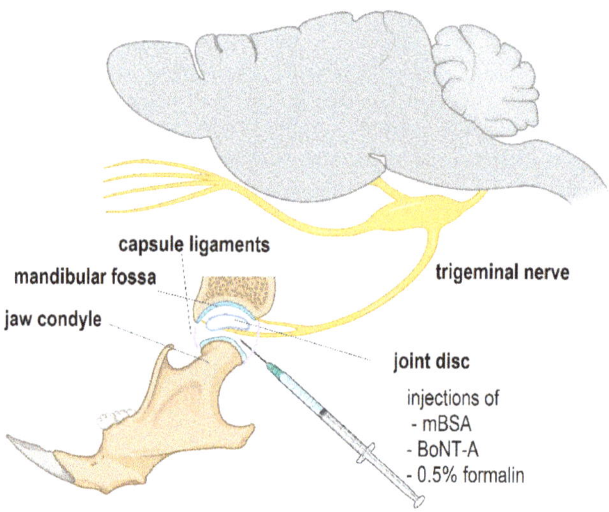

Figure 1. Schematic diagram of the structure of the temporomandibular (TMJ) joint innervated by the trigeminal nerve mandibular branch, as well as the site of injection of experimental substances, such as methylated bovine serum albumin (mBSA), abobotulinumtoxinA, or onabotulinumtoxinA (BoNT-A), and subsequent low-dose formalin challenge. The sizes of structures and their relative positions are not to scale. The illustration was generated by using BioRender© (Biorender.com, assessed on 16 February 2022).

In animals exposed to PIH and challenged with a low-dose formalin in the TMJ, we previously demonstrated that a treatment with an intra-TMJ injection of botulinum neurotoxin type A1 (BoNT-A), a potent neurotoxin naturally produced by *Clostridium botulinum*, reduced spontaneous behavioral facial grooming and head flinching responses 24h and 14 days after its administration [8]. Additionally, BoNT-A reduced the upregulated levels of substance P and calcitonin gene-related peptide (CGRP) in the trigeminal ganglion and interleukin 1 beta (IL-1β) in the TMJ of exposed rats [8]. Despite the apparent association of BoNT-A antinociceptive action with the mentioned peripheral effects, numerous studies in different pain models have suggested a necessity for an axonal transport and a direct central analgesic action of the toxin [15,16]. Therefore, we examined the occurrence of cleaved SNAP-25 (cSNAP-25), along with nociceptive neuronal and astroglial activation, on the trigeminal brainstem sensory areas of rats induced to PIH and treated with BoNT-A. In addition, spontaneous and evoked nociceptive behaviors were also assessed, and the efficacy of two different commercially available pharmaceutical preparations of BoNT-A: abobotulinumtoxinA (aboBoNT-A) and onabotulinumtoxinA (onaBoNT-A) administered at equiefficatious doses were compared.

2. Results

2.1. BoNT-A Effects on Rat Grimace Scale (RGS) and Spontaneous Nocifensive Responses (Flinching, Sctratching)

Previously in the PIH model, the number of spontaneous motor responses (flinches and scratching) after the formalin stimulation was shown to be reduced by onaBoNT-A [8]. Herein, we extended the measurement of spontaneous nociception by examining facial grimacing related to pain (RGS) and examined if the spontaneous nociceptive responses are present both before and after TMJ stimulation with low-dose formalin 0.5%). In the RGS test, we found a lack of painful facial expression (score = 0) and a low number of behavioral nocifensive responses in all groups at the pre-formalin assessments. Conversely, the post-formalin assessments showed higher values of RGS (Figure 2A,B) and a significant increase in the nocifensive responses (Figure 2C) on induced animals compared to noninduced.

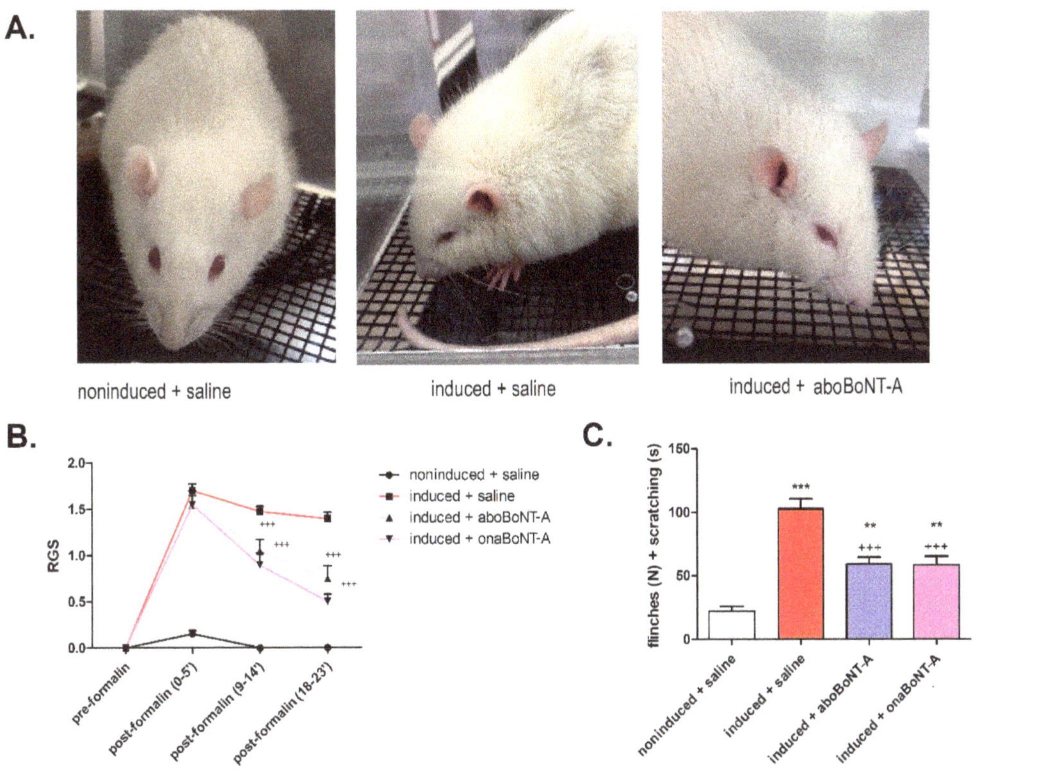

Figure 2. Effect of botulinum toxin type A1 (abobotulinumtoxinA—aboBoNT-A and onabotulinumtoxinA—onaBoNT-A) on spontaneous behavioral responses and the rat grimace scale

(RGS) prior to and after low-dose (0.5%) formalin stimulation of the temporomandibular joint (TMJ) of hypernociceptive rats. (**A**) Representative photographs showing rat facial expression in noninduced + saline (noninduced), induced + saline (induced), and induced + abobotulinumtoxinA (aboBoNT-A) animals after formalin stimulation. In the photos, the best visible features of nociceptive grimacing are orbital tightening, as well as characteristic spread-out ear positioning in animals experiencing more intensive pain (induced + saline vs. induced + onaBoNT/A), compared to no observable painful grimacing (noninduced + saline). (**B**) Both BoNT-A pharmaceutical preparations reduced the nociceptive facial grimacing evoked by 0.5% formalin, as assessed by RGS. No face grimacing was observed at the pre-formalin assessments. The individual RGS are calculated as the average of 4 distinctive facial features (orbital tightening, nose/cheek bulging, ear positioning, and whisker positioning) on a scale from 0 to 2. Mean ± SEM; +++ $p < 0.001$ vs. induced + saline (two-way repeated measures ANOVA followed by Bonferroni´s post hoc test). (**C**) Both BoNT-A intra-articular (i.a.) treatments (14 U/kg aboBoNT-A or 7 U/kg onaBoNT-A) reduced the spontaneous nocifensive behaviors (total sum of the number of flinches—N, plus the duration of facial grooming and rubbing—s) evoked by low-dose (0.5%) i.a. formalin in rats exhibiting persistent immunogenic hypersensitivity compared to noninduced animals. N(animals/group) = 10. Mean ± SEM, *** $p < 0.001$ and ** $p < 0.01$ vs. noninduced + saline and +++ $p < 0.001$ vs. induced + saline ($F_{3,36} = 49.59$, one-way ANOVA followed by Tukey's post hoc test).

Compared to induced animals treated with saline, both BoNT-A groups exhibited a significant reduction of the nocifensive responses triggered by formalin (Figure 2C). Furthermore, both BoNT-A treatments resulted in a similar reduction ($p < 0.01$) of the RGS score after formalin injection (Figure 2B).

2.2. *Effects of BoNT-A on Mechanically-Evoked Responses at the TMJ Area*

By employing the von Frey filaments, we examined the possible occurrence of facial mechanical allodynia over the skin covering the stimulated TMJ area both before and after the stimulation with 0.5% formalin. All animals showed a lack of allodynic responses at the pre-formalin evaluation. Following the formalin challenge, PIH-induced animals developed bilateral mechanical allodynia over the area of TMJ, measured with von Frey filaments (Figure 3A,B). The after-formalin assessments showed that BoNT-A (aboBoNT-A and onaBoNT-A) reduced the mechanically evoked TMJ bilateral allodynia of the hypersensitive rats (Figure 2A,B).

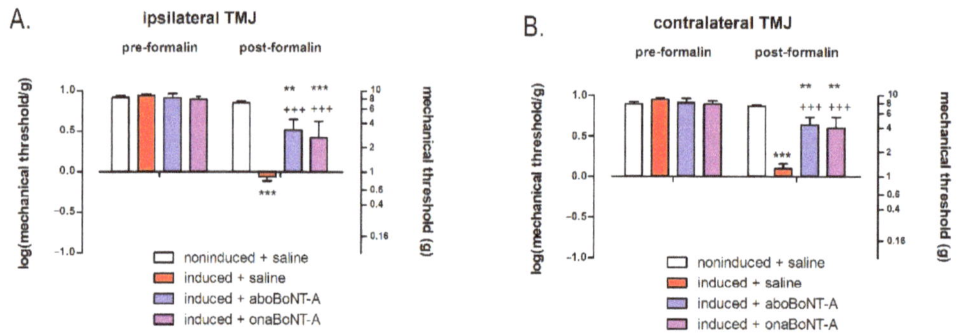

Figure 3. Effects of abobotulinumtoxinA (aboBoNT-A, 14 U/kg) and onabotulinumtoxinA (onaBoNT-A, 7 U/kg) on mechanical allodynia assessed by von Frey filaments in the (**A**) ipsilateral and (**B**) contralateral temporomandibular joint (TMJ) of PIH rats prior to and after i.a. stimulation with 0.5% low-dose formalin. The values were shown as the log-transformed data ($\log_{10}$(value in grams)). The values on the right y-axis represent the dose range of filaments used (in grams) on the corresponding logarithmic scale. N(animals/group) = 10; *** $p < 0.001$ and ** $p < 0.01$ vs. noninduced + saline and +++ $p < 0.001$ vs. induced + saline (two-way repeated measures ANOVA followed by Bonferroni's post hoc test).

2.3. Effect of BoNT-A on Neuronal and Astrocyte Activation in the TNC

We further examined the possible BoNT-A effects on neuronal and astrocyte activation associated with nociception in the trigeminal nucleus caudalis (TNC), the first-order pain processing sensory nucleus of the cranial area. We assessed the post-formalin neuronal activation of second-order nociceptive neurons in the TNC by employing c-Fos immunohistochemistry (Figure 3A,B) and the counting of c-Fosexpressing neuronal profiles (Figure 2C). After a normally nonpainful formalin injection, induced animals presented high levels of c-Fos-positive nuclei in both ipsilateral and contralateral TNC. Unilateral applications of BoNT-A (aboBoNT-A or onaBoNT-A) significantly decreased post-formalin c-Fos expression in both ipsilateral and contralateral TNC (Figure 4B,C).

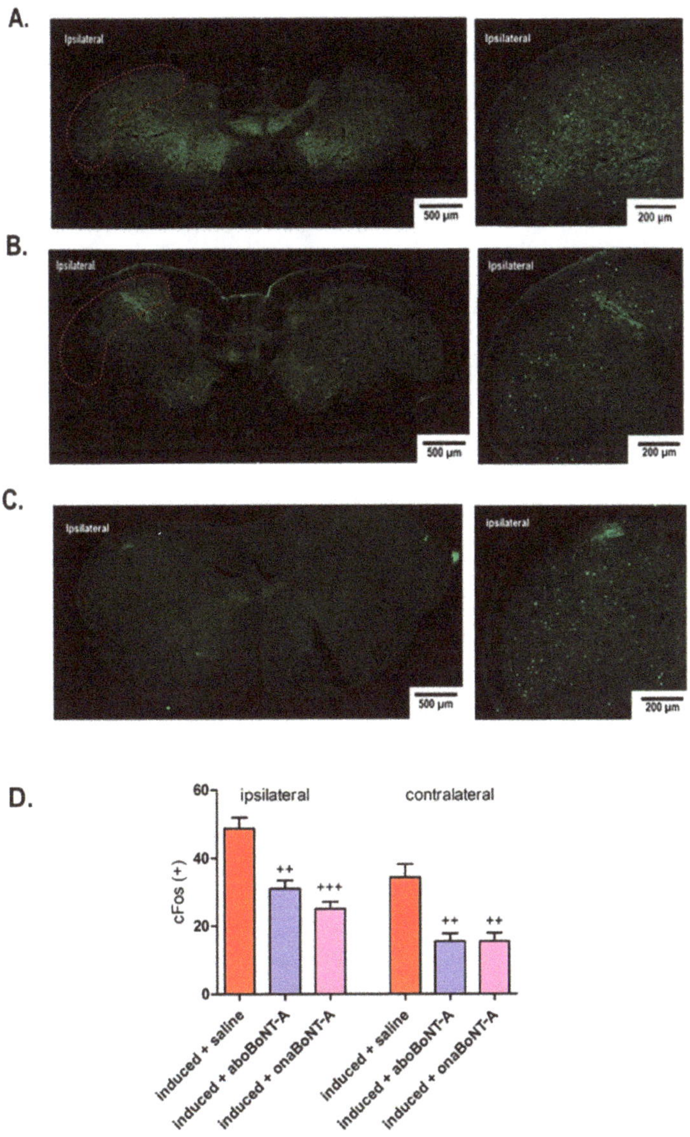

Figure 4. Bilateral effects of intra-articular (i.a.) abobotulinumtoxinA (aboBoNT-A, 14 U/kg) or

onabotulinumtoxinA (onaBoNT-A, 7 U/kg) on neuronal activation (c-Fos green punctate immunoreactivity) in the trigeminal nucleus caudalis (TNC) after low-dose (0.5%) i.a. formalin stimulation of rats exposed to the persistent immunogenic hypersensitivity model in the temporomandibular joint. Immunohistochemical staining of c-Fos activation in the TNC of (**A**) induced + saline, (**B**) induced + aboBoNT-A-injected animals, and (**C**) induced + onaBoNT-A animals. The red dotted line indicates the analyzed TNC area. (**D**) Quantification of c-Fos-expressing neuronal profiles (5 sections/animal, N = 5 animals/treatment group). Mean ± SEM; +++ = $p < 0.001$ and ++ = $p < 0.01$ vs. induced + saline ($F_{5,24} = 20.1$, one-way ANOVA followed by Tukey's post hoc test).

Analysis of TNC showed an ipsilateral increase in astrocytic marker glial fibrillary acidic protein (GFAP) immunoreactivity on induced rats compared to noninduced animals (Figure 5A). BoNT-A reduced the GFAP-immunoreactive area 14 days after treatment (Figure 5B,C). The levels of calcitonin gene-related peptide (CGRP) expression were not increased in the hypernociceptive rats or altered by BoNT-A (aboBoNT-A or onaBoNT-A) treatments (Figure 6).

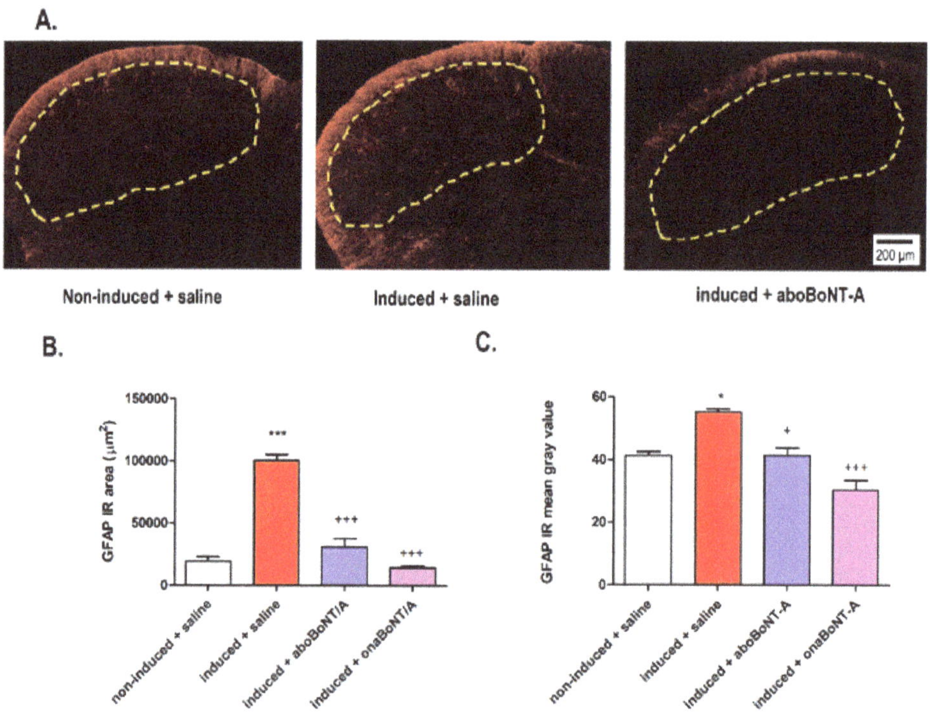

Figure 5. Effects of abobotulinumtoxinA (aboBoNT-A, 15 U/kg) and onabotulinumtoxinA (onaBoNT-A, 7 U/kg) on glial fibrillary acidic protein (GFAP) expression in the trigeminal nucleus caudalis (TNC) of rats exposed to the persistent immunogenic hypersensitivity model in the temporomandibular joint. (**A**) The representative microphotographs of TNC immunostained for GFAP (representative of 5 animals per group). Astrocyte activation was evident as (**B**) an increase in the GFAP-immunoreactive (IR) area and (**C**) increase in the mean intensity (mean gray value) of arthritis-induced animals and prevented by aboBoNT-A and onaBoNT-A. The analysis was performed on 5 randomly selected slices per animal (n = 5 animals/group). The yellow dotted lines indicate the analyzed TNC area. Mean ± SE; *** $p < 0.001$ and * = $p < 0.05$ vs. noninduced + saline and +++ = $p < 0.001$ and + = $p < 0.05$ vs. induced + saline ($F_{3,16}$ (GFAP area) = 73.14; $F_{3,16}$ (GFAP gray value) = 23.21; one-way ANOVA followed by Tukey's multiple comparisons test).

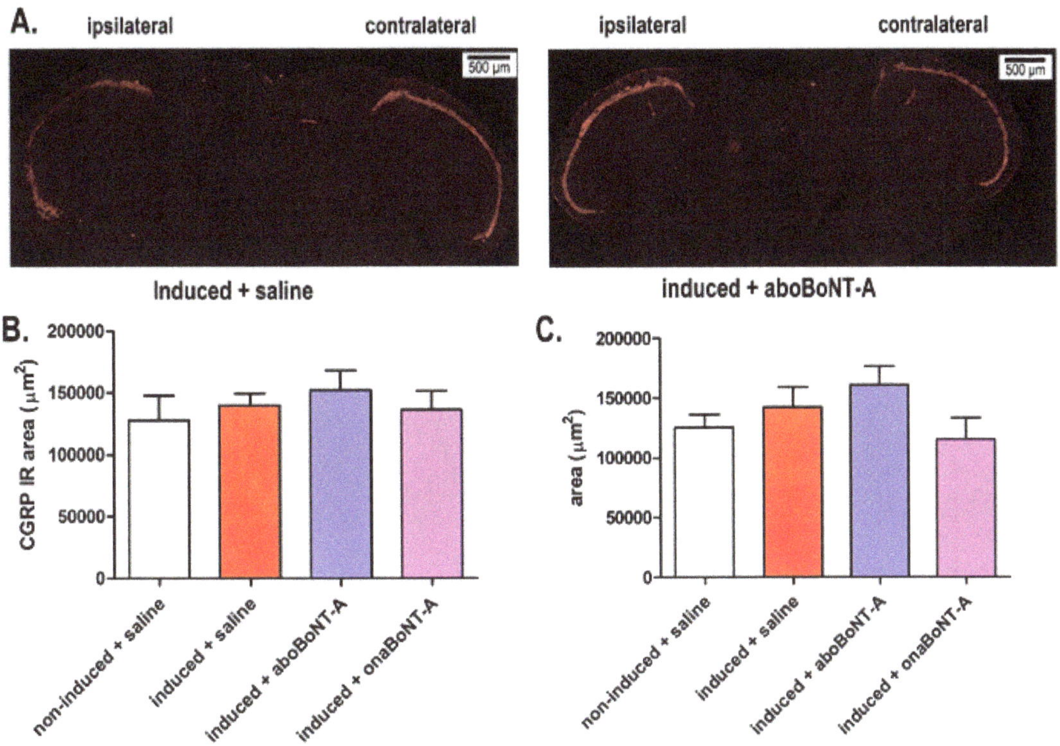

Figure 6. Lack of increase of CGRP immunoreactivity in the trigeminal nucleus caudalis (TNC) of rats exposed to the persistent immunogenic hypersensitivity model in the temporomandibular joint (**A**). Representative images of TNC sections immunostained for CGRP (representative of 5 animals/group). Quantification of the pixel threshold area of CGRP immunoreactivity in the ipsilateral (**B**) and contralateral (**C**) TNC demonstrated lack of changes of CGRP expression. The analysis was performed on 5 randomly selected slices per animal (n (animals) = 5/group). The yellow dotted lines indicate the analyzed TNC area. Mean ± SE; ($F_{3,16}$ (ipsilateral) = 0.3819; F $F_{3,16}$ (contralateral) = 1.267; one-way ANOVA).

2.4. Immunohistochemical Localization of cSNAP-25 in the Brain

Previously, we discovered that the BoNT/A antinociceptive action in the craniofacial area is associated with the occurrence of cleaved SNAP-25 (cSNAP-25) in the TNC following axonal transport [17]. Thus, we further examined if the antinociceptive actions of aboBoNT/A and onaBoNT/A are similarly associated with the direct toxin activity in the TNC. Fourteen days after BoNT-A (aboBoNT-A or onaBoNT-A) i.a. injections into the left TMJ, BoNT-A cSNAP-25-positive nerve fibers were found in the ipsilateral trigeminal sensory regions (TNC). Similar cSNAP-25 staining was observed in both aboBoNT-A and onaBoNT-A-treated animals. cSNAP-25 on the contralateral (nontreated) side was not observed (Figure 7).

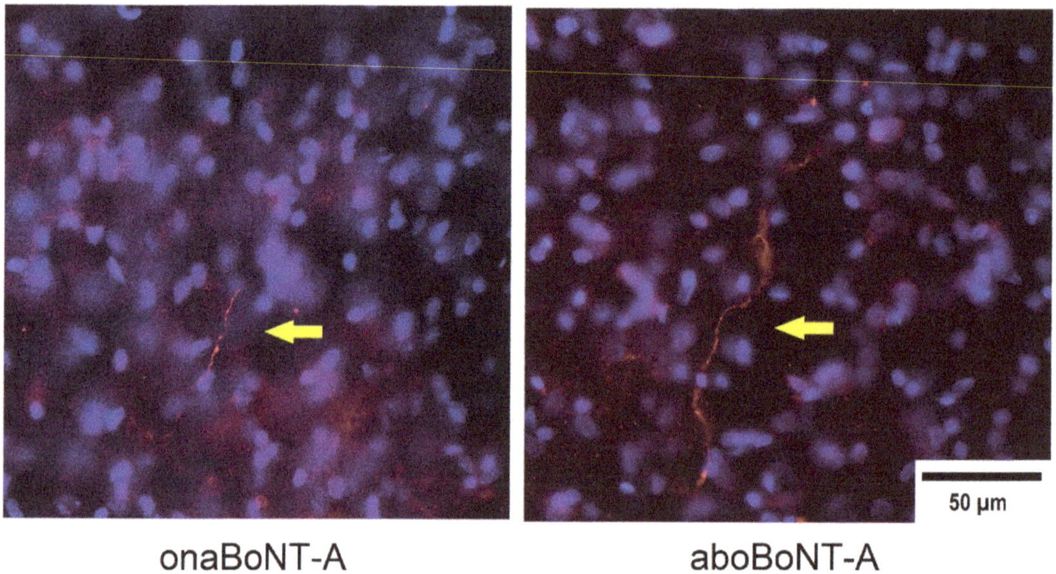

Figure 7. Immunohistochemical staining of cleaved SNAP-25 (cSNAP-25) in the trigeminal nucleus caudalis (TNC) ipsilateral to BoNT-A injections in the temporomandibular joint (TMJ) on day 14 after 7-U/kg onabotulinumtoxinA (onBoNT-A) or 14-U/kg abobotulinumtoxinA (aboBoNT-A) injection. In the epifluorescent microphotographs, yellow arrows indicate cSNAP-25-positive individual fibers (in bright red), while the blue counterstain represents the cell nuclei stained by 4′,6-diamidino-2-phenylindole (DAPI). The sections shown are representative of 10 sections per animal (N = 3/treatment group).

3. Discussion

In rats induced to PIH of the TMJ, we demonstrated that the analgesic activity of BoNT-A, expressed as a reduction of spontaneous and evoked nocifensive behaviors after a low-dose formalin stimulation, is associated with a decrease of neuronal and astroglial activity in the TNC. Moreover, in the same arthritis model, we previously demonstrated that BoNT-A reduced the nocifensive behaviors, as well as the expression of proinflammatory mediators and peptides in the TMJ and trigeminal ganglion [8], in accordance with other studies [18–20]. Notwithstanding, a causal role between BoNT-A antinociceptive activity and its peripheral effects has not been established. Furthermore, it was suggested that the BoNT-A analgesic effect necessarily involves an axonal transport of the toxin to central nervous system (CNS), where it may interact with neuronal and/or non-neuronal cells [15,21].

3.1. Peripherally Injected BoNT-A Reduces Spontaneous and Mechanically Evoked Nocifensive Behaviors in Rats with PIH

Despite the existence of numerous tests for animals' behavior assessment [22], spontaneous behavioral assessments may be more relevant than stimulus-evoked assessments for future clinical translations, increasing the validity of the model [7]. However, the orofacial region remains an underrepresented area in spontaneous pain behavioral assessments. Persistent scratching and rubbing of the facial region after chemical or inflammatory stimulation (e.g., the orofacial formalin test) is used as a sign of pain in freely moving rodents [23,24]. For this reason, we also assessed the spontaneous nociceptive behaviors by employing RGS. Since pain is a well-characterized emotion based on the facial action coding system [25], the RGS is capable of detecting spontaneous nociception [22] and measuring the analgesic efficacy of drugs [26].

The arthritic rats showed no visible signs of evoked or spontaneous pain prior to formalin TMJ stimulation. However, low-dose formalin (0.5%), known not to induce pain-related behaviors in normal animals [27], induced a significant increase of facial grimacing, head flinching, facial rubbing, and bilateral mechanical allodynia (Figures 2 and 3), in accordance with previous studies [8,28]. This suggests that the mBSA-induced monoarthritis is nonpainful per se; however, it induces a profound chemical hypersensitivity to normally nonpainful concentrations of inflammatory substances such as formalin [12]. This exacerbation of nociceptive responses by normally nonpainful chemical stimuli suggests a quick triggering of central sensitization processes [10].

BoNT-A decreased the flinching and rubbing nocifensive responses (Figure 2C) and reduced the facial nociceptive grimacing of rats exposed to PIH in the TMJ. Importantly, the RGS was quantified at three different periods after formalin stimulation in the TMJ. Shortly after the formalin injection (0–5 min), BoNT-A was not effective in RGS reduction, similar to the observed lack of BoNT-A action on immediate formalin-evoked pain sensation (phase I of formalin test) [17,19]. However, at later time points (9–14 and 18–23 min), both BoNT-A groups significantly reduced the spontaneous pain grimacing score (Figure 2B), demonstrating BoNT-A action in the tonic pain phase when central sensitization takes place.

In addition, we assessed the mechanical allodynia by von Frey monofilaments prior and after 0.5% formalin stimulation of the exposed TMJ. Allodynia over the TMJ area in PIH-induced rats was observed only after the 0.5% formalin challenge at both sides (ipsi- and contralateral TMJ). This bilateral allodynia was reduced by a unilateral BoNT-A treatment resembling the toxin's bilateral action on experimental trigeminal neuropathic pain [29] and polyneuropathic pain [16].

3.2. *Effects of BoNT-A on Nociceptive Neuronal and Glial Activation*

Apart from BoNT-A behavioral antinociceptive activity, we examined its action at the level of central sensory nociceptive nuclei by exploiting c-Fos activation as a reliable marker of neuronal activity [23,30]. Interestingly, we found an occurrence of c-Fos activated neurons in the ipsilateral and contralateral TNC after 0.5% formalin stimulation of the TMJ area (Figure 4), which was not observed in the orofacial or hind paw formalin tests [31–33]. The unilateral stimulation of mouse masseter with complete Freund's adjuvant (CFA) similarly induces bilateral expression of c-Fos in the TNC [23]. This suggests that pain in the TMJ region and related muscles may be associated with the bilateral activation of trigeminal nociception, similar to referred pain in humans [34,35]. All these results indicate that the pain triggered by formalin activates contralateral trigeminal nociceptive nuclei, explaining the occurrence of bilateral mechanical allodynia (Figure 3A,B). Importantly, both BoNT-A preparations (aboBoNT-A and onaBoNT-A) injected into the ipsilateral TMJ reduced both bilateral neuronal activation and the bilateral allodynia in the TMJ area.

In addition, we examined the possible changes in the activation of glial cells in the CNS, since they are key to initiate and maintain central sensitization after peripheral tissue/nerve injury [10,36,37]. Astrocytes represent the majority of glial cells and play a significant role in the persistence of pain [38]. Both BoNT-A preparations reduced GFAP (astrocyte marker) immunoreactivity measured by surface area immunoreactivity and grey intensity in the TNC of arthritic rats (Figure 5). An in vitro study showed a slight or no direct effect of BoNT-A on astrocytes [39]. Another in vivo study detected the presence of cSNAP-25 in spinal cord astrocytes of a mouse with induced neuropathy and treated in the hind paw with high-dose (15 pg/mouse) BoNT-A devoid of complexing proteins [40]. The reduced astroglial expression supports the hypothesis that BoNT-A may have a modulatory action on glial cells and their role in central nociceptive sensitization [20]. However, BoNT-A effects on glial cells (microglia and astrocytes) should be more carefully explored under pharmacologically relevant moderate doses.

Finally, we assessed possible changes in the peptidergic content in the central afferent terminals by examining the expression of CGRP in the TNC, which was previously found to be increased in the more acute CFA-evoked unilateral monoarthritis of the TMJ [41]. Herein, we did not observe changes in the CGRP expression in the TNC (Figure 6).

3.3. Localization of cSNAP-25 in TNC

Formerly, it was assumed that BoNT-A actions resided peripherally, and its antinociceptive activity was linked to a local reduction of the synaptic transmitter/proinflammatory peptide release [18,42]. However, its analgesic effects were shown to be mediated by axonal transport and associated with toxin's enzymatic activity in the CNS [15,17,32]. Previously, we discovered that inhibition of the axonal transport within the peripheral nerve or in the sensory ganglion prevent the antinociceptive effect of peripherally injected BoNT-A, while the toxin injections in the peripheral nerve or in the trigeminal ganglion at doses lower than the ones effective in the periphery produce an antinociceptive effect [15,17,29]. Herein, we confirmed the occurrence of BoNT-A (aboBoNT-A and onaBoNT-A) in the brainstem after i.a. injections aimed at the TMJ (Figure 7). cSNAP-25 appeared as fiber-like processes in the ipsilateral TNC, which is the primary pain-processing area for the orofacial nociceptive input [43]. In accordance with previous studies, we observed no cSNAP-25 in the contralateral TNC, in line with our observation that the toxin is axonally transported primarily via sensory neurons ipsilateral to the BoNT-A injection site and their central projections [32].

A possible transcytosis of the toxin within the central neurons has also been proposed [40,44]. However, this was not corroborated in previous studies employing low to moderate non-systemic onaBoNT-A doses (5-15 U/kg), which suggests that BoNT-A enzymatic activity resides in central afferent terminals of transient receptor potential vanilloid 1-expressing neurons [32]. A lack of immunohistochemical evidence for contralateral occurrence of the toxin's enzymatic activity might suggest an unknown indirect analgesic effect of the toxin on the contralateral side. The exact mechanism of the BoNT-A action on contralateral pain processing remains unknown.

3.4. Preclinical Comparison of the Analgesic Efficacy of Different BoNT-A Pharmaceutical Preparations

Currently, there is an intense debate regarding the differences in efficacy of various BoNT-A formulations [45–47]. Although both aboBoNT-A and onaBoNT-A are produced from identical Hall A strain synthesizing the A1 toxin subtype; they differ in the composition of complexing auxiliary proteins (onaBoNT-A is composed of the entire 900-kDa complex, while aboBoNT-A has a variable composition of complexing proteins per neurotoxin molecule), as well as the relative potency of the different products [46]. Despite the possibility that these differences also depend on the amount of active toxins in individual preparations [47], they have rarely been studied in both animals and humans. To compare the effect of aboBoNT-A vs. onaBoNT-A, we employed a 2:1 dose conversion ratio according to Scaglione et al. [45]. Following a study from Field et al., who calculated the picogram content of each international unit (U) (5.38 pg of 150-kDa neurotoxin protein for aboBoNT-A, while 1 U of onaBoNT-A contains 9 pg of 150-kDa protein) [47], approx. 400-g rats were herein injected with similar 150-kDa BoNT-A amounts (30.1 pg/animal of aboBoNT-A and 25.2 pg/animal of onaBoNT-A). Despite some possible differences in the total amount of the neurotoxin injected, we found no significant differences in the antinociceptive activity of any investigated outcome between the two BoNT-A preparations (Figures 2–6). This suggests that both products may have similar analgesic effects [48,49].

3.5. Limitations

As a limitation of the study, we must consider that this experiment is restricted to the trigeminal region, so differences in the mechanism of action of BoNT-A applied to other extracranial joints (e.g., knee and ankle) and actions at the spinal levels cannot be excluded. Additionally, during the period after BoNT-A treatment, the aboBoNT-A- or onaBoNT-A-treated rats did not exhibit a significantly slower weight gain compared to the saline-treated rat groups (p = 0.070, one-way ANOVA, results not shown). This excludes possible systemic effects or a local muscular BoNT-A action (e.g., mastication and feeding) interfering with the animal weight gain.

4. Materials and Methods

Animal procedures were approved by the Ethics Committee of the University of Zagreb School of Medicine (permit NoEP 108/2017) and conducted according to the European Communities Council Directive and the International Association for the Study of Pain guidelines [50]. All efforts were made to reduce the number of animals. All data was presented in accordance with the ARRIVE guidelines [51].

4.1. Persistent Immunogenic Hypersensitivity (PIH) in the TMJ

Forty male Sprague–Dawley rats (300–400 g, age 6–8 weeks at the start of experiment, University of Zagreb School of Medicine, Zagreb, Croatia) were housed in plastic cages with a maximum of 3 animals per cage. Animals were kept in a room with a controlled constant temperature (22 ± 0.5 °C) and a 12 h:12 h light:dark cycle (lights on from 07:00 to 19:00. Food and water were available ad libitum.

Animals were induced to PIH in the TMJ according to previous experiments [8,10] (summarized experimental design on Figure 8). On day 0, the rats were immunized systemically with methylated bovine serum albumin (mBSA, 500 µg; Sigma, St. Louis, MO, USA) used as the antigen, diluted in an emulsion containing immunologic response enhancer complete Freund´s adjuvant (CFA, 100 µL; Sigma, St. Louis, MO, USA) containing inactivated *Mycobacterium tuberculosis* in mineral oil, and 100-µL phosphate-buffered saline (PBS) injected subcutaneously (s.c.) into their back (day 0). On days 7 and 14, animals received further booster s.c. injections (aimed at other parts of the back) containing additional 500-µg doses of mBSA, in which CFA was replaced by incomplete Freund´s adjuvant (IFA, 100 µL; Sigma, St. Louis, MO, USA) devoid of inactivated Mycobacterium. After systemic immunization with mBSA and CFA/IFA, to induce a more localized TMJ monoarthritis, the animals further received unilateral (left) intra-TMJ injections of the low-dose antigen (mBSA; 10 µg, i.a.) dissolved in PBS (15 µL) on days 21, 28, and 35.

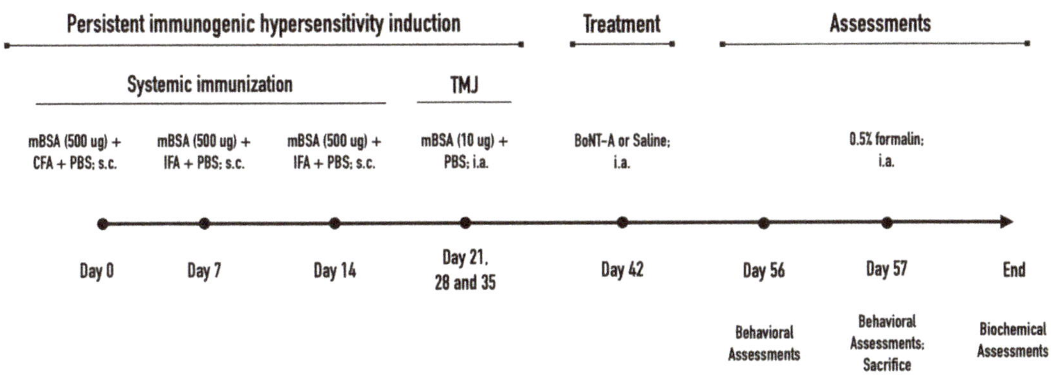

Figure 8. Experimental design indicating the time points of the persistent immunogenic hypersensitivity model, applied treatments, and assessments. mBSA, methylated bovine serum albumin; CFA, complete Freund's adjuvant; IFA, incomplete Freund's adjuvant; PBS, phosphate-buffered solution; BoNT-A, botulinum toxin type A1 (abobotulinumtoxinA or onabotulinumtoxinA); TMJ, temporomandibular joint; s.c., subcutaneous; i.a., intra-articular.

In comparison to PIH-induced animals, the control animals (noninduced to PIH) were similarly treated with mBSA; however, they were not subjected to systemic immunologic adjuvants CFA and IFA. These rats received a s.c. injection of mBSA (500 μg) diluted only in phosphate-buffered saline (PBS, 100 μL) on days 0, 7, and 14. Then, similar to PIH rats, on days 21, 28, and 35, the animals received a unilateral (left) intra-TMJ injection of sterile mBSA (10 μg; i.a.) dissolved in PBS (15 μL).

4.2. Study Design

Animals were randomly allocated into four different experimental groups (n = 10). The first group included animals noninduced to PIH that received i.a. injection of saline on day 42. The second group included animals induced to PIH and treated with saline solution into the TMJ on day 42. The third and the fourth groups included animals induced to PIH and treated with either abobotulinumtoxinA or onabotulinumtoxinA into the TMJ on day 42.

AbobotulinumtoxinA (aboBoNT-A; Dysport®, Ipsen, Wrexham, UK) or onabotulinumtoxinA (onaBoNT-A; Botox®, Allergan Inc., Irvine, CA, USA) was injected i.a. into the ipsilateral TMJ (left side) of anesthetized animals (20 μL, saline-diluted, using a 30-gauge needle) on day 42 of the experiment (Figure 1). The applied doses (14 U/kg for aboBoNT-A, equivalent to 30.1 pg/400 g animal and 7 U/kg for onaBoNT-A, equivalent to 25.2 pg/400 g animal) were chosen according to previously employed doses in our laboratory [8] and a conventional dose conversion ratio between aboBoNT-A and onaBoNT-A [45–47]. Sterile saline solution (NaCl 0.9%) was used as a control treatment for noninduced and induced groups.

4.3. Spontaneous and Mechanically Evoked Nocifensive Behaviors

All behavioral evaluations were conducted before (day 13; pre-formalin) and after a low dose of intra-TMJ formalin injection (day 14, post-formalin) in alignment with the time points used in our previous experiment (Figure 1) [8]. Animals were allowed to accommodate for 10 min inside the testing cages before each evaluation in a quiet environment with the appropriate lighting (between 9 a.m. and 4 p.m.). For the pre-formalin assessments, animals were placed directly in the testing cage. For the post-formalin assessment, animals were briefly anesthetized with isoflurane (5% induction; 2.5% maintenance), injected i.a. with low-dose formalin (0.5%; 15 μL), and immediately returned to the testing cage to recover (recovery time: 1 to 2 min) and to perform the behavioral evaluations in the following order: (1) behavioral nocifensive responses and rat grimace scale evaluated simultaneously and (2) mechanical sensitivity using von Frey filaments. The experimenter who performed the behavioral assessments was blinded to all treatments.

4.3.1. Behavioral Nocifensive Responses

The spontaneous nocifensive responses were assessed for 45 min, as described by previous studies [8,12]. The assessment of the behavioral nocifensive responses was calculated by the sum of seconds the animal spent rubbing or scratching the orofacial region (assessed live by a stopwatch) plus the total number of head flinches (1 flinch = 1 s).

4.3.2. The Rat Grimace Scale (RGS)

A video camera was placed in the upper part of the testing cage in a position that allowed a complete view of each animal. Rats were continuously recorded during the evaluation of their behavioral nocifensive responses (starting after awakening from isoflurane anesthesia) divided across 3 blocks of 5 min with a 4-min interval between each block (0–5′, 9–14′, and 18–23′). After recording, the videos were analyzed by an experimenter blinded to the treatment. A 0–2 score (0 = not present, 1 = moderate, and 2 = obvious pain) was assigned to each facial parameter (orbital tightening, nose/cheek flattening, ear changes, and whisker changes) observed in each 5-min block, as previously described [52]. A mean of the scores in all blocks was considered the total RGS score.

4.3.3. Mechanical Sensitivity to von Frey Filaments

Mechanical allodynia was assessed immediately after the completion of the assessment of spontaneous behavioral nociceptive responses by using von Frey monofilaments (Stoeling Co., Wood Dale, IL, USA). The assessment of mechanical allodynia lasted for 5–10 min. The test was performed as previously described [29]. The filaments (flat contact area and weights in grams: 0.16, 0.4, 0.6, 1, 1.4, 2, 4, 6, 8, and 10) were applied bilaterally over the TMJ area, starting with the contralateral nontreated TMJ (right TMJ). In each session, each filament was applied three times (starting with 0.16-g filament and continuing in ascending order) until a repeated positive response (escape and/or defensive movement) after the stimulation of the evaluated area was elicited. The maximum weight (10 g) was assigned when no response was observed.

4.4. Immunohistochemistry

On day 57, immediately after behavioral assessments, all animals were deeply anesthetized (ketamine 70 mg/kg and xylazine 7 mg/kg, intraperitoneally (i.p.)) and transcardially perfused with saline followed by buffered 4% paraformaldehyde (PFA) in PBS (pH 7.4). The brain was dissected and placed in the PFA solution overnight (18–20 h) at +4 °C. Then, the samples were transferred to 30% sucrose dissolved in PBS for 48h at +4 °C. Then, all samples were stored at −80 °C prior to further use. The caudal part of the medulla oblongata containing the trigeminal nucleus caudalis (TNC; from −5.28 mm to −6.72 interaural, based on Paxinos and Watson [53]) was cut into 35-μm coronal cross-sections using a cryostat. The position of the sections relative to the bregma/interaural line was determined based on visible landmarks such as the obex, position of the central canal, and shape of the white and gray matter, according to the rat brain atlas [53]. For each immunohistochemical analysis, 5 sections per animal were randomly chosen.

4.4.1. Localization of cSNAP-25 in the Brainstem

Sections corresponding to the TNC region of the brainstem were randomly selected, washed in PBS with 0.25% Triton X-100 (PBST), and exposed to 50 μL of 3% peroxide diluted in 200-μL PBS for 1h. Sections were washed again and blocked with 10% normal goat serum (NGS) in PBST. A well-characterized rabbit polyclonal antibody that specifically cleaves the BoNT-A-truncated SNAP-25 (nonaffinity-purified rabbit antiserum (anti-SNAP-25_{197}, National Institute for Biological Standards and Control, Potters Bar, UK, a generous gift from Dr. Thea Sesardic) was diluted in 1% NGS (1:4000), and samples were incubated overnight at room temperature. The next day, the sections were washed, and 200 μL of Alexa Fluor 555 secondary antibody (1:400, Molecular Probes, Invitrogen, Carlsbad, CA, USA) was used to enhance the cSNAP-25 signal. Samples were then washed 3 times before mounting into the glass slide and coverslipped with an antifading agent. All samples were visualized with an epifluorescent microscope (Olympus BX-51, Tokyo, Japan). Fiber-shaped staining was searched, and pictures were taken using a digital camera (DP-70, Olympus, Tokyo, Japan).

4.4.2. Quantification of Neuronal and Astrocyte Activation in the Brainstem

The brainstem sections of each animal containing the TNC were randomly selected for immunohistochemical analyses. The sections were placed in free-floating plate wells in PBS with PBST and washed and blocked with 10% NGS in PBST. Sections of TNC were incubated overnight at room temperature in 1% NGS with either anti-c-Fos (1:500, Santa Cruz Biotechnology, Dallas, TX, USA), anti-calcitonin gene-related peptide (CGRP, 1:10,000, Sigma, St Louis, MO, USA), or anti-glial fibrillary acidic protein (anti-GFAP; 1:600, Sigma, St. Louis, MO, USA) antibodies. The next day, the sections were incubated with fluorescent secondary antibody (goat anti-rabbit Alexa Fluor 555; Molecular Probes, Invitrogen, Carlsbad, CA, USA/goat anti-rabbit Alexa Fluor 488; Invitrogen, Carlsbad, CA, USA). Sections were washed with PBS mounted on glass slides, and coverslipped with an antifading agent.

Immunostained sections were visualized with an Olympus BX-51 epifluorescent microscope coupled to a DP-70 digital camera (Olympus, Tokyo, Japan), and brain regions were identified using the rat stereotaxic atlas [53]. All images were processed and quantified by using cellSens Dimension Software (Olympus, Tokyo, Japan). The c-Fos-positive neurons were automatically counted in the ipsilateral and contralateral side of the dorsal horns of each section (TNC). For the acquisition of GFAP, microphotographs of each coronal section containing the TNC was taken ipsilaterally at a 10× objective magnification and 0.5× camera adapter, while for CGRP quantification, we employed 4× objective magnification and a 0.5× camera adapter. A constant exposure was used to acquire individual microphotographs for each subsequent quantification, as well as the manual thresholds used for calculation of the object number and immunoreactivity areas. The immunohistochemical quantification of the area of immunoreactivity, as well as the number of objects within the area of interest, were based on manual pixel thresholding of grey images obtained from separated RGB channels: red or green based on the fluorophore used (pixel threshold range 115–256 for GFAP and c-Fos images taken at 10× objective magnification; range 62–256 for CGRP microphotographs taken at 4× objective magnification).

4.5. Statistical Analysis

Quantitative data were presented as the mean ± standard error of the mean (SEM). Data from repeated behavioral measurements were analyzed using the two-way repeated measures analysis of variance (RM ANOVA), followed by Bonferroni's post hoc test to compare the intertreatment effects. Single time observations were analyzed by one-way ANOVA, followed by Tukey's post hoc test. The data representing mechanical sensitivity, assessed by Von Frey filaments, were log-transformed prior to statistical analysis, according to the fact that mechanical sensation is perceived on a logarithmic scale (Weber's law) [54].

All statistical analyses were performed with GraphPad Prism version 5 (GraphPad Software, Inc., San Diego, CA, USA). A p-value < 0.05 was considered significant.

5. Conclusions

Based on the present data and previous results [8], the analgesic activity of BoNT-A in rats exposed to a PIH in the TMJ seems to be associated with peripheral and central sites of BoNT-A action. The present results suggest that BoNT-A action is associated with a reduction of spontaneous and evoked nociceptive pain measures suggestive of central sensitization (reduction of pain grimacing in the later assessment phase similar to the BoNT-A action on phase II of the formalin test and reduction of bilateral mechanical allodynia). In addition, we found that the antinociceptive effect of BoNT-A is accompanied by reduction of neuronal and astrocytic activation in the TNC, both effects associated with central pain processing. The mentioned actions did not significantly differ with respect to different pharmacological preparations of the toxin (aboBoNT-A vs. onaBoNT-A). In addition, the antinociceptive activity of aboBoNT-A and onaBoNT-A was accompanied by the occurrence of central cSNAP-25, suggestive of the toxin's direct central action. However, the relative importance of each site of action (peripheral vs. central) for this effect remains to be further investigated in the present pain model.

Author Contributions: V.R.M.M.-L.: substantial contributions to conception, design, acquisition, analysis, and interpretation of the data; drafting the article; and final approval of the version to be published. A.D.O. substantial contributions to acquisition of the data. I.M.: substantial contributions to conception, design, acquisition, analysis, and interpretation of the data; revising critically the article; and final approval of the version to be published. A.A.D.B.C.: substantial contributions to conception, design, analysis, and interpretation of the data; revising critically the article; and final approval of the version to be published. M.K.: substantial contributions to revising critically the article, and final approval of the version to be published. Z.L.: substantial contributions to conception, design, acquisition, analysis, and interpretation of the data; revising the article critically for important intellectual content; general supervision of the research group; and final approval of the version to be published. All authors have read and agreed to the published version of the manuscript.

Funding: This study was funded by Ipsen Innovation, 91940 Les Ulis, France. V.R.M. Muñoz-Lora was a guest Ph.D. student at the University of Zagreb School of Medicine and received a scholarship from the São Paulo Research Foundation (FAPESP; BEPE scholarship 2018/13575-5, regular scholarship 2017/07741-7).

Institutional Review Board Statement: Animal procedures were approved by the Ethics Committee of the University of Zagreb School of Medicine and Croatian Ministry of Agriculture (permit No EP 108/2017) and conducted according to the European Communities Council Directive and the International Association for the Study of Pain guidelines. All efforts were made to reduce the number of animals. All data were presented in accordance with the ARRIVE guidelines.

Informed Consent Statement: Not applicable.

Data Availability Statement: The data presented in this study are available on request from the corresponding author.

Acknowledgments: The authors express thanks to Thea Sesardic from the National Institute for Biological Standards and Control (NIBSC; Hertfordshire, UK) for the kind gift of the cSNAP-25 antibody used in this study.

Conflicts of Interest: This study was funded by Ipsen. Mikhail Kalinichev is an employee of Ipsen Innovation, France. All other authors declare no conflicts of interest.

References

1. Alstergren, P.; Pigg, M.; Kopp, S. Clinical diagnosis of temporomandibular joint arthritis. *J. Oral Rehabil.* **2018**, *45*, 269–281. [CrossRef] [PubMed]
2. Ahmed, N.; Catrina, A.I.; Alyamani, A.O.; Mustafa, H.; Alstergren, P. Deficient cytokine control modulates temporomandibular joint pain in rheumatoid arthritis. *Eur. J. Oral Sci.* **2015**, *123*, 235–241. [CrossRef] [PubMed]
3. Ruparelia, P.B.; Shah, D.S.; Ruparelia, K.; Sutaria, S.P.; Pathak, D. Bilateral TMJ involvement in rheumatoid arthritis. *Case Rep. Dent.* **2014**, *2014*, 262430. [CrossRef]
4. Romero-Reyes, M.; Uyanik, J.M. Orofacial pain management: Current perspectives. *J. Pain Res.* **2014**, *7*, 99–115. [CrossRef]
5. International Headache Society. International Classification of Orofacial Pain Committee International Classification of Orofacial Pain, 1st edition (ICOP). *Cephalalgia* **2020**, *40*, 129–221. [CrossRef] [PubMed]
6. Sodhi, A.; Naik, S.; Pai, A.; Anuradha, A. Rheumatoid arthritis affecting temporomandibular joint. *Contemp. Clin. Dent.* **2015**, *6*, 124–127. [CrossRef]
7. Mogil, J.S. Animal models of pain: Progress and challenges. *Nat. Rev. Neurosci.* **2009**, *10*, 283–294. [CrossRef] [PubMed]
8. Muñoz-Lora, V.R.M.; Clemente-Napimoga, J.T.; Ballassini, H.; Gomes-Macedo, C.; De la Torre Canales, G.; Rizzatti-Barbosa, C.M. Botulinum toxin type A reduces inflammatory hypernociception induced by arthritis in the temporomadibular joint of rats. *Toxicon* **2017**, *129*, 52–57. [CrossRef]
9. Leuchtweis, J.; Segond von Banchet, G.; Eitner, A.; Ebbinghaus, M.; Schaible, H.-G. Pain-related behaviors associated with persistence of mechanical hyperalgesia after antigen-induced arthritis in rats. *Pain* **2020**, *161*, 1571–1583. [CrossRef]
10. Bonfante, R.; Napimoga, M.H.; Macedo, C.G.; Abdalla, H.B.; Pieroni, V.; Clemente-Napimoga, J.T. The P2X7 receptor, cathepsin S and fractalkine in the trigeminal subnucleus caudalis signal persistent hypernociception in temporomandibular rat joints. *Neuroscience* **2018**, *391*, 120–130. [CrossRef]
11. de Sousa, L.M.; dos Santos Alves, J.M.; da Silva Martins, C.; Pereira, K.M.A.; Goes, P.; Gondim, D.V. Immunoexpression of canonical Wnt and NF-κB signaling pathways in the temporomandibular joint of arthritic rats. *Inflamm. Res.* **2019**, *68*, 889–900. [CrossRef] [PubMed]
12. Roveroni, R.C.; Parada, C.A.; Cecília, M.; Veiga, F.A.; Tambeli, C.H. Development of a behavioral model of TMJ pain in rats: The TMJ formalin test. *Pain* **2001**, *94*, 185–191. [CrossRef]
13. Atkinson, S.M.; Nansen, A. Pharmacological value of murine delayed-type hypersensitivity arthritis: A robust mouse model of rheumatoid arthritis in C57BL/6 mice. *Basic Clin. Pharmacol. Toxicol.* **2017**, *120*, 108–114. [CrossRef] [PubMed]
14. Billiau, A.; Matthys, P. Modes of action of Freund's adjuvants in experimental models of autoimmune diseases. *J. Leukoc. Biol.* **2001**, *70*, 849–860. [CrossRef]
15. Bach-Rojecky, L.; Lacković, Z. Central origin of the antinociceptive action of botulinum toxin type A. *Pharmacol. Biochem. Behav.* **2009**, *94*, 234–238. [CrossRef]
16. Favre-Guilmard, C.; Chabrier, P.-E.; Kalinichev, M. Bilateral analgesic effects of abobotulinumtoxinA (Dysport®) following unilateral administration in the rat. *Eur. J. Pain* **2017**, *21*, 927–937. [CrossRef]
17. Matak, I.; Bach-Rojecky, L.; Filipović, B.; Lacković, Z. Behavioral and immunohistochemical evidence for central antinociceptive activity of botulinum toxin A. *Neuroscience* **2011**, *186*, 201–207. [CrossRef]
18. Rapp, D.E.; Turk, K.W.; Bales, G.T.; Cook, S.P. Botulinum toxin type a inhibits calcitonin gene-related peptide release from isolated rat bladder. *J. Urol.* **2006**, *175*, 1138–1142. [CrossRef]

19. Cui, M.; Khanijou, S.; Rubino, J.; Aoki, K.R. Subcutaneous administration of botulinum toxin A reduces formalin-induced pain. *Pain* **2004**, *107*, 125–133. [CrossRef]
20. Chuang, Y.-C.; Yoshimura, N.; Huang, C.-C.; Wu, M.; Chiang, P.-H.; Chancellor, M.B. Intravesical botulinum toxin A administration inhibits COX-2 and EP4 expression and suppresses bladder hyperactivity in cyclophosphamide-induced cystitis in rats. *Eur. Urol.* **2009**, *56*, 159–167. [CrossRef]
21. Rojewska, E.; Piotrowska, A.; Popiolek-Barczyk, K.; Mika, J. Botulinum toxin type A—A modulator of spinal neuron–glia interactions under neuropathic pain conditions. *Toxins* **2018**, *10*, 145. [CrossRef]
22. Deuis, J.R.; Dvorakova, L.S.; Vetter, I. Methods used to evaluate pain behaviors in rodents. *Front. Mol. Neurosci.* **2017**, *10*, 284. [CrossRef]
23. Romero-Reyes, M.; Akerman, S.; Nguyen, E.; Vijjeswarapu, A.; Hom, B.; Dong, H.-W.; Charles, A.C. Spontaneous behavioral responses in the orofacial region: A model of trigeminal pain in mouse. *Headache J. Head Face Pain* **2013**, *53*, 137–151. [CrossRef] [PubMed]
24. Vos, B.P.; Hans, G.; Adriaensen, H. Behavioral assessment of facial pain in rats: Face grooming patterns after painful and non-painful sensory disturbances in the territory of the rats infraorbital nerve. *Pain* **1998**, *76*, 173–178. [CrossRef]
25. Ekman, P.; Friesen, W.V. *Facial Action Coding System: A Technique for the Measurement of Facial Movement*; Consulting Psychologists Press: Palo Alto, CA, USA, 1978.
26. Langford, D.J.; Bailey, A.L.; Chanda, M.L.; Clarke, S.E.; Drummond, T.E.; Echols, S.; Glick, S.; Ingrao, J.; Klassen-Ross, T.; Lacroix-Fralish, M.L.; et al. Coding of facial expressions of pain in the laboratory mouse. *Nat. Methods* **2010**, *7*, 447–449. [CrossRef] [PubMed]
27. Clavelou, P.; Dallel, R.; Orliaguet, T.; Woda, A.; Raboisson, P. The orofacial formalin test in rats: Effects of different formalin concentrations. *Pain* **1995**, *62*, 295–301. [CrossRef]
28. Quinteiro, M.S.; Napimoga, M.H.; Mesquita, K.P.; Clemente-Napimoga, J.T. The indirect antinociceptive mechanism of 15d-PGJ2 on rheumatoid arthritis-induced TMJ inflammatory pain in rats. *Eur. J. Pain* **2012**, *16*, 1106–1115. [CrossRef]
29. Filipović, B.; Matak, I.; Bach-Rojecky, L.; Lacković, Z. Central action of peripherally applied botulinum toxin type A on pain and dural protein extravasation in rat model of trigeminal neuropathy. *PLoS ONE* **2012**, *7*, e29803. [CrossRef]
30. Bullitt, E. Expression ofC-fos-like protein as a marker for neuronal activity following noxious stimulation in the rat. *J. Comp. Neurol.* **1990**, *296*, 517–530. [CrossRef]
31. Drinovac, V.; Bach-Rojecky, L.; Matak, I.; Lacković, Z. Involvement of μ-opioid receptors in antinociceptive action of botulinum toxin type A. *Neuropharmacology* **2013**, *70*, 331–337. [CrossRef]
32. Matak, I.; Rossetto, O.; Lacković, Z. Botulinum toxin type A selectivity for certain types of pain is associated with capsaicin-sensitive neurons. *Pain* **2014**, *155*, 1516–1526. [CrossRef] [PubMed]
33. Matak, I.; Tékus, V.; Bölcskei, K.; Lacković, Z.; Helyes, Z. Involvement of substance P in the antinociceptive effect of botulinum toxin type A: Evidence from knockout mice. *Neuroscience* **2017**, *358*, 137–145. [CrossRef] [PubMed]
34. Exposto, F.G.; Udagawa, G.; Naganawa, T.; Svensson, P. Comparison of masseter muscle referred sensations after mechanical and glutamate stimulation. *Pain* **2018**, *159*, 2649–2657. [CrossRef]
35. Masuda, M.; Iida, T.; Exposto, F.; Baad-Hansen, L.; Kawara, M.; Komiyama, O.; Svensson, P. Referred pain and sensations evoked by standardized palpation of the masseter muscle in healthy participants. *J. Oral Facial Pain Headache* **2018**, *32*, 159–166. [CrossRef] [PubMed]
36. Ji, R.-R.; Berta, T.; Nedergaard, M. Glia and pain: Is chronic pain a gliopathy? *Pain* **2013**, *154*, S10–S28. [CrossRef]
37. Scholz, J.; Woolf, C.J. The neuropathic pain triad: Neurons, immune cells and glia. *Nat. Neurosci.* **2007**, *10*, 1361–1368. [CrossRef] [PubMed]
38. Li, T.; Chen, X.; Zhang, C.; Zhang, Y.; Yao, W. An update on reactive astrocytes in chronic pain. *J. Neuroinflamm.* **2019**, *16*, 140. [CrossRef]
39. Piotrowska, A.; Kwiatkowski, K.; Rojewska, E.; Makuch, W.; Mika, J. Maraviroc reduces neuropathic pain through polarization of microglia and astroglia—Evidence from in vivo and in vitro studies. *Neuropharmacology* **2016**, *108*, 207–219. [CrossRef]
40. Marinelli, S.; Vacca, V.; Ricordy, R.; Uggenti, C.; Tata, A.M.; Luvisetto, S.; Pavone, F. the analgesic effect on neuropathic pain of retrogradely transported botulinum neurotoxin A involves Schwann cells and astrocytes. *PLoS ONE* **2012**, *7*, e47977. [CrossRef]
41. Lacković, Z.; Filipović, B.; Matak, I.; Helyes, Z. Activity of botulinum toxin type A in cranial dura: Implications for treatment of migraine and other headaches. *Br. J. Pharmacol.* **2016**, *173*, 279–291. [CrossRef]
42. Aoki, K.R. Pharmacology and immunology of botulinum toxin serotypes. *J. Neurol.* **2001**, *248*, I3–I10. [CrossRef] [PubMed]
43. Casatti, C.A.; Frigo, L.; Bauer, J.A. Origin of sensory and autonomic innervation of the rat temporomandibular joint: A retrograde axonal tracing study with the fluorescent dye fast blue. *J. Dent. Res.* **1999**, *78*, 776–783. [CrossRef] [PubMed]
44. Antonucci, F.; Rossi, C.; Gianfranceschi, L.; Rossetto, O.; Caleo, M. Long-distance retrograde effects of botulinum neurotoxin A. *J. Neurosci.* **2008**, *28*, 3689–3696. [CrossRef] [PubMed]
45. Scaglione, F. Conversion ratio between Botox®, Dysport®, and Xeomin® in clinical practice. *Toxins* **2016**, *8*, 65. [CrossRef] [PubMed]
46. Walker, T.J.; Dayan, S.H. Comparison and overview of currently available neurotoxins. *J. Clin. Aesthet. Dermatol.* **2014**, *7*, 31–39.

47. Field, M.; Splevins, A.; Picaut, P.; van der Schans, M.; Langenberg, J.; Noort, D.; Foster, K. AbobotulinumtoxinA (Dysport®), OnabotulinumtoxinA (Botox®), and IncobotulinumtoxinA (Xeomin®) neurotoxin content and potential implications for duration of response in patients. *Toxins* **2018**, *10*, 535. [CrossRef]
48. Wohlfarth, K.; Schwandt, I.; Wegner, F.; Jürgens, T.; Gelbrich, G.; Wagner, A.; Bogdahn, U.; Schulte-Mattler, W. Biological activity of two botulinum toxin type A complexes (Dysport and Botox) in volunteers: A double-blind, randomized, dose-ranging study. *J. Neurol.* **2008**, *255*, 1932–1939. [CrossRef]
49. Kollewe, K.; Mohammadi, B.; Dengler, R.; Dressler, D. Hemifacial spasm and reinnervation synkinesias: Long-term treatment with either Botox or Dysport. *J. Neural Transm.* **2010**, *117*, 759–763. [CrossRef]
50. Zimmermann, M. Ethical guidelines for investigations of experimental pain in conscious animals. *Pain* **1983**, *16*, 109–110. [CrossRef]
51. Kilkenny, C.; Browne, W.J.; Cuthi, I.; Emerson, M.; Altman, D.G. Improving bioscience research reporting: The ARRIVE guidelines for reporting animal research. *Vet. Clin. Pathol.* **2012**, *41*, 27–31. [CrossRef]
52. Sotocinal, S.G.; Sorge, R.E.; Zaloum, A.; Tuttle, A.H.; Martin, L.J.; Wieskopf, J.S.; Mapplebeck, J.C.S.; Wei, P.; Zhan, S.; Zhang, S.; et al. The Rat Grimace Scale: A partially automated method for quantifying pain in the laboratory rat via facial expressions. *Mol. Pain* **2011**, *7*, 55. [CrossRef] [PubMed]
53. Paxinos, G.; Watson, C. *The Rat Brain in Stereotaxic Coordinates*, 5th ed.; Elsevier Academic: Burlington, MA, USA, 2005.
54. Mills, C.; Leblond, D.; Joshi, S.; Zhu, C.; Hsieh, G.; Jacobson, P.; Meyer, M.; Decker, M. Estimating efficacy and drug ED50's using von Frey thresholds: Impact of weber's law and log transformation. *J. Pain* **2012**, *13*, 519–523. [CrossRef] [PubMed]

toxins

MDPI

Article

Intramuscular Neural Distribution of the Serratus Anterior Muscle: Regarding Botulinum Neurotoxin Injection for Treating Myofascial Pain Syndrome

Kyu-Ho Yi [1,2,†], Ji-Hyun Lee [2,†] and Hee-Jin Kim [2,3,*]

1 COVID-19 Division, Wonju City Public Health Center, Wonju-si 26417, Korea; kyuho90@daum.net
2 Division in Anatomy and Developmental Biology, Department of Oral Biology, Human Identification Research Institute, BK21 FOUR Project, College of Dentistry, Yonsei University, Seoul 03722, Korea; jhlee119@yuhs.ac
3 Department of Materials Science & Engineering, College of Engineering, Yonsei University, Seoul 03722, Korea
* Correspondence: hjk776@yuhs.ac
† These authors contributed equally to this work.

Abstract: The serratus anterior muscle is commonly involved in myofascial pain syndrome and is treated with many different injective methods. Currently, there is no definite injection point for the muscle. This study provides a suggestion for injection points for the serratus anterior muscle considering the intramuscular neural distribution using the whole-mount staining method. A modified Sihler method was applied to the serratus anterior muscles (15 specimens). The intramuscular arborization areas were identified in terms of the anterior (100%), middle (50%), and posterior axillary line (0%), and from the first to the ninth ribs. The intramuscular neural distribution for the serratus anterior muscle had the largest arborization patterns in the fifth to the ninth rib portion of between 50% and 70%, and the first to the fourth rib portion had between 20% and 40%. These intramuscular neural distribution-based injection sites are in relation to the external anatomical line for the frequently injected muscles to facilitate the efficiency of botulinum neurotoxin injections. Lastly, the intramuscular neural distribution of serratus anterior muscle should be considered in order to practice more accurately without the harmful side effects of trigger-point injections and botulinum neurotoxin injections.

Keywords: myofascial pain syndrome; Sihler's method; serratus anterior; trigger point injection

Key Contribution: The research provides intramuscular distribution of the serratus anterior regarding botulinum neurotoxin injection for serratus anterior muscles.

Citation: Yi, K.-H.; Lee, J.-H.; Kim, H.-J. Intramuscular Neural Distribution of the Serratus Anterior Muscle: Regarding Botulinum Neurotoxin Injection for Treating Myofascial Pain Syndrome. *Toxins* **2022**, *14*, 271. https://doi.org/10.3390/toxins14040271

Received: 16 February 2022
Accepted: 31 March 2022
Published: 11 April 2022

1. Introduction

Myofascial pain syndrome (MPS) is extremely common, occurring in up to 95% of individuals [1]. Repetitive movements and incorrect posturing habits contribute to the advancement of MPS by triggering overload on a particular muscle; the serratus anterior (SA) muscle is the most commonly involved [2]. As a part of MPS, serratus anterior myofascial pain syndrome (SAMPS) is separately named due to its frequency [2]. Points with taut banded parts and pinched tenderness of the muscle belly are termed myofascial trigger points (MTrPs). SAMPS occurs with deep respiratory distress while running, repetitive coughing due to respiratory disease, lifting heavy loads, and other psychological stresses [3].

The cause of SAMPS is hyperactivated SA muscle contractions [4–6]. Pathological findings indicate an increase in the release of acetylcholine by the neuromuscular junction under relaxing conditions. Elevated and prolonged acetylcholine release generates persistent depolarization of the muscle fiber, which causes sarcomere shortening and involuntary muscle contraction [2]. This point is anatomically known to be the thickest muscle belly, with the most intramuscular neural arborization [7–11].

The therapeutic options for MPS include releasing MTrPs using injective agents such as botulinum neurotoxin (BoNT), lidocaine, steroids, normal saline, and combinations of agents. BoNT blocks neural transmission by stalling the release of acetylcholine at the neuromuscular junction and impedes muscle contraction [12]. In myofascial pain control, BoNT injection is renowned for offering better consequences than oral medications in terms of pain management and functional movement [13–15]. Therefore, injection of BoNT is widely used as a treatment option for MPS, especially SAMPS [16–21].

At present, BoNT injection is acknowledged as the most secure and effective treatment for inactivating the muscle [22–25]. The consequences of BoNT depend on uptake by the presynaptic membranes at the neuromuscular junction; thus, injections should be directed into the neuromuscular junction area where most neuromuscular junctions exist [12,26,27]. The significance of utilizing neuromuscular arborization-directed BoNT injections has been verified by clinical trials in the iliopsoas and bicep brachii muscles. These injections resulted in higher pain reduction as well as volume reduction compared to conventional injections [26,27].

However, intramuscular neural distribution of the muscle for accurate injection points is necessary for BoNT, as excessive amounts of BoNT may potentially cause the toxin to spread to the neighboring muscles, resulting in paralysis [28,29]. The adverse effect of paralyzed muscle is reported in cases of overdose of BoNT [30–32]. Moreover, repetitive and overdose of BoNT injections build up antibodies that will result in an insufficient treatment effect [30,31,33,34]. Consequently, BoNT should be injected into the arborized regions to enhance efficacy and decrease adverse effects. To direct the injection points while preventing these adverse effects, numerous studies have revealed the intramuscular neural arborization of various muscles, but not the SA [14,33–43]. This study aimed to reveal the intramuscular neural arborization of the SA and provide anatomical information of the SA muscle.

2. Results

2.1. *Running of the Thoracic Nerve Trunk*

The long thoracic nerve runs superficial to the SA muscle and pierces the muscle at each level until the seventh rib. Thirteen of the fifteen specimens had a trunk of the long thoracic nerve running at 30 to 50% throughout the level of the first to the seventh rib. The other two had the long thoracic nerve running down at 40 to 50% at the level of the first to the fourth rib and 30 to 40% at the level of the fifth to the seventh rib.

2.2. *Intramuscular Arborization Patterns of the SA Muscle*

Twelve of the fifteen SA muscles had two regions in which the arborization patterns were the largest: the sixth to the ninth rib portion had between 50 and 70% and the first to the fifth rib portion had between 20 and 40%, following three anatomical lines: the anterior (100%), middle (50%), and posterior axillary line (0%) (Figure 1B). The other two had the largest patterns in the fourth to the ninth rib portion, between 50% and 60%; the first to the third rib portion had between 20% and 30%. The last muscle had the largest patterns in the fourth to the ninth rib portion, between 50% and 70%; the first to the third rib portion had between 30 to 40%.

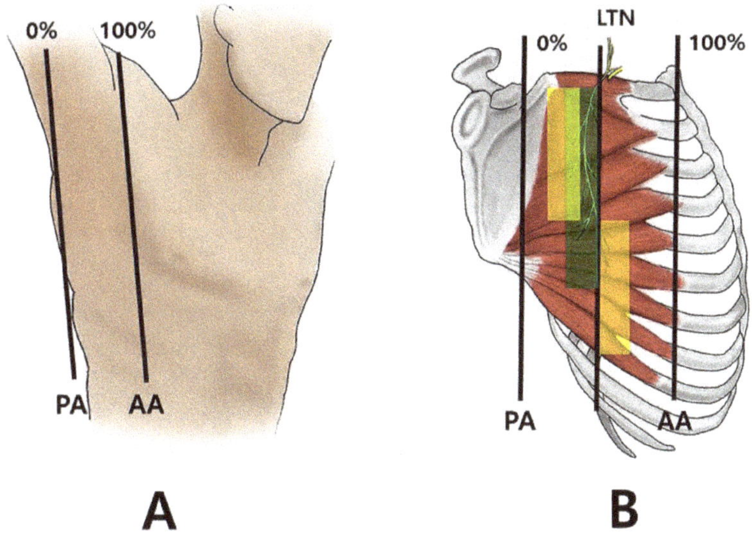

Figure 1. Specimens were harvested according to two anatomical lines: the anterior axillary line (AA) and posterior axillary line (PA). The AA was 100% to PA of 0%, respectively (**A**). The long thoracic nerve runs superficial to the SA muscle and pierces into the muscle at each level until the 7th rib. The specimens had a trunk of the long thoracic nerve running on 30 to 40% throughout the level of 1st to 7th rib (green shaded). The intramuscular arborization patterns were the largest: the 6th to 9th rib portion had between 50% to 70% and the 1st to 5th rib portion had between 20 to 40%. The injection should be guided to these arborized areas (**B**).

3. Discussion

3.1. Anatomy of the SA Muscle

The SA muscle is a flat and wide muscle covering the lateral ribs; it is anatomically divided into three muscle bellies [2]. It consists of an upper, middle, and lower muscle belly, each of which contribute to the movement of the scapular bone during upper extremity actions [44]. The upper belly of the SA lies parallel to the first rib and inserts into the superior angle of the scapula [45]. The middle belly of the SA originates from the second, third, and fourth ribs and inserts into the medial scapular border. The lower muscle belly of the SA is where the MTrPs frequently exist, originating from the fifth to the ninth ribs and inserting into the inferior angle of the scapula [45,46]. The SA muscle is innervated by the long thoracic nerve, which originates from the anterior rami of spinal nerves C5–C7 [47,48]. The long thoracic nerve runs superficially over the SA muscle along the anterior axillary line. The SA muscle is mostly involved in upper extremity movements; however, it is the prime stabilizer of the shoulder girdle and acts on shoulder flexion, abduction, and upward rotation [44].

3.2. MPS in SA Muscle

MPS is a chronic pain disorder caused by MTrPs situated at the muscle belly; it has been recognized as the main cause of pain in 85% of patients attending pain clinics [48,49]. SA muscle MTrPs may be triggered by muscle strain during excessive running, overloaded weightlifting, or repetitive coughing, especially susceptible to torsional stresses. Another cause of MTrPs initiation in the SA muscle is breast surgery due to cancer or esthetic purposes [50].

Studies have revealed that sarcomere shortening is related to MTrP etiology, and the shortening is due to an increase in activation of the neuromuscular junction and its over-release of acetylcholine. In addition, a large quantity of calcium released at the sarcoplasmic reticulum over a dysfunctional ryanidine receptor causes prolonged muscle contraction [51]. Therefore, to release muscle contraction, BoNT is currently frequently used as an injective agent for MPS [51–53]. The primary known therapeutic effects are releasing muscular contractions and alleviating the vicious pain cycle [54–56]. Injection treatment of MPS with local anesthetics is reported to be highly effective and currently represents the gold standard [57]. The local anesthetics are thought to bring relief from muscle tightness. Additionally, the injection of BoNT is another treatment option inhibiting the diffusion of neurotransmitters in the peripheral nerve, avoiding peripheral sensitization [58,59].

In the study of Kamanli et al., lidocaine injection is more practical, since it causes less disturbance than dry needling and is more cost effective than BoNT injection, and it seems to be the treatment of choice in MPS [60]. However, they have proposed that BoNT could be selectively used in MPS patients resistant to conventional treatments. In many of the assessment scores with lidocaine, dry needling and BoNT injection, depression and anxiety scores significantly improved only in the BoNT-injected group [60].

Neuromuscular junctions are the underlying causes of MPS; therefore, injecting BoNT and other injective treatments such as lidocaine, steroids, and normal saline are frequently performed to target the neuromuscular junctions [61–63]. Unlike oral medications and lidocaine injections that have short-term effects, the effectiveness of BoNT treatment in MPS has been known to continue for up to 4 months [62,63].

As BoNT acts on the neuromuscular junction, accurate anatomical knowledge of the neuromuscular arborization patterns of the SA muscles is vital for achieving the highest relief with the smallest possible dose of BoNT. Although BoNT procedures are minimally invasive compared to surgical intervention, there is a probability of damaging the nerve trunks that are not present near the neural arborized area. Therefore, precise knowledge of the anatomical features of the SA muscle should be considered. In this study, we carried out Sihler's staining, which is a whole-mount staining procedure that stains myelin sheaths and is effective in tracing the nerve endings without destroying the nerves [14,33–35,64]. The application of Sihler's staining to the SA muscle will enable an accurate and thorough understanding of the neural distribution.

Moreover, identifying the neural arborization area of the SA muscle is important in diagnosing long thoracic nerve palsy [65]. Surface electromyography in the SA muscle is challenging because multiple thin digitations make it difficult to place the electrode for recording [66,67]. When detecting long thoracic nerve palsy, the technical limitations of electromyography are interrupted signals from the neighboring muscles and difficulty with accurate electrode placement since the SA is not a bulky muscle.

At present, there is no anatomical guidance for the injection or EMG of the SA muscle. The authors acknowledge the following limitations in the current study. The results are solely based on the analysis from Sihler staining of cadavers' SA muscles. Additionally, the cadavers are from elderly people with an average age of 76.6, and from a single race (Korean).

In this study, we have revealed intramuscular distribution of the SA muscle that might help clinicians guiding electromyography and injective treatments including BoNT, lidocaine, normal saline, and steroids. In the anatomical aspect, clinicians should be able to carefully target the three regions in the middle portion, between the sixth to ninth rib portion and the first to the fifth rib (Figure 1B).

4. Materials and Methods

This study was performed in accordance with the principles outlined in the Declaration of Helsinki. Informed consent and approval were obtained from the families of the cadavers before the dissections were performed and approved by the Institutional Review Board of Yonsei University College of Medicine (approval number 20-006, approved date: 26 February 2020). A total of 15 SA muscles from Korean cadavers (5 men and 4 women with a mean age of 76.6 years; range, 73–95 years) were dissected in Yonsei University medical center from May 2020 to October 2020, and modified Sihler staining was applied to clarify the intramuscular neural arborization patterns.

Before dissection, the SA muscles were aligned in their anatomical positions (Figure 2). The arborizing patterns of the SA muscles were tracked according to the three anatomical lines: anterior (100%), middle (50%), posterior axillary line (0%), and from the first to the ninth ribs (Figure 1A).

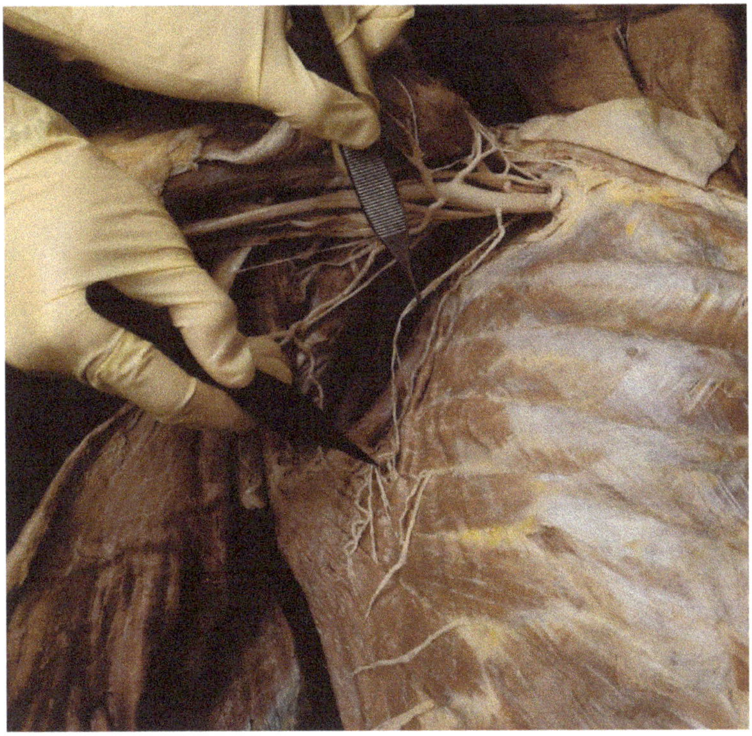

Figure 2. The serratus anterior muscle the long thoracic nerve running over the muscle. The long thoracic nerve has been pointed out by the forceps.

The SA muscles underwent Sihler staining, as modified by Liem and Douwe van Willingen (Figure 3) [67].

This technique involves several steps to acquire the visual representation of the intramuscular neural arborization pattern. The changes over Sihler's method of the SA specimens are shown in Figure 4.

Following Sihler staining, the SA muscles were divided into 10 sections according to the vertical lines from the anterior and posterior axillary lines and the curved lines of the first to ninth ribs.

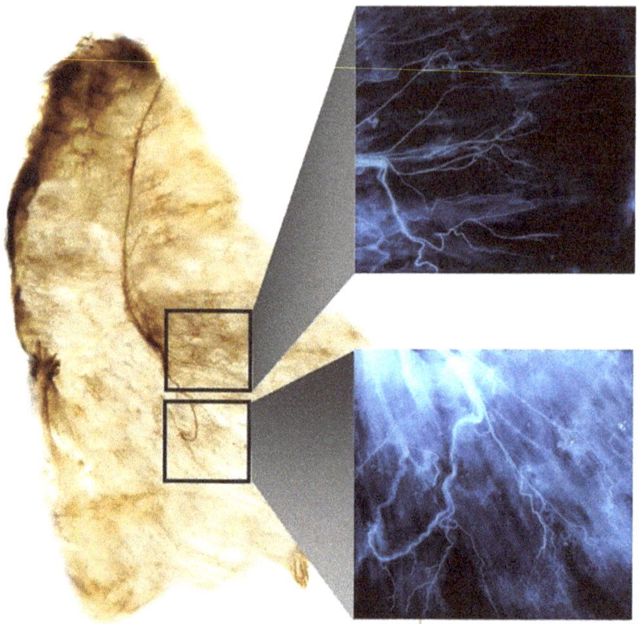

Figure 3. The result of Sihler's staining of the serratus anterior muscle. The intramuscular neural distribution of the serratus anterior muscle is observed with enlarged views.

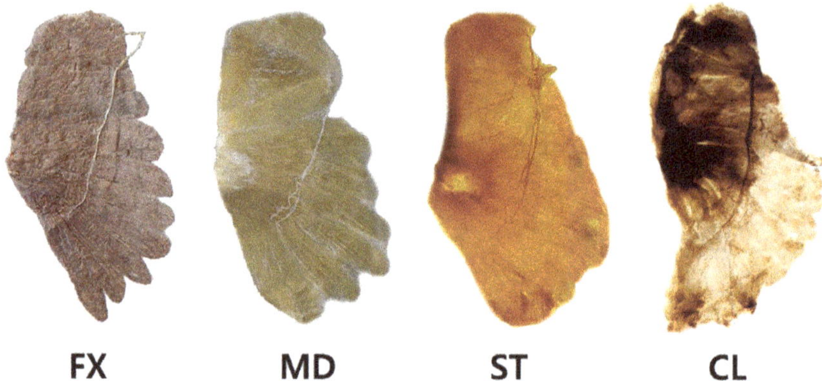

Figure 4. The serratus anterior muscle underwent modified Sihler's method. The method consists of stages of fixation (FX), maceration and depigmentation (MD), decalcification, staining (ST), and clearing (CL).

Modified Sihler Staining

Fixation: The SA muscles were stored for one month in a container filled with 10% un-neutralized formalin. The solution was replaced with fresh solution whenever it turned cloudy.

Maceration and depigmentation: The fixed SA specimens were washed in running water for an hour. Then, they were placed for one month in a container filled with 3% aqueous potassium hydroxide and hydrogen peroxide solution.

Decalcification: The depigmented SA specimens were then placed in Sihler I solution, a compound of glycerin, glacial acetic acid, and aqueous chloral hydrate.

Staining: The decalcified SA specimens were then stained with the Sihler II solution, a compound of glycerin, aqueous chloral hydrate, and acetic acid. The staining process takes 30–35 days for intramuscular nerve visualization.

De-staining: The stained SA specimens were cleansed in a container filled with Sihler I solution. This step is used to de-stain the SA muscle fibers so that only the intramuscular nerve distributions are visualized.

Neutralization: The de-stained SA specimens were neutralized in clean water for half an hour. Consequently, the SA specimens were placed in a solution of 0.05% lithium carbonate.

Clearing: Finally, the neutralized SA specimens were taken into the clearing stage with glycerin by increasing the concentrations from 20% to 100%. This stage took nearly 4–5 h.

Author Contributions: Conceptualization, K.-H.Y.; Writing—Original Draft Preparation, K.-H.Y. and J.-H.L.; Writing—Review and Editing, K.-H.Y. and J.-H.L.; Visualization, J.-H.L. and K.-H.Y.; Supervision, H.-J.K. All authors have read and agreed to the published version of the manuscript.

Funding: This work was supported by the Korea Medical Device Development Fund grant funded by the Korea government (the Ministry of Science and ICT, the Ministry of Trade, Industry and Energy, the Ministry of Health & Welfare, the Ministry of Food and Drug Safety) (Project Number: 1711138194, KMDF_PR_20200901_0109-01).

Institutional Review Board Statement: The study was conducted according to the guidelines of the Declaration of Helsinki and approved by the Institutional Review Board of Yonsei University College of Medicine (approval number 20-006, approved date: 26 February 2020). Informed Consent Statement: Informed consent was obtained from all subjects involved in the study.

Informed Consent Statement: Informed consent was obtained from all subjects involved in the study.

Data Availability Statement: Not applicable.

Acknowledgments: The authors thank Eun-Byul Yi from Eonbuk elementary school for the illustrations.

Conflicts of Interest: The authors declare no conflict of interest.

References

1. Alvarez, D.J.; Rockwell, P.G. Trigger points: Diagnosis and management. *Am. Fam. Physician* **2002**, *65*, 653–660. [PubMed]
2. Bautista, A.; Webb, C.; Rosenquist, R. Serratus Anterior Muscle Pain Syndrome: A Diagnostic Conundrum. *Pain Med.* **2017**, *18*, 1600–1602. [CrossRef] [PubMed]
3. Donnelly, J.M.; Simons, D.G. *Travell, Simons & Simons' Myofascial Pain and Dysfunction: The Trigger Point Manual*; Lippincott Williams & Wilkins: Philadelphia, PA, USA, 2019.
4. Lee, S.T.; Moon, J.; Lee, S.H.; Cho, K.H.; Im, S.H.; Kim, M.; Min, K. Changes in Activation of Serratus Anterior, Trapezius and Latissimus Dorsi With Slouched Posture. *Ann. Rehabil. Med.* **2016**, *40*, 318–325. [CrossRef] [PubMed]
5. Manfredini, D.; Cocilovo, F.; Stellini, E.; Favero, L.; Guarda-Nardini, L. Surface electromyography findings in unilateral myofascial pain patients: Comparison of painful vs. non painful sides. *Pain Med.* **2013**, *14*, 1848–1853. [CrossRef]
6. Szyszka-Sommerfeld, L.; Machoy, M.; Lipski, M.; Wozniak, K. The Diagnostic Value of Electromyography in Identifying Patients with Pain-Related Temporomandibular Disorders. *Front. Neurol.* **2019**, *10*, 180. [CrossRef]
7. Saxena, A.; Chansoria, M.; Tomar, G.; Kumar, A. Myofascial pain syndrome: An overview. *J. Pain Palliat. Care Pharmacother.* **2015**, *29*, 16–21. [CrossRef]
8. Lluch, E.; Arguisuelas, M.D.; Coloma, P.S.; Palma, F.; Rey, A.; Falla, D. Effects of deep cervical flexor training on pressure pain thresholds over myofascial trigger points in patients with chronic neck pain. *J. Manipulative Physiol. Ther.* **2013**, *36*, 604–611. [CrossRef]
9. Javanshir, K.; Ortega-Santiago, R.; Mohseni-Bandpei, M.A.; Miangolarra-Page, J.C.; Fernandez-de-Las-Penas, C. Exploration of somatosensory impairments in subjects with mechanical idiopathic neck pain: A preliminary study. *J. Manipulative Physiol. Ther.* **2010**, *33*, 493–499. [CrossRef]
10. Iglesias-Gonzalez, J.J.; Munoz-Garcia, M.T.; Rodrigues-de-Souza, D.P.; Alburquerque-Sendin, F.; Fernandez-de-Las-Penas, C. Myofascial trigger points, pain, disability, and sleep quality in patients with chronic nonspecific low back pain. *Pain Med.* **2013**, *14*, 1964–1970. [CrossRef]
11. Duyur Cakit, B.; Genc, H.; Altuntas, V.; Erdem, H.R. Disability and related factors in patients with chronic cervical myofascial pain. *Clin. Rheumatol.* **2009**, *28*, 647–654. [CrossRef]

12. Childers, M.K. Targeting the neuromuscular junction in skeletal muscles. *Am. J. Phys. Med. Rehabil.* **2004**, *83*, S38–S44. [CrossRef]
13. Comella, C.L. The treatment of cervical dystonia with botulinum toxins. *J. Neural Transm.* **2008**, *115*, 579–583. [CrossRef]
14. Yi, K.H.; Choi, Y.J.; Cong, L.; Lee, K.L.; Hu, K.S.; Kim, H.J. Effective botulinum toxin injection guide for treatment of cervical dystonia. *Clin. Anat.* **2020**, *33*, 192–198. [CrossRef]
15. Vasileiadis, G.I.; Sakellariou, V.I.; Papagelopoulos, P.J.; Zoubos, A.B. Posttraumatic focal dystonia of the shoulder. *Orthopedics* **2012**, *35*, e977–e980. [CrossRef]
16. Climent, J.M.; Kuan, T.S.; Fenollosa, P.; Martin-Del-Rosario, F. Botulinum toxin for the treatment of myofascial pain syndromes involving the neck and back: A review from a clinical perspective. *Evid. Based Complement. Alternat. Med.* **2013**, *2013*, 381459. [CrossRef]
17. Kwanchuay, P.; Petchnumsin, T.; Yiemsiri, P.; Pasuk, N.; Srikanok, W.; Hathaiareerug, C. Efficacy and Safety of Single Botulinum Toxin Type A (Botox(R)) Injection for Relief of Upper Trapezius Myofascial Trigger Point: A Randomized, Double-Blind, Placebo-Controlled Study. *J. Med. Assoc. Thai* **2015**, *98*, 1231–1236.
18. Xie, P.; Qin, B.; Yang, F.; Yu, T.; Yu, J.; Wang, J.; Zheng, H. Lidocaine Injection in the Intramuscular Innervation Zone Can Effectively Treat Chronic Neck Pain Caused by MTrPs in the Trapezius Muscle. *Pain Physician* **2015**, *18*, E815–E826.
19. Mor, N.; Tang, C.; Blitzer, A. Temporomandibular Myofacial Pain Treated with Botulinum Toxin Injection. *Toxins* **2015**, *7*, 2791–2800. [CrossRef]
20. Chaurand, J.; Pacheco-Ruiz, L.; Orozco-Saldivar, H.; Lopez-Valdes, J. Efficacy of botulinum toxin therapy in treatment of myofascial pain. *J. Oral. Sci.* **2017**, *59*, 351–356. [CrossRef]
21. Layeeque, R.; Hochberg, J.; Siegel, E.; Kunkel, K.; Kepple, J.; Henry-Tillman, R.S.; Dunlap, M.; Seibert, J.; Klimberg, V.S. Botulinum toxin infiltration for pain control after mastectomy and expander reconstruction. *Ann. Surg.* **2004**, *240*, 608–613; discussion 613–604. [CrossRef]
22. Rosales, R.L.; Kong, K.H.; Goh, K.J.; Kumthornthip, W.; Mok, V.C.; Delgado-De Los Santos, M.M.; Chua, K.S.; Abdullah, S.J.; Zakine, B.; Maisonobe, P.; et al. Botulinum toxin injection for hypertonicity of the upper extremity within 12 weeks after stroke: A randomized controlled trial. *Neurorehabil. Neural Repair* **2012**, *26*, 812–821. [CrossRef] [PubMed]
23. Bhakta, B.B.; Cozens, J.A.; Bamford, J.M.; Chamberlain, M.A. Use of botulinum toxin in stroke patients with severe upper limb spasticity. *J. Neurol. Neurosurg. Psychiatry* **1996**, *61*, 30–35. [CrossRef] [PubMed]
24. Brashear, A.; Gordon, M.F.; Elovic, E.; Kassicieh, V.D.; Marciniak, C.; Do, M.; Lee, C.H.; Jenkins, S.; Turkel, C.; Botox Post-Stroke Spasticity Study, G. Intramuscular injection of botulinum toxin for the treatment of wrist and finger spasticity after a stroke. *N. Engl. J. Med.* **2002**, *347*, 395–400. [CrossRef] [PubMed]
25. Hesse, S.; Jahnke, M.T.; Luecke, D.; Mauritz, K.H. Short-term electrical stimulation enhances the effectiveness of Botulinum toxin in the treatment of lower limb spasticity in hemiparetic patients. *Neurosci. Lett.* **1995**, *201*, 37–40. [CrossRef]
26. Van Campenhout, A.; Verhaegen, A.; Pans, S.; Molenaers, G. Botulinum toxin type A injections in the psoas muscle of children with cerebral palsy: Muscle atrophy after motor end plate-targeted injections. *Res. Dev. Disabil.* **2013**, *34*, 1052–1058. [CrossRef]
27. Gracies, J.M.; Lugassy, M.; Weisz, D.J.; Vecchio, M.; Flanagan, S.; Simpson, D.M. Botulinum toxin dilution and endplate targeting in spasticity: A double-blind controlled study. *Arch. Phys. Med. Rehabil.* **2009**, *90*, 9–16.e12. [CrossRef]
28. Kinnett, D. Botulinum toxin A injections in children: Technique and dosing issues. *Am. J. Phys. Med. Rehabil.* **2004**, *83*, S59–S64. [CrossRef]
29. Hsu, T.S.; Dover, J.S.; Arndt, K.A. Effect of volume and concentration on the diffusion of botulinum exotoxin A. *Arch. Dermatol.* **2004**, *140*, 1351–1354. [CrossRef]
30. Peng, H.P.; Peng, J.H. Complications of botulinum toxin injection for masseter hypertrophy: Incidence rate from 2036 treatments and summary of causes and preventions. *J. Cosmet. Dermatol.* **2018**, *17*, 33–38. [CrossRef]
31. Song, J.H.; Cho, E.S.; Kim, S.T.; Ahn, H.J. Change of distribution and timing of bite force after botulinum toxin type A injection evaluated by a computerized occlusion analysis system. *Yonsei Med. J.* **2014**, *55*, 1123–1129. [CrossRef]
32. Rafferty, K.L.; Liu, Z.J.; Ye, W.; Navarrete, A.L.; Nguyen, T.T.; Salamati, A.; Herring, S.W. Botulinum toxin in masticatory muscles: Short- and long-term effects on muscle, bone, and craniofacial function in adult rabbits. *Bone* **2012**, *50*, 651–662. [CrossRef]
33. Yi, K.H.; Cong, L.; Bae, J.H.; Park, E.S.; Rha, D.W.; Kim, H.J. Neuromuscular structure of the tibialis anterior muscle for functional electrical stimulation. *Surg. Radiol. Anat.* **2017**, *39*, 77–83. [CrossRef]
34. Yi, K.H.; Rha, D.W.; Lee, S.C.; Cong, L.; Lee, H.J.; Lee, Y.W.; Kim, H.J.; Hu, K.S. Intramuscular nerve distribution pattern of ankle invertor muscles in human cadaver using sihler stain. *Muscle Nerve* **2016**, *53*, 742–747. [CrossRef]
35. Rha, D.W.; Yi, K.H.; Park, E.S.; Park, C.; Kim, H.J. Intramuscular nerve distribution of the hamstring muscles: Application to treating spasticity. *Clin. Anat.* **2016**, *29*, 746–751. [CrossRef]
36. Yi, K.-H.; Lee, H.-J.; Seo, K.K.; Kim, H.-J. Intramuscular Neural Arborization of the Latissimus Dorsi Muscle: Application of Botulinum Neurotoxin Injection in Flap Reconstruction. *Toxins* **2022**, *14*, 107. [CrossRef]
37. Yi, K.H.; Lee, H.J.; Seo, K.K.; Kim, H.J. Botulinum neurotoxin injection guidelines regarding flap surgeries in breast reconstruction. *J. Plast. Reconstr. Aesthet. Surg.* **2021**, *75*, 503–505. [CrossRef]
38. Yi, K.H.; Lee, J.H.; Kim, H.M.; Kim, H.J. The botulinum neurotoxin for pain control after breast reconstruction: Neural distribution of the pectoralis major muscle. *Reg. Anesth. Pain Med.* **2022**. [CrossRef]
39. Yi, K.H.; Lee, H.J.; Lee, J.H.; Lee, K.L.; Kim, H.J. Effective botulinum neurotoxin injection in treating iliopsoas spasticity. *Clin. Anat.* **2021**, *34*, 431–436. [CrossRef]

40. Yi, K.H.; Lee, H.J.; Choi, Y.J.; Lee, K.; Lee, J.H.; Kim, H.J. Anatomical guide for botulinum neurotoxin injection: Application to cosmetic shoulder contouring, pain syndromes, and cervical dystonia. *Clin. Anat.* **2021**, *34*, 822–828. [CrossRef]
41. Yi, K.H.; Lee, K.L.; Lee, J.H.; Hu, H.W.; Lee, K.; Seo, K.K.; Kim, H.J. Guidelines for botulinum neurotoxin injections in piriformis syndrome. *Clin. Anat.* **2021**, *34*, 1028–1034. [CrossRef]
42. Yi, K.H.; Lee, H.J.; Lee, J.H.; Seo, K.K.; Kim, H.J. Application of Botulinum Neurotoxin Injections in TRAM Flap for Breast Reconstruction: Intramuscular Neural Arborization of the Rectus Abdominis Muscle. *Toxins* **2021**, *13*, 269. [CrossRef]
43. Yi, K.H.; Lee, J.H.; Lee, D.K.; Hu, H.W.; Seo, K.K.; Kim, H.J. Anatomical locations of the motor endplates of sartorius muscle for botulinum toxin injections in treatment of muscle spasticity. *Surg. Radiol. Anat.* **2021**, *43*, 2025–2030. [CrossRef] [PubMed]
44. Hamada, J.; Igarashi, E.; Akita, K.; Mochizuki, T. A cadaveric study of the serratus anterior muscle and the long thoracic nerve. *J. Shoulder Elbow Surg.* **2008**, *17*, 790–794. [CrossRef] [PubMed]
45. Moore, K.L.; Agur, A.M.R.; Dalley, A.F. *Clinically Oriented Anatomy*; Wolters Kluwer India Pvt, Ltd.: Gurugram, India, 2019.
46. Webb, A.L.; O'Sullivan, E.; Stokes, M.; Mottram, S. A novel cadaveric study of the morphometry of the serratus anterior muscle: One part, two parts, three parts, four? *Anat. Sci. Int.* **2018**, *93*, 98–107. [CrossRef] [PubMed]
47. Nasu, H.; Yamaguchi, K.; Nimura, A.; Akita, K. An anatomic study of structure and innervation of the serratus anterior muscle. *Surg. Radiol. Anat.* **2012**, *34*, 921–928. [CrossRef] [PubMed]
48. Fishbain, D.A.; Goldberg, M.; Meagher, B.R.; Steele, R.; Rosomoff, H. Male and female chronic pain patients categorized by DSM-III psychiatric diagnostic criteria. *Pain* **1986**, *26*, 181–197. [CrossRef]
49. Tough, E.A.; White, A.R.; Richards, S.; Campbell, J. Variability of criteria used to diagnose myofascial trigger point pain syndrome—Evidence from a review of the literature. *Clin. J. Pain.* **2007**, *23*, 278–286. [CrossRef]
50. Torres Lacomba, M.; Mayoral del Moral, O.; Coperias Zazo, J.L.; Gerwin, R.D.; Goni, A.Z. Incidence of myofascial pain syndrome in breast cancer surgery: A prospective study. *Clin. J. Pain* **2010**, *26*, 320–325. [CrossRef]
51. McPartland, J.M. Travell trigger points—Molecular and osteopathic perspectives. *J. Am. Osteopath Assoc.* **2004**, *104*, 244–249.
52. Shah, J.P.; Gilliams, E.A. Uncovering the biochemical milieu of myofascial trigger points using in vivo microdialysis: An application of muscle pain concepts to myofascial pain syndrome. *J. Bodyw. Mov. Ther.* **2008**, *12*, 371–384. [CrossRef]
53. Hsieh, R.L.; Lee, W.C. Are the effects of botulinum toxin injection on myofascial trigger points placebo effects or needling effects? *Arch. Phys. Med. Rehabil.* **2008**, *89*, 792–793; author reply 793. [CrossRef]
54. Tepper, S.J. Treatment of headache pain with botulinum neurotoxins. *Pain Pract.* **2004**, *4* (Suppl. 1), S38–S46. [CrossRef]
55. Ondo, W.G.; Vuong, K.D.; Derman, H.S. Botulinum toxin A for chronic daily headache: A randomized, placebo-controlled, parallel design study. *Cephalalgia* **2004**, *24*, 60–65. [CrossRef]
56. Troost, B.T. Botulinum toxin type A (Botox) in the treatment of migraine and other headaches. *Expert Rev. Neurother.* **2004**, *4*, 27–31. [CrossRef]
57. Giamberardino, M.A.; Affaitati, G.; Fabrizio, A.; Costantini, R. Myofascial pain syndromes and their evaluation. *Best Pract. Res. Clin. Rheumatol.* **2011**, *25*, 185–198. [CrossRef]
58. Robertson, C.E.; Garza, I. Critical analysis of the use of onabotulinumtoxinA (botulinum toxin type A) in migraine. *Neuropsychiatr. Dis. Treat* **2012**, *8*, 35–48. [CrossRef]
59. Ramachandran, R.; Yaksh, T.L. Therapeutic use of botulinum toxin in migraine: Mechanisms of action. *Br. J. Pharmacol.* **2014**, *171*, 4177–4192. [CrossRef]
60. Kamanli, A.; Kaya, A.; Ardicoglu, O.; Ozgocmen, S.; Zengin, F.O.; Bayik, Y. Comparison of lidocaine injection, botulinum toxin injection, and dry needli.ing to trigger points in myofascial pain syndrome. *Rheumatol. Int.* **2005**, *25*, 604–611. [CrossRef]
61. Gobel, H.; Heinze, A.; Heinze-Kuhn, K.; Jost, W.H. Evidence-based medicine: Botulinum toxin A in migraine and tension-type headache. *J. Neurol.* **2001**, *248* (Suppl. 1), 34–38. [CrossRef]
62. Dodick, D.W. Botulinum neurotoxin for the treatment of migraine and other primary headache disorders: From bench to bedside. *Headache* **2003**, *43* (Suppl. 1), S25–S33. [CrossRef]
63. Chan, V.W.; McCabe, E.J.; MacGregor, D.L. Botox treatment for migraine and chronic daily headache in adolescents. *J. Neurosci. Nurs.* **2009**, *41*, 235–243. [CrossRef] [PubMed]
64. Yi, K.H.; Lee, H.J.; Choi, Y.J.; Lee, J.H.; Hu, K.S.; Kim, H.J. Intramuscular Neural Distribution of Rhomboid Muscles: Evaluation for Botulinum Toxin Injection Using Modified Sihler's Method. *Toxins* **2020**, *12*, 289. [CrossRef] [PubMed]
65. Wiater, J.M.; Flatow, E.L. Long thoracic nerve injury. *Clin. Orthop. Relat. Res.* **1999**, *368*, 17–27. [CrossRef]
66. Helgadottir, H.; Kristjansson, E.; Einarsson, E.; Karduna, A.; Jonsson, H., Jr. Altered activity of the serratus anterior during unilateral arm elevation in patients with cervical disorders. *J. Electromyogr. Kinesiol.* **2011**, *21*, 947–953. [CrossRef] [PubMed]
67. Liem, R.S.; Douwe van Willigen, J. In toto staining and preservation of peripheral nervous tissue. *Stain Technol.* **1988**, *63*, 113–120. [CrossRef]

toxins

MDPI

Review

Multidimensional Effectiveness of Botulinum Toxin in Neuropathic Pain: A Systematic Review of Randomized Clinical Trials

Lorenzo Lippi [1,2], Alessandro de Sire [3], Arianna Folli [1], Francesco D'Abrosca [1], Elisa Grana [4], Alessio Baricich [1,5], Stefano Carda [4] and Marco Invernizzi [1,2,*]

1 Physical and Rehabilitative Medicine, Department of Health Sciences, University of Eastern Piedmont "A. Avogadro", 28100 Novara, Italy; lorenzolippi.mt@gmail.com (L.L.); arianna.folli23@gmail.com (A.F.); fradabro@gmail.com (F.D.); alessio.baricich@med.uniupo.it (A.B.)
2 Translational Medicine, Dipartimento Attività Integrate Ricerca e Innovazione (DAIRI), Azienda Ospedaliera SS. Antonio e Biagio e Cesare Arrigo, 15121 Alessandria, Italy
3 Physical Medicine and Rehabilitation Unit, Department of Medical and Surgical Sciences, University of Catanzaro Magna Graecia, 88100 Catanzaro, Italy; alessandro.desire@unicz.it
4 Neuropsychology and Neurorehabilitation Service, Department of Clinical Neuroscience, Lausanne University Hospital, 1004 Lausanne, Switzerland; granaelisa@gmail.com (E.G.); stefano.carda@gmail.com (S.C.)
5 Physical and Rehabilitation Medicine, "Ospedale Maggiore della Carità" University Hospital, 28100 Novara, Italy
* Correspondence: marco.invernizzi@med.uniupo.it; Tel.: +39-(0)3-2137-34800

Abstract: Although botulinum toxin (BoNT) has been suggested as a treatment to counter neuropathic pain, no previous systematic reviews investigated the multidimensional effects of BoNT on pain relief and Health-Related Quality of Life (HR-QoL). The aim of this systematic review is to summarize the current evidence on the effectiveness of BoNT treatment for neuropathic pain, and to characterize its multidimensional effectiveness in order to guide physicians in clinical practice. Five databases were systematically searched up to 4 April 2022, to identify randomized controlled trials satisfying the following criteria: adults suffering from neuropathic pain, BoNT administration, any comparator, multidimensional assessment of pain as primary outcome, HR-QoL, physical function, anxiety and depression, and sleep quality as secondary outcomes. Twelve studies were included. The multidimensional pain scales used were short-form McGill Pain Questionnaire, Neuropathic pain scale, Neuropathic Pain Symptom Inventory, International SCI Pain Basic Data Set, West Haven-Yale Multidimensional Pain Inventory, Brief Pain Inventory, and Douleur Neuropathique 4. These scales highlighted the positive effects of BoNT administration. According to the Jadad scale, all the RCTs included were high-quality studies. BoNT administration might be effectively introduced in the comprehensive management of neuropathic pain. Further research should focus on optimal and cost-effective therapeutic protocols.

Keywords: botulinum toxin (BoNT); neuropathic pain; pain management; quality of life; rehabilitation

Key Contribution: BoNT administration might be effectively introduced in the comprehensive management of neuropathic pain.

Citation: Lippi, L.; de Sire, A.; Folli, A.; D'Abrosca, F.; Grana, E.; Baricich, A.; Carda, S.; Invernizzi, M. Multidimensional Effectiveness of Botulinum Toxin in Neuropathic Pain: A Systematic Review of Randomized Clinical Trials. *Toxins* **2022**, *14*, 308. https://doi.org/10.3390/toxins14050308

Received: 9 April 2022
Accepted: 24 April 2022
Published: 27 April 2022

Publisher's Note: MDPI stays neutral with regard to jurisdictional claims in published maps and institutional affiliations.

1. Introduction

Pain is currently defined by the International Association for the Study of Pain (IASP) as 'an unpleasant sensory and emotional experience associated with, or resembling that associated with, actual or potential tissue damage' [1]. Among the different pain types, neuropathic pain is characterized by increased pain sensitivity and/or spontaneous pain and is defined by the presence of neuropathy, a lesion or disease affecting the somatosensory nervous system [1]. It is currently considered a challenge in the clinical setting due to its chronic course and poor responsiveness to medications [2–5]. In further detail, the recent systematic review by van Hecke et al. [6] reported that approximately 6.9–10% of the European population suffer from neuropathic pain, with detrimental consequences in terms of physical and psychosocial wellbeing, health-related quality of life (HR-QoL) and economic burden [3,7,8].

Pathophysiology of neuropathic pain has been widely investigated and the present evidence underlines the role of independent mechanisms triggered by various damages to an afferent pathway [5,9–12]. However, the exact pathophysiological mechanisms underpinning neuropathic pain are far from being fully understood [13]. In this context, the downregulation of sodium channels [14], the dysregulation of Transient Receptor Potential Vanilloid 1 (TRPV1) and Transient Receptor Potential Member 8 (TRPM8) receptors have been proposed to have a role in this complex framework [15–17]. Interestingly, it has recently been highlighted that abnormal sensory messages characterizing neuropathic pain might stimulate the cortex, promoting the excitation of neurons in the limbic areas related to anxiety, depression, and sleep problems, frequently accompanying neuropathic pain [5]. A deeper assessment of these pathological mechanisms might play a key role in optimizing a multidimensional treatment, selecting a precise pathophysiological pathway [3,18]. In this complex framework, multimodal therapeutic interventions targeting specific structures involved in the neuropathic pain circuits are crucial to promoting the optimal response in pain relief [5]. Furthermore, several pharmacological and non-pharmacological approaches have been proposed for the complex management of neuropathic pain and growing evidence recommends a comprehensive patient-centered approach in order to improve pain management and minimize the side effects of single therapies [3,19–22].

In the last decade, botulinum toxin (BoNT) has been proposed as a therapeutic option to treat neuropathic pain [23], although the antinociceptive effects of BoNT have been widely ascribed to the muscle relaxation effects alone [24,25]. However, several studies reported positive results of BoNT treatment in the management of neuropathic pain [26–28]. Park et al. [28], in a neuropathic pain animal model, demonstrated the dissociation between the duration of muscle relaxation and pain relief after BoNT treatment, suggesting a pure antinociceptive role.

Despite these findings, the recent GRADE classification by Finnerup et al. [29] reported BoNT as a third-line pharmacological treatment in general neuropathic pain. They suggested gabapentin, pregabalin, SNRIs (duloxetine/venlafaxine), and tricyclic antidepressants as first-line treatments, followed by capsaicin patches, lidocaine patches and tramadol. Similarly, the guidelines for neuropathic pain published by Moisset et al. [30] in 2020 recommended the use of BoNT as second-line therapy for peripheral neuropathic pain, while lidocaine plasters or Transcutaneous Electrical Nerve Stimulation (TENS) therapy were the first-line interventions. On the other hand, SNRI drugs, gabapentin, or tricyclic antidepressant were recommended for the treatment of central neuropathic pain, while pregabalin, tramadol, or combination therapy were recommended as second-line therapy.

Although BoNT has been suggested as an effective treatment to counter neuropathic pain [29–35], evidence in the literature is mainly focused on the unidimensional evaluation of pain, with different systematic reviews assessing the Visual Analogue Scale (VAS) or Numerical Rating Scale (NRS). Conversely, given the psychosocial and functional burden of neuropathic pain, a multidimensional assessment of this condition is needed, in order to promote a patient-centered approach.

However, to the best of our knowledge, no previous systematic reviews investigated the multidimensional effectiveness of BoNT on pain relief and quality of life in patients suffering from neuropathic pain.

Therefore, this systematic review of randomized controlled trials (RCTs) aimed at summarizing the current evidence on the efficacy of BoNT treatment for neuropathic pain, characterizing the multidimensional effectiveness of BoNT related to different neuropathic pain etiologies to improve the complex management of this burdensome condition.

2. Results

The search strategy performed on 4 April 2022 identified 1688 records from the five databases and six records from the reference lists of the included studies. Figure 1 shows the PRISMA flow diagram of the search process. After duplication removal, 1269 studies were assessed for eligibility and screened for title and abstract. After the exclusion of 1187 records, 82 full-text records were assessed for eligibility. Due to inconsistency with the eligibility criteria, 70 articles were excluded (36 were not RCTs, 3 studies involved animals, 4 were in a language other than English, 1 was retired for plagiarism, 11 did not assess patients with neuropathic pain, 1 was an ongoing trial, 5 were congress abstracts, 5 were registered protocols not published, and 4 did not assess appropriate outcomes). The Supplementary Table S2 shows the lists of excluded studies assessed in full text and the reasons for exclusions. As a result, 12 studies were included in the present work [26,36–46]. The studies included in this systematic review were published between 2006 [46] and 2020 [43]. Among these, two studies were conducted in the USA [37,46]; three studies were conducted in Iran [39,42,43]; one study was conducted in Taiwan [45]; one was conducted in Greece [36]; one was conducted in Canada [38]; one was conducted in South Korea [40]; one was conducted in France [41]; and one was conducted in China [44]. The remaining study was an international collaboration (France and Brazil, n = 1 [26]).

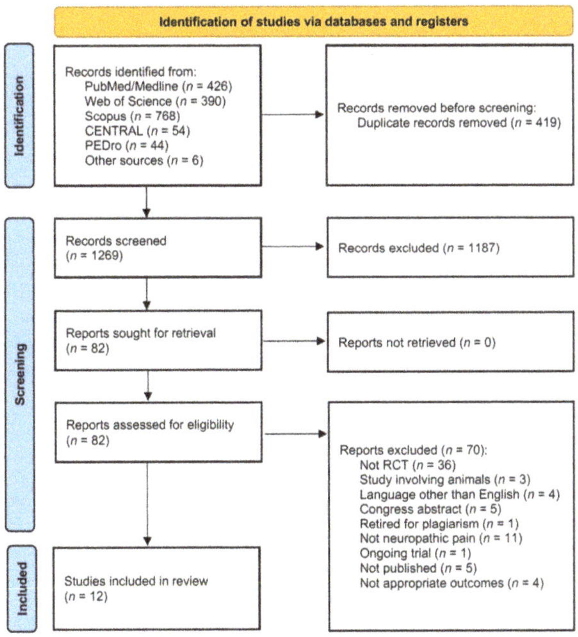

Figure 1. PRISMA 2020 flow chart.

A total of 522 subjects were included in the present systematic review (90 with postherpetic neuropathic pain [36,44], 48 with spinal cord injuries [37,40], 29 with posttraumatic/postoperative nerve lesion or post-herpetic neuropathy [41], 66 with peripheral nerve lesion [26], 231 with diabetic neuropathy [39,42,43,45], 38 with thoracic outlet syndrome [38], and 20 with carpal tunnel syndrome [46]). The mean age of the subjects included ranged from 36.8 ± 8.9 years [38] to 77.5 ± 8.2 years [36], while 230 patients were males and 272 females. However, it should be noted that Breuer et al. [46] did not report the mean age and sex of the study participants.

The intervention was compared to placebo or other treatments. In particular, the control group in each study was composed of patients suffering from neuropathic pain with the same etiological cause of the intervention group [26,36–46]. Control groups were treated with normal saline injections in 11 studies [26,36–43,45]. Only Xiao et al. [44] compared BoNT-A injection to both an active control group (treated with 0.5% lidocaine injection) and a placebo group (treated with saline injection). Two studies were crossover studies [37,45].

The time of follow-up varied somewhat among the studies included with five studies reporting a total duration of 24 weeks [26,36,38,41,45], one study of 20 weeks [37], two studies of 12 weeks [42,44], one study of 13 weeks [46], one study of 8 weeks [40], one study of 4 weeks [43], and one study of 3 weeks [39]. The sample characterization of each study included has been summarized in detail in Table 1.

In conclusion, it should be noted that among the 12 RCTs included in the present review, 3 studies [36,42,43] did not report any funding, while 9 studies [26,37–41,44–46] received external funding and, in particular, 3 studies [38,40,46] received funding by pharmaceutical companies. Lastly, 2 studies [37,41] declared that they were supplied with the toxin by pharmaceutical companies.

2.1. BoNT Intervention

A high heterogeneity of BoNT type, source, amount, injection sites, number of injections and injection technique was reported in the studies included in the present work. Out of the 12 RCTs assessed, 11 (91.7%) [26,36–45] utilized BoNT-A, while 1 (8.3%) [46] administered BoNT-B.

Two studies assessed patients with postherpetic neuropathy with an amount of BoNT-A in the study protocols ranging between 100 and 200 units [36,44]. Both study protocols assessed the effects of BoNT-A injected subcutaneously in the painful area. In further detail, one of the two administration protocols was characterized by a chessboard distribution of 40 different injections, with a minimum distance of 1 cm between injection sites [36]. The other study assessed injections every 1.0–2.0 cm radius of skin [44].

Only one study protocol has been proposed in patients with peripheral nerve lesions and was characterized by two subcutaneous injections of up to 300 units of OnabotulinumtoxinA performed after 12 weeks [26]. The maximum number of injections was 60, all performed at a distance of 1.5–2 cm [26].

Similarly, the study by Ranoux et al., assessed the effects of subcutaneous injections (maximum 200 units) of OnabotulinumtoxinA in patients with post-traumatic/postoperative nerve lesion or postherpetic neuropathy [41]. The authors performed up to 40 different subcutaneous injections at a distance of 1.5 cm [41].

One study assessed the effects of 75 units of OnabotulinumtoxinA in the anterior and middle scalene muscles in patients with neuropathic pain due to thoracic outlet syndrome [38]. The intramuscular injections were performed under EMG guidance [38].

The effects of BoNT-A in patients with diabetic neuropathy have been studied by four different RCTs with different injection protocols. All the studies included administered BoNT-A subcutaneously. One study performed an injection of 50 units of OnabotulinumtoxinA [45] while two studies injected 100 units of AbobotulinumtoxinA [39,42]. BoNT-A was administered in the sole, dorsum or entire surface of the foot. The injections number ranged between 12–20 *sites* per patient [39,42,43,45].

In patients with neuropathic pain due to Spinal Cord Injury (SCI), two studies assessed different protocols of subcutaneous administration of BoNT-A in the painful area. The amount of BoNT-A ranged between 200 and 400 units [37,40]. One study performed 5 units per injection site [37] and the other did not report the single injection amount of BoNT-A [40].

Lastly, one study injected BoNT-B in patients with carpal tunnel syndrome [46], dividing 2500 units of RimabotulinumtoxinB into three intramuscular injections (one for each muscle), under EMG guidance to identify opponens digiti minimi and flexor digiti minimi, and anatomically locating palmaris brevis muscle [46].

All the BoNT administration protocols are summarized in detail in Table 2.

Table 1. Characteristics of the RCTs included.

Article	Design	Intervention Characteristics	Comparison	Sample Size		Outcomes Measures	Follow-Up	Main Findings	Indications
				Intervention	*Comparator*				
Apalla et al. [36]	Randomized, double-blind, placebo-controlled clinical trial	Forty injections of Onabotulinumtoxin-A (100 units) in the painful area	Placebo (normal saline) injections	Patients with post-herpetic neuropathy *n*: 15 male/female: 8/7 Mean age: 73.2 ± 10.5	Patients with post-herpetic neuropathy *n*: 15 male/female: 10/5 Mean age: 77.5 ± 8.2	Quality of sleep, assessed by five-item questionnaire with a score ranging from 0 to 3	24 weeks	*Intervention Group* Sleep scores: significant improvements at week 2, which remained stable between weeks 2 and 4; after the initial decline, sleep scores remain unchanged until week 12 *Comparator Group* Sleep scores: no significant improvement at week 2, which remained unchanged between weeks 2 and 4 *Between groups* Sleep scores: significant differences at week 2	BoNT-A administration significantly improves quality of sleep at 2 weeks.
Xiao et al. [44]	Randomized, double-blind, placebo-controlled clinical trial	BoNT-A group: administrations (comprising several injections) of Lanbotulinum-toxinA (up to 200 units).	Lidocaine (active control) group: administrations (comprising several injections) of 0.5% lidocaine. Placebo group: placebo (normal saline) injections.	BoNT-A group: patients with postherpetic neuropathy *n*: 20 male/female: 11/9 Mean age: 70 ± 15.4	Lidocaine group: patients with postherpetic neuropathy *n*: 20 male/female: 8/12 Mean age: 65 ± 14.2 Placebo group: patients with postherpetic neuropathy *n*: 20 male/female: 9/11 Mean age: 67 ± 12.1	Sleep time (hours)	3 months	*Intervention Group* Sleep time: significant improvement on day 7 and after 3 months *Comparator Group* Sleep time: significant improvement on day 7 and after 3 months in both lidocaine group and placebo group *Between groups* Sleep time: improvement of IG was significantly greater compared with lidocaine and placebo groups	BoNT-A administration significantly improves sleep time.

Table 1. *Cont.*

Article	Design	Intervention Characteristics	Comparison	Sample Size		Outcomes Measures	Follow-Up	Main Findings	Indications
				Intervention	*Comparator*				
Attal et al. [26]	Randomized, double-blind, placebo-controlled, parallel-group clinical trial	Two administrations of Onabotulinumtoxin A (up to 300 units), 12 weeks apart.	Two administrations of saline, 12 weeks apart.	Patients with peripheral nerve lesion *n*: 34 male/female: 17/17 Mean age: 51.6 ± 16.7	Patients with peripheral nerve lesion *n*: 32 male/female: 20/12 Mean age: 52.3 ± 15.8	BPI, NPSI, EQ5D VAS, HADS, Sleep Problem Index	24 weeks	*Intervention Group* BPI VAS: significant reduction NPSI burning pain and paroxysmal pain subitem: significant improvement HADS, EQ5D VAS, Sleep Problem Index: p = NS *Comparator Group* BPI, NPSI, HADS, EQ5D VAS, Sleep Problem Index: p = NS *Between groups* BPI VAS: significant differences NPSI subscales: significant differences in paroxysmal pain and allodynia HADS: significant differences in anxiety Sleep Problem Index—6 items: significant differences Sleep Problem Index—9 items: significant differences	BoNT-A administration significantly improves of BPI VAS, NPSI burning pain and paroxysmal pain subitem, HADS, and Sleep Problem Index
Ranoux et al. [41]	Randomized, double-blind, placebo-controlled, parallel-group clinical trial	Administrations of Onabotulinumtoxin A (up to 200 units).	Administrations of saline.	Patients with posttraumatic/postoperative or postherpetic neuropathy *n*: 15 male/female: 6/9 Mean age: 53.8 ± 13.9	Patients with posttraumatic/postoperative or postherpetic neuropathy *n*: 14 male/female: 4/10 mean age: 49.7 ± 15.9	NPSI, BPI, HADS	24 weeks	*Intervention Group* NPSI subitems (burning, paroxysmal pain, allodynia): significant improvement at 12 weeks General activity and mood: improvement without significance HADS anxiety: slight improvement without significance *Comparator Group* NPSI, BPI, HADS: p = NS *Between groups* NPSI subscales (burning, electric shock and evoked pain to cold): significant differences at 12 weeks, without significant differences at week 24 BPI pain intensity: significant differences HADS: NS at 24 weeks	BoNT-A administration significantly improves NPSI subscales (burning, electric shock, and evoked pain to cold), and BPI pain intensity
Finlayson et al. [38]	Randomized, double-blind, placebo-controlled clinical trial	Seventy-five units of OnabotulinumtoxinA injected in the anterior and middle scalene muscles under EMG guidance.	Saline injected in the anterior and middle scalene muscles under EMG guidance.	Patients with thoracic outlet syndrome *n*: 20 male/female: 3/17 Mean age: 36.8 ± 8.9	Patients with thoracic outlet syndrome *n*: 18 male/female: 4/14 Mean age: 38.7 ± 7.0	DASH, SF-36	6 months	*Intervention Group* DASH, SF-36: not significant *Comparator Group* DASH, SF-36: not significant *Between groups* DASH, SF-36: not significant differences	BoNT-A administration did not improve DASH and SF-36

Table 1. *Cont.*

Article	Design	Intervention Characteristics	Comparison	Sample Size		Outcomes Measures	Follow-Up	Main Findings	Indications
				Intervention	*Comparator*				
Ghasemi et al. [39]	Randomized, double-blind, placebo-controlled clinical trial	100 units of Abobotulinum-toxinA in 0.9% saline were injected, each injection approximately 8–10 units	Placebo (normal saline) injections	Patients with diabetic neuropathy *n*: 20 male/female: 9/11 Mean age: 62.7 ± 9.9	Patients with diabetic neuropathy *n*: 20 male/female: 13/7 Mean age: 59.3 ± 9.6	NPS and DN4 questionnaire.	3 weeks	*Intervention Group* NPS subitems: significant differences, except for cold sensation DN4 questionnaire subitems: significant improvement (electric shocks, burning, pins and needle, and brushing) *Comparator Group* NPS subitems: NS DN4 questionnaire subitems: NS *Between groups* NR	BoNT-A administration significantly improves NPS (except for cold sensation) and DN4 questionnaire subitems (electric shocks, burning, pins and needles, and brushing subitems)
Salehi et al. [42]	Randomized double-blind, placebo-controlled clinical trial	Twelve injections of Abobotulinum-toxinA (8.33 units each point) in the dorsal foot surface	Placebo (normal saline) injections	Patients with diabetic neuropathy *n*:16 male/female: 6/10 Mean age: 58.3 ± 5.3	Patients with diabetic neuropathy *n*:16 male/female: 6/10 Mean age: 56.7 ± 7.5	NPS, SF-36, and PSQI questionnaires.	12 weeks	*Intervention Group* NPS, SF-36 subitems: significant improvement PSQI: significant decrease *Comparator Group* NR *Between groups* SF-36: significant differences PSQI: significant differences NPS subitems: significant differences, except for sharp sensation, sensory sensation, and deep sensation	BoNT-A administration significantly improve SF-36, PSQI and NPS subitems (except for sharp sensation, sensory sensation, and deep sensation)

Table 1. *Cont.*

Article	Design	Intervention Characteristics	Comparison	Sample Size		Outcomes Measures	Follow-Up	Main Findings	Indications
				Intervention	*Comparator*				
Taheri et al. [43]	Randomized, double-blind, placebo-controlled clinical trial	Group 1: twenty injections of BoNT-A (for a total of 150 units) in the sole of the right foot (7.5 units each injection); in the other feet, same procedure with saline placebo. Group 2: twenty injections of BoNT-A (for a total of 75 units) in the sole of both feet (3.75 units each injection, for a total of 150 units).	Placebo group: both feet with placebo (normal saline) injections.	Group 1: Patients with diabetic neuropathic pain *n*: 47 male/female: 16/31 Mean age: 54.5 ± 7.6 Group 2: Patients with diabetic neuropathic pain *n*: 47 male/female: 20/27 Mean age: 56.9 ± 6.2	Group N: Patients with diabetic neuropathic pain *n*: 47 male/female: 19/28 Mean age: 54.3 ± 8.2	NPS	4 weeks	*Intervention Group* Group 1: NPS subitems: significant improvements, except for dull sensation and cold sensation Group 2: NPS subitems: significant improvements, except for dull sensation ($p = 0.622$) and cold sensation *Comparator Group* NPS subitems: significant improvements, except for dull sensation, cold sensation ($p = 0.067$), unpleasant sensation and surface pain. *Between groups* NPS subitems: pain intensity, sharp sensation, hot sensation, sensitive sensation, unpleasant sensation, deep pain, and surface pain improved significantly after IGs vs. CG. Hot sensation subitem showed a significant difference between Group 1 vs. Group 2. Dull and cold sensations improvement did not show a significant difference between Group 2 and N.	BoNT-A administration significantly improve NPS (pain intensity, sharp sensation, hot sensation, sensitive sensation, unpleasant sensation, deep pain, and surface pain subitems)
Yuan et al. [45]	Randomized, double-blind, placebo-controlled, crossover clinical trial	OnabotulinumtoxinA injection of 50 units into each foot (4 units per injection); then crossover after 12 weeks.	Saline injection into each foot; then crossover after 12 weeks.	Patients with diabetic neuropathy *n*: 9 male/female: 6/12 Mean age: 65.6 ± 9.2	Patients with diabetic neuropathy *n*: 9 male/female: 6/12 Mean age: 65.6 ± 9.2	CPSQI, and SF-36	24 weeks	*Intervention Group* CPSQI: NS at week 12 *Comparator Group* NR *Between groups* CPSQI: significant improvements at 4 weeks SF-36: NS	BoNT-A administration significantly improve CPSQI

Table 1. *Cont.*

Article	Design	Intervention Characteristics	Comparison	Sample Size		Outcomes Measures	Follow-Up	Main Findings	Indications
				Intervention	*Comparator*				
Chun et al. [37]	Randomized, double-blind, placebo-controlled, crossover clinical trial	Injection of up to 400 units Onabotulinum-toxinA (phase 1). After 12 weeks of follow up, cross-over of participants was performed and subcutaneous injection of normal saline was administered (phase 2, P2)	Injection of normal saline (placebo) (phase 1). After 12 weeks of follow up, crossover of participants was performed and subcutaneous injection of up to 400 units Onabotulinum-toxin A were administered (phase 2, P2)	Patients with SCI *n*: 8 male/female: 6/2 Mean age: 45 (32–61)	Patients with SCI *n*: 8 male/female: 6/2 Mean age: 45 (32–61)	ISCIPBDS and QOL	20 weeks	*Intervention Group* ISCIPBDS subitems: change in pain intensity at 8 and 12 weeks in 33% of patients. QOL: 33% of patients reported at least moderate improvements at 2, 4, and 12 weeks, 17% at 8 weeks *Comparator Group* ISCIPBDS subitems: no patient reported a change in pain intensity at 8 and 12 weeks. 17% reported decreased pain interference with daily activities at 2 and 4 weeks QOL: no changes at 2, 4, 8 and 12 weeks *Between groups* NR	BoNT-A administration improve ISCIPBDS subitems
Han et al. [40]	Randomized, double-blind, placebo-controlled clinical trial	200 units Leti-botulinumtoxin A in 4 mL saline solution 1-time injection in painful area	Placebo (normal saline) injections	Patients with SCI *n*: 20 male/female: 15/5 Mean age: 53.1 ± 9.1	Patients with SCI *n*: 20 male/female: 14/6 Mean age: 48.9 ± 14.2	SF-MPQ and WHOQOL-BREF.	8 weeks	*Intervention Group* NR *Comparator Group* NR *Between groups* SF-MPQ: significant differences at 4 weeks and 8 weeks WHOQOL-BREF: NS	BoNT-A administration significantly improves SF-MPQ
Breuer et al. [46]	Randomized, double-blind, placebo-controlled clinical pilot trial	2500 units of rimabotulinum-toxin B in 0.5 mL of normal saline divided in 3 intramuscular under EMG guidance for opponens digiti minimi and flexor digiti minimi, and anatomically located for palmaris brevis muscle	Placebo (normal saline) intramuscular under EMG guidance for opponens digiti minimi and flexor digiti minimi, and anatomically located for palmaris brevis muscle	Patients with carpal tunnel syndrome *n*: 11 male/female: NR Mean age: NR	Patients with carpal tunnel syndrome *n*: 9 male/female: NR Mean age: NR	WHYMPI, Quality of sleep	13 weeks	*Intervention Group* WHYMPI quality of life indicators: improvements with statistical or borderline significance at different time points Pain interference with sleep (assessed with diary report): improved for weeks 2 through 8 *Comparator Group* Pain interference with sleep (assessed with phone report): improved for weeks 2 through 8 *Between groups* WHYMPI, Quality of sleep: no significant differences	BoNT-B administration did not show differences between groups in WHYMPI, and Quality of sleep

Abbreviations: BoNT-A: Botulinum Neurotoxin type A; BoNT: Botulinum Neurotoxin; BPI: Brief Pain Inventory; CG: comparator group; CPSQI: Chinese version of the Pittsburgh Sleep Quality Index; DASH: Disabilities of the Arm, Shoulder, and Hand; DN4: Douleur Neuropathique en 4 questions; EMG: electromyography; HADS: Hospital Anxiety and Depression Scale; IG: intervention group; ISCIBODS: International SCI Pain Basic Data Set; LSB: lumbar sympathetic block; NPS: Neuropathic Pain Scale; NPSI: Neuropathic Pain Symptom Inventory; NS: not significant; SCI: Spinal Cord Injury; SF-36: 36-Item Short Form Health Survey; SF-MPQ: short-form McGill Pain Questionnaire; PSQI: Pittsburgh Sleep Quality Index; WHOQOL-BREF: World Health Organization quality of life questionnaire; WHYMPI: West Haven-Yale Multidimensional Pain Inventory.

Table 2. BoNT administration protocols of the RCTs included in the present systematic review.

Study	Type of BoNT	Source of BoNT	Amount of BoNT	Injection Sites	Number of Injections	Rout of Injection
Post Herpetic Neuropathy						
Apalla et al. [36]	Onabotulinumtoxin A	Botox, Allergan	100 units	Painful area	Chessboard distribution, with a minimum distance of 1 cm between injections' sites, 40 injections in total.	Subcutaneous
Xiao et al. [44]	NA	BoNT-A (Lanzhou Institute of Biological Products, Lanzhou, China)	Total maximum administration of 200 units	Painful area	Over the affected area, injections every 1.0–2.0 cm radius of skin.	Subcutaneous
Peripheral nerve lesion						
Attal et al. [26]	Onabotulinumtoxin A	Botox; Allergan	Total maximum administration of 300 units, 5 units per injection	Painful area	Up to 60 injections, at sites 1.5–2 cm apart.	Subcutaneous
Posttraumatic/Postoperative Nerve Lesion or Post Herpetic Neuropathy						
Ranoux et al. [41]	Onabotulinumtoxin A	Botox, Allergan	Total maximum administration of 200 units, 5 units per injection	Painful area	Up to 40 injections, at sites 1.5 cm apart in the area mapped with a pen.	Subcutaneous
Thoracic Outlet Syndrome						
Finlayson et al. [38]	Onabotulinumtoxin A	Botox, Allergan	75 units	Anterior and middle scalene muscles	1 injection	Intramuscular under EMG guidance
Diabetic neuropathy						
Ghasemi et al. [39]	Abobotulinumtoxin A	Dysport, Ipsen	100 units; each injection approximately 8–10 units	Dorsum of the foot	Grid distribution pattern covering a total of 12 (3 × 4) sites.	Subcutaneous
Salehi et al. [42]	Abobotulinumtoxin A	Dysport, Ipsen	100 units; 0.1 mL (8.33 units) injection per site	Foot surface	Grid pattern of 12 points (3 × 4).	Subcutaneous
Taheri et al. [43]	NA	NA	150 units total; Group D1 each injection 7.5 U, Group D2 each injection 3.75 units.	Sole of the foot	Twenty points at distance of 1 cm from each other (a 5 × 4 grid).	Subcutaneous
Yuan et al. [45]	Onabotulinumtoxin A	Botox, Allergan	50 units per foot; each injection 4 units	Dorsum of the foot	Grid distribution pattern covering a total of 12 (3 × 4) sites.	Subcutaneous
Spinal Cord Injury						
Chun et al. [37]	Onabotulinumtoxin A	Botox, Allergan	Total maximum administration of 400 units, 5 units per injection	Painful area	Up to 80 injections; the area of pain was marked using a skin marker and a plastic cut-out template for injection sites separated from each other by a 1 cm radius.	Subcutaneous
Han et al. [40]	Letibotulinumtoxin A	Meditoxin (Medytox, Seoul, Korea)	200 units	Painful area	Checkerboard pattern over the maximally affected area.	Subcutaneous
Carpal Tunnel Syndrome						
Breuer et al. [46]	Rimabotulinumtoxin B	Myobloc, Supernus Pharmaceuticals	2500 units divided in 3 injections	Opponens digiti minimi, flexor digiti minimi, palmaris brevis muscle	3 injections (one for each muscle)	Intramuscular under EMG guidance for opponens digiti minimi and flexor digiti minimi, and anatomically located for palmaris brevis muscle

Abbreviations: BoNT: Botulinum Neurotoxin; EMG: electromyography; cm: centimeters.

2.2. *Main Findings*

2.2.1. Primary Outcome-Multidimensional Pain Assessment

The primary outcome assessed in this review was the effectiveness of BoNT injections in terms of multidimensional pain scales and were assessed in eight studies [26,37,39–43,46].

In further detail, three studies [39,42,43] assessed pain with the Neuropathic Pain Scale (NPS), showing significant improvements in most of the subitems considered [39,42,43]. The Neuropathic Pain Symptom Inventory (NPSI) has been used in two RCTs, reporting significant improvements in specific subitem scales (burning [26,41], paroxysmal [26], electric shock [41] and evoked pain to cold [41]). However, Ranoux et al. [41] did not show significant NPSI modifications 24 weeks after BoNT-A treatment. Similarly, these studies [26,41] assessed the Brief Pain Inventory (BPI) scale, reporting significant differences ($p < 0.05$) compared to placebo, but only in terms of pain intensity subitems [26,41].

Interestingly, Chun et al. [37] assessed pain intensity with the International SCI Pain Basic Data Set. The authors reported that 33% of patients assessed showed a significant change in pain intensity at 8 and 12 weeks, 50% showed a decreased pain interference with daily activities at 2 and 4 weeks, 50% reported a reduced pain interference with mood at 2 weeks, 33% at 4 and 8 weeks, and 50% reported a reduced pain interference with sleep at 2 and 4 weeks, 17% at 8 and 12 weeks. On the other hand, it should be noted that the statistical analysis was based on a descriptive approach [37].

In contrast, Han et al. [40] investigated the effectiveness of BoNT-A administration with the short-form McGill Pain Questionnaire, reporting significant differences between groups at 4 weeks ($p < 0.05$) and 8 weeks ($p < 0.05$) [40]. The Douleur Neuropathique 4 question scale has been used to assess pain in patients with diabetic neuropathy, showing significant improvement in electric shocks ($p = 0.01$), burning ($p < 0.01$), pins and needles ($p = 0.03$) and brushing ($p < 0.001$) subitems [39].

Lastly, Breuer et al. [46] assessed pain intensity with the West Haven-Yale Multidimensional Pain Inventory, highlighting improvements in quality-of-life indicators, reaching significance in some of the different time points assessed ($p < 0.05$). However, no significant differences between groups were reported (p = NS) [46].

2.2.2. Secondary Outcomes

The secondary outcomes assessed in the present systematic review were HR-QoL, physical function, anxiety and depression, and sleep quality.

In particular, HR-QoL has been assessed in six studies [26,37,38,40,42,45]. SF-36 has been assessed in three RCTs [38,42,45] reporting controversial results. In further detail, in patients with diabetic neuropathy, the RCT by Salehi et al. [42] reported significant improvement of SF-36 ($p = 0.007$) [42], while Yuan et al. [45] did not demonstrate significant differences between groups (p = NS) [45]. On the other hand, the study by Finlayson et al. [38] did not report significant improvement in SF-36 (p = NS) after BoNT-A treatment [38]. The World Health Organization Quality of Life questionnaire (WHOQOL-BREF) has been proposed by Han et al. [40] to assess pain relief in patients with SCI undergoing BoNT-A treatment; however, no significant differences were reported in the four domains of WHOQOL-BREF after the BoNT-A intervention [40]. On the other hand, Attal et al. [26] assessed EQ5D VAS scale, without underlining significant differences between groups (p = NS). Lastly, Chun et al. [37] reported at least moderate improvements in QoL in 33% of patients assessed at 2, 4, and 12 weeks, 17% at 8 weeks. Unfortunately, descriptive statistics was performed, without assessing the significance of the reported results [37]. Interestingly, physical function was assessed by Finlayson et al. [38] by the Disabilities of the Arm, Shoulder, and Hand scale; however, the authors did not report significant improvement (p = NS) after BoNT-A treatment [38].

On the other hand, anxiety and depression have been assessed through the Hospital Anxiety and Depression Scale (HADS) in two studies [26,41]: there were no significant differences between the BoNT-A and the placebo group (p = NS) in both studies [26,41].

Sleep quality has been specifically studied in six RCTs [26,36,42,44–46] assessing patients with herpetic neuropathy [36,44], peripheral nerve lesions [26], diabetic neuropathy [42,45] and carpal tunnel syndrome [46]. Thus, a high heterogeneity of the outcome measures was reported. The Pittsburgh Sleep Quality Index (PSQI) has been used in two studies [42,45] performed on patients with diabetic neuropathy. In further detail, the study by Salehi et al. [42] reported significant differences in the PSQI after the BoNT-A intervention compared to the placebo group [42]. On the contrary, in the RCT by Yuan et al. [45], the difference in the improvement in sleep quality between the BoNT-A group and the placebo group reached significance (1.72 ± 1.82 vs. -0.11 ± 2.78, $p < 0.05$) exclusively at 4 weeks after intervention. [45]. The Sleep Problem Index has been used by Attal et al. [26], reporting significant differences (Sleep Problem Index I, six items: 43.9 ± 21.4 vs. 40.6 ± 20.6; $p = 0.02$; Sleep Problem Index I, nine items: 45.3 ± 19.3 vs. 41.7 ± 19.6; $p = 0.03$) in the intergroup analysis [26]. Similarly, the study by Apalla et al. [36] assessed sleep quality with a five-item questionnaire in 14 patients with post-herpetic neuropathy treated with BoNT-A, showing significant improvements at week 2 ($p < 0.001$) and week 4 ($p < 0.001$), compared to placebo [36]. Sleep time has been also assessed by Xiao et al. [44], showing a significant improvement at day 7 and after 3 months from the BoNT-A treatment ($p < 0.01$) in patients with postherpetic neuropathy compared with lidocaine and placebo groups [44].

Lastly, the RCT by Breuer et al. [46] assessed sleep interference by pain in patients with carpal tunnel syndrome, reporting significant improvements ($p < 0.05$) in some of the time points assessed. On the other hand, the authors did not find a statistically significant difference between groups (p = NS) [46].

Table 1 reported further detail of the main results of the RTCs included in the present review.

2.3. Study Quality

According to the Jadad scale [47], all the RCTs included (n = 12, 100%) were high quality studies [26,36–46]. Table 3 showed in detail the score of each subitem of the Jadad scale for the RCTs included.

Table 3. Quality assessment of the studies included in the present systematic review.

Articles	Domain					Score
	Random Sequence Generation	Appropriate Randomization	Blinding of Participants or Personnel	Blinding of Outcome Assessors	Withdrawals and Dropouts	
Apalla et al. [36]	1	1	1	1	1	5
Attal et al. [26]	1	1	1	1	1	5
Breuer et al. [46]	1	0	1	1	0	3
Chun et al. [37]	1	1	1	1	1	5
Finlayson et al. [38]	1	1	1	1	1	5
Ghasemi et al. [39]	1	1	1	1	1	5
Han et al. 2016 [40]	1	1	1	1	1	5
Ranoux et al. [41]	1	1	1	1	1	5
Salehi et al. [42]	1	1	1	0	1	4
Taheri et al. [43]	1	1	1	1	0	4
Xiao et al. [44]	1	0	1	1	1	4
Yuan et al. [45]	1	0	1	1	1	4

Points were awarded as follows: study described as randomized, 1 point; appropriate randomization, 1 point; subjects blinded to intervention, 1 point; evaluator blinded to intervention, 1 point; description of withdrawals and dropouts, 1 point.

The risk of bias assessed by RoBv.2 [48] showed that 10 studies (83.3%) [26,36–43,45] ensured correct randomization, while 4 studies (33.3%) [42–45] showed some concerns in the second domain due to the lack of details about the blinding of the study participants. One study (6.7%) resulted in high risk of bias because it did not reach the target sample size [37]. All studies (n = 12, 100%) [26,36–46] showed low risk of bias in missing outcome data and outcome assessment, and 11 studies (91.7%) showed low risk of bias in selection of the reported results. See Figure 2 for further details.

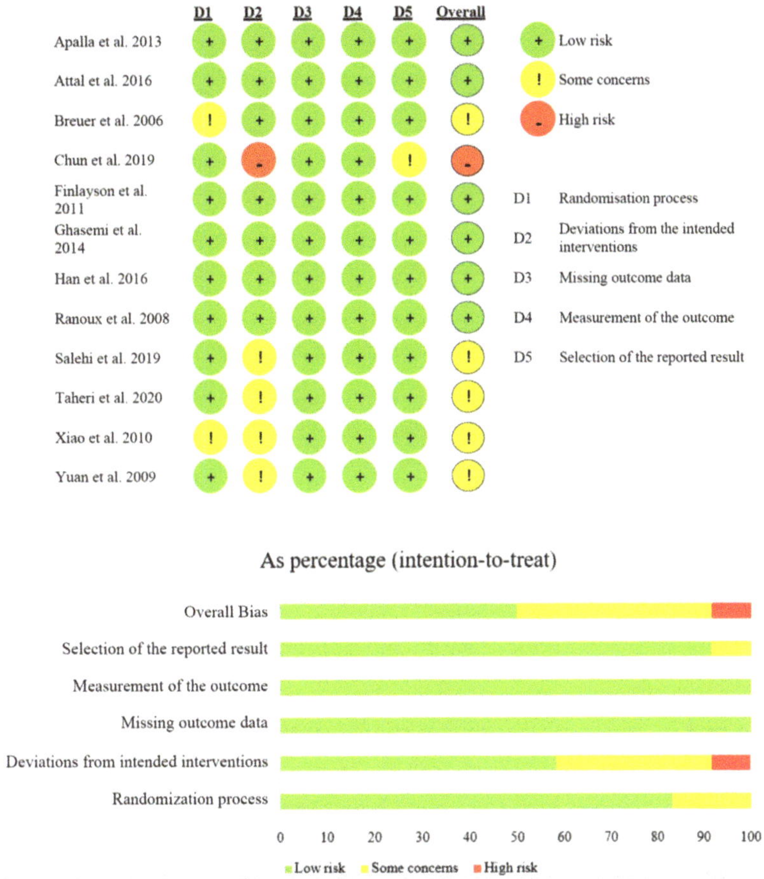

Figure 2. Risk of bias summary of the included studies.

3. Discussion

To date, there is a lack of consensus about the multidimensional effectiveness of BoNT in neuropathic pain and the optimal BoNT administration protocols are still debated [49–52].

Our findings showed a significant effect of BoNT administration in patients suffering from neuropathic pain due to postherpetic neuralgia [36,44], SCI [37,40], peripheral nerve lesion [26], diabetic neuropathy [39,42,43,45], post-traumatic/postoperative neuropathies [41], and carpal tunnel syndrome [46]. Our study results are in accordance with other evidence reporting positive effects of BoNT on pain management of non-specific neuropathic pain [51–53]. On the contrary, the etiological cause of neuropathic pain seems strictly related to the treatment effectiveness. In particular, Finlayson et al. [38] did not show a statistically significant improvement in patients with thoracic outlet syndrome pain [38]. Accordingly, Breuer et al. [46] did not reveal any significant differences between BoNT-B administration and placebo group in carpal tunnel syndrome, suggesting that BoNT might not provide additional benefits in the management of neuropathic pain with nerve compression etiology [38,46]. In contrast, most of the studies included highlighted positive results in the multidimensional management of neuropathic pain in several pathological conditions [26,36,37,39–45]. However, it should be noted that all these studies [26,36,37,39–45] assessed the effectiveness of BoNT-A administration, while the RCT by Breuer et al. [46]

was the only study that assessed BoNT-B; therefore, the role of BoNT-B in neuropathic pain management is far from being fully characterized.

As a result, clinicians should be aware of the evidence supporting BoNT use in specific conditions and the therapeutic intervention should be based on a precise diagnosis in order to select the patients more suitable to achieve better pain relief. Interestingly, our data showed positive long terms results of BoNT compared to lidocaine injections [44] in patients with post-herpetic neuropathy. These controversial data might be related to the characteristics of neuropathic pain and the BoNT administration protocols that were often heterogeneous in the studies included in the present review [26,36–46].

In this scenario, the current literature underlines a large gap of knowledge regarding the optimal BoNT therapeutic strategy, and this might be related to the lack of standardized BoNT administration protocols and injective techniques [54–56]. On the other hand, the French Recommendation for Neuropathic Pain of 2020 [30] provided a general indication of dosage from 50 to 300 units (onabotulinumtoxinA) every 3 months, without fully characterizing the intervention protocols or without suggesting any differences based on patients' characteristics. Our findings showed that although the maximum amount of BoNT-A injection might reach 400 units [37] and BoNT-B might reach 2500 units [46], specific subgroup analysis based on neuropathic pain should be considered and a wide difference in the dosage injected based on patient characteristics has been reported [26,36–45]. Similarly, in the past few years, different narrative and systematic reviews assessed the effects of BoNT administration characterizing patients with different types of neuropathic pain [49,51,52].

In particular, the systematic review by Hary et al. [57] assessed the effectiveness of BoNT-A administration in terms of pain relief in patients with peripheral neuropathic pain [57]. The authors reported significant effects in VAS scores, reporting better results in patients with diabetic polyneuropathy compared to patients with postherpetic, posttraumatic, or postsurgical neuralgia at 1 and 3 months post injection [57]. However, the authors mainly focused on unidimensional pain assessment and sleep improvement [57].

On the other hand, it should be noted that unidimensional scales lack the ability to characterize pain as a complex personal experience: these measurements heavily weight not only patient treatment satisfaction but also physician decision making [58]. In this context, multidimensional pain scales might better characterize pain intensity, nature, and location, and its consequences in function or mood, producing a quantitative description aiming at becoming the preferential assessment in a holistic approach [58]. Therefore, to the best of our knowledge, the present work represents the first systematic review of RCTs summarizing the current evidence on specific BoNT administration protocols providing data about the multidimensional effectiveness based on etiological cause of neuropathic pain to guide physicians in effective and safe therapeutic interventions in clinical practice.

Moreover, the large heterogeneity of administration protocols raises questions about the need to identify the lowest effective dose, not only to minimize the risk of adverse events but also from a cost effectiveness standpoint [59–61]. In particular, given the high prevalence of neuropathic pain and the strictly related sanitary costs [6,62], cost-effective therapies are mandatory in large-scale interventions aiming at improving quality of life and well-being of these patients. None of the studies included in this systematic review provided a precise cost analysis to better address the critical issue of the sanitary costs of pain management [26,36–46]. Therefore, future research is needed to better address this emerging issue in the clinical scenario.

Although unidimensional assessment has been proposed in the current literature to provide relevant quantitative data about BoNT efficacy [3,8,63], to date, clear evidence in multidimensional assessment for neuropathic pain is still lacking. In the present work, we sought to assess the effects of BoNT-induced pain relief in HR-QoL of patients with neuropathic pain. Unfortunately, our results revealed conflicting evidence regarding supporting BoNT efficacy on the overall well-being of patients suffering from neuropathic pain [26,36–46]. Nevertheless, it should be noted that recent research is now focused on a multidisciplinary framework with growing evidence emphasizing the need for comprehensive and synergic treatments to improve outcomes of patients with chronic neuropathic pain [3,20,64,65].

In this complex scenario, several therapeutic approaches have been investigated and might be proposed to improve pain management in patients with chronic neuropathic pain, including mini-invasive interventions [22,66–68], pharmacological drugs [67,69,70], nutraceuticals [71,72], physical exercise [21,73,74] and instrumental rehabilitative techniques [20,75]. In addition, recent advances in understanding the pathophysiological mechanisms underpinning pain chronification reveals that specific peripheral [11,12,76] and central circuits might be involved [12,77]. In further detail, it has been reported that after peripheral nerve damage, sodium channels increase in quantity in both the involved fibers and surrounding ones, which might lower the action potential threshold of the stimulus [14]. Hence, pain in the absence of an external stimulus might be due to an ectopic signal generated along this pathway [78–80]. In the hyperalgesia state, some receptors like TRPV1, involved in the noxious heat pathway [15], and the receptor TRPM8 involved in the cold pathway are upregulated [16,17]. Furthermore, it has been proposed that central sensitization mechanisms might affect continuous discharge of peripheral afferent fibers in the dorsal horn of the spinal cord inducing structural modifications in postsynaptic neurons [9,11]. Other contributors to pain hypersensitivity after a nerve lesion are inflammation, loss of inhibitory GABAergic interneurons in the spinal horn and enhanced sympathetic activity [9]. Even if the precise mechanism of action of BoNT is far from being understood in detail [77], recent research suggested a possible role in the nociceptive peripheral pathway, inflammation and even in central activities related to retrograde axonal transport to the spinal cord [81]. BoNT might implement HR-QoL in patients affected by neuropathic pain and might be considered as a part of a comprehensive management strategy including both pharmacological and non-pharmacological approaches [82,83]. However, to date, the effects of combined interventions to treat neuropathic pain have not been studied in the current literature; indeed, the role of BoNT injections in integrated multitarget interventions still remains unknown.

Taken together, the findings of this systematic review of RCTs might improve the knowledge about the possible role of BoNT treatment in chronic neuropathic pain [84]. In addition, the data reported by the RCTs included in the present review might support previous evidence suggesting positive effects of BoNT in patients with neuropathic pain [54–56], underlining the effectiveness of specific administration protocols tailored to patient characteristics.

Despite these considerations, we are aware that the present systematic review is not free from limitations. In particular, the lack of meta-analysis represents the main limitation of the present work. Unfortunately, the large heterogeneity of participants, intervention and outcomes assessed did not allow performance of a quantitative analysis, in accordance with the Cochrane Handbook for Systematic Review of Intervention (Ver, 6.1, 2020) [85]. Moreover, our search strategy might not include all records in these field and other sources have been searched in accordance with the PRISMA 2020 guidelines [86] in order to cover the relevant literature related to this topic. Lastly, 5 out of 12 studies included were directly supplied by pharmaceutical companies distributing the BoNT. Therefore, potential conflicts of interests in the studies included should be considered before making strong conclusions.

4. Conclusions

To date, the mechanisms underpinning the therapeutic role of BoNT in neuropathic pain are not completely understood, but the RCTs included in the present systematic review showed promising results in terms of pain relief, suggesting that BoNT-A might effectively improve symptoms in patients with neuropathic pain.

Although our findings provided evidence about the current BoNT protocols for specific neuropathic pain treatment, this systematic review of RCTs underlined the need for high-quality studies to better elucidate the optimal and cost-effective therapeutic strategies of BoNT administration.

Therefore, further evidence with standardized BoNT protocols in deeply characterized populations is needed to provide strong conclusions aiming to guide clinicians to implement precise and tailored treatment to improve the management of neuropathic pain.

5. Materials and Methods

5.1. Registration

This systematic review of randomized controlled trials (RCTs) was performed ethically in accordance with the Preferred Reporting Items for Systematic Reviews and Meta-analyses (PRISMA) statement [86]. The study protocol was realized before study initiation and submitted to PROSPERO (https://www.crd.york.ac.uk/prospero; accessed on 4 April 2022) with registration number CRD42022299703.

5.2. Search Strategy

We systematically searched PubMed/Medline, Scopus, Cochrane Central Register of Controlled Trials (CENTRAL), Physiotherapy Evidence Database (PEDro), and Web of Science for RCTs published up to 9 December 2021. Two investigators independently searched the databases. The search strategy is reported in Supplementary Table S1.

5.3. Selection Criteria

In accordance with the PICO model [87], we considered eligible RCTs satisfying the following criteria:

1. (P) Participants: adults suffering from neuropathic pain.
2. (I) Intervention: BoNT type A (BoNT-A) or BoNT type B (BoNT-B) administration.
3. (C) Comparator: any comparator, including placebo, other pharmacological treatment, non-pharmacological treatment or no treatment.
4. (O) Outcome: the primary outcome was self-reported pain relief in terms of multi-dimensional pain scales. The secondary outcomes were HR-QoL, physical function, anxiety and depression, and sleep quality.

We included RCTs published in peer-reviewed international journals in the English language. The exclusion criteria were the following: (i) studies involving animals; (ii) language other than English; (iii) participants with pregnancy; (iv) conference abstracts.

After duplication removal, two investigators independently reviewed the title and abstracts of retrieved articles to choose relevant articles. Any discordance was resolved by collegial discussion. If consensus was not achieved, a third reviewer was asked.

5.4. Data Extraction and Synthesis

All data were assessed and extracted by two authors independently from full-text documents into Microsoft Excel (IBM, New York, NY, USA). Missing data was directly requested from corresponding authors. Any disagreement between the two reviewers was solved by collegial discussion among the authors. In case of disagreement, a third author was asked.

The following data were extracted: (1) title; (2) authors; (3) publication year; (4) nationality; (5) participants (number, mean age and age range, gender); (6) interventions' characteristics; (7) comparator; (8) outcomes; (9) main findings; (10) funding.

The data extracted were summarized in tables. Subgroup analysis was performed based on neuropathic pain characteristics and by the BoNT administration modalities.

5.5. Quality Assessment and Risk of Bias

The quality of the studies included was assessed independently by two of the authors, according to the Jadad scale [47]. Discordances were solved by discussion between the authors or by asking a third reviewer. A Jadad score between 3 to 5 points was considered high quality.

The risk of bias was assessed by the Cochrane risk-of-bias tool for randomized trials (RoBv.2) [48]. Bias was classified as low, high, or unclear based on the item of RoBv.2. In particular, the domains assessed by RoBv.2 were: (i) random process; (ii) deviation from the intended interventions; (iii) missing outcome data; (iv) measurement of the outcome; and (v) selection of the reported result.

Supplementary Materials: The following supporting information can be downloaded at: https://www.mdpi.com/article/10.3390/toxins14050308/s1, Table S1. Search strategy; Table S2. Records excluded.

Author Contributions: Conceptualization, S.C. and M.I.; methodology, L.L., A.d.S., S.C. and M.I.; databases searching, L.L. and A.d.S.; data screening, L.L., A.d.S. and M.I.; data extraction, L.L., A.d.S. and M.I.; data synthesis and interpretation, L.L., A.d.S. and M.I.; manuscript drafting, L.L. and A.F.; critical revision, A.d.S., S.C. and M.I.; visualization, F.D., E.G. and A.B.; study supervision, S.C. and M.I.; study submission, M.I. All authors have read and agreed to the published version of the manuscript.

Funding: This research received no external funding.

Institutional Review Board Statement: The study was conducted according to the guidelines of the Declaration of Helsinki and approved by the Institutional Review Board.

Informed Consent Statement: Not applicable.

Data Availability Statement: The datasets generated during the current study are available from the corresponding author on reasonable request.

Acknowledgments: The authors would like to thank Alessio Turco for his support in this work.

Conflicts of Interest: The authors declare no conflict of interest.

Abbreviations

BoNT	botulinum toxin
BoNT-A	BoNT type A
BoNT-B	BoNT type B
CENTRAL	Cochrane Central Register of Controlled Trials
HADS	Hospital Anxiety and Depression Scale
HR-QoL	health-related quality of life
NRS	Numerical Rating Scale
NS	non significant
PEDro	Physiotherapy Evidence Database
RCTs	randomized controlled trials
SNRIs	Serotonin and norepinephrine reuptake inhibitors
TENS	Transcutaneous Electrical Nerve Stimulator
TRPM8	Transient Receptor Potential Member 8
TRPV1	Transient Receptor Potential Vanilloid 1
VAS	Visual Analogue Scale
WHOQOL-BREF	World Health Organization Quality of Life questionnaire

References

1. Scholz, J.; Finnerup, N.B.; Attal, N.; Aziz, Q.; Baron, R.; Bennett, M.I.; Benoliel, R.; Cohen, M.; Cruccu, G.; Davis, K.D.; et al. The IASP classification of chronic pain for ICD-11: Chronic neuropathic pain. *Pain* **2019**, *160*, 53–59. [CrossRef] [PubMed]
2. Doth, A.H.; Hansson, P.T.; Jensen, M.P.; Taylor, R.S. The burden of neuropathic pain: A systematic review and meta-analysis of health utilities. *Pain* **2010**, *149*, 338–344. [CrossRef] [PubMed]

3. Bernetti, A.; Agostini, F.; de Sire, A.; Mangone, M.; Tognolo, L.; Di Cesare, A.; Ruiu, P.; Paolucci, T.; Invernizzi, M.; Paoloni, M. Neuropathic Pain and Rehabilitation: A Systematic Review of International Guidelines. *Diagnostics* **2021**, *11*, 74. [CrossRef] [PubMed]
4. Langley, P.C.; Van Litsenburg, C.; Cappelleri, J.C.; Carroll, D. The burden associated with neuropathic pain in Western Europe. *J. Med. Econ.* **2013**, *16*, 85–95. [CrossRef]
5. Colloca, L.; Ludman, T.; Bouhassira, D.; Baron, R.; Dickenson, A.H.; Yarnitsky, D.; Freeman, R.; Truini, A.; Attal, N.; Finnerup, N.B.; et al. Neuropathic pain. *Nat. Rev. Dis. Primers* **2017**, *3*, 17002. [CrossRef]
6. Van Hecke, O.; Austin, S.K.; Khan, R.A.; Smith, B.H.; Torrance, N. Neuropathic pain in the general population: A systematic review of epidemiological studies. *Pain* **2014**, *155*, 654–662. [CrossRef]
7. Deng, Y.; Luo, L.; Hu, Y.; Fang, K.; Liu, J. Clinical practice guidelines for the management of neuropathic pain: A systematic review. *BMC Anesth.* **2016**, *16*, 12. [CrossRef]
8. Bates, D.; Schultheis, B.C.; Hanes, M.C.; Jolly, S.M.; Chakravarthy, K.V.; Deer, T.R.; Levy, R.M.; Hunter, C.W. A Comprehensive Algorithm for Management of Neuropathic Pain. *Pain Med.* **2019**, *20*, S2–S12. [CrossRef]
9. Baron, R.; Binder, A.; Wasner, G. Neuropathic pain: Diagnosis, pathophysiological mechanisms, and treatment. *Lancet Neurol.* **2010**, *9*, 807–819. [CrossRef]
10. Braz, J.; Solorzano, C.; Wang, X.; Basbaum, A.I. Transmitting pain and itch messages: A contemporary view of the spinal cord circuits that generate gate control. *Neuron* **2014**, *82*, 522–536. [CrossRef]
11. Tsuda, M.; Koga, K.; Chen, T.; Zhuo, M. Neuronal and microglial mechanisms for neuropathic pain in the spinal dorsal horn and anterior cingulate cortex. *J. Neurochem.* **2017**, *141*, 486–498. [CrossRef] [PubMed]
12. Nelson, T.S.; Fu, W.; Donahue, R.R.; Corder, G.F.; Hokfelt, T.; Wiley, R.G.; Taylor, B.K. Facilitation of neuropathic pain by the NPY Y1 receptor-expressing subpopulation of excitatory interneurons in the dorsal horn. *Sci. Rep.* **2019**, *9*, 7248. [CrossRef] [PubMed]
13. Papuc, E.; Rejdak, K. The role of neurostimulation in the treatment of neuropathic pain. *Ann. Agric. Environ. Med.* **2013**, *1*, 14–17.
14. Wu, G.; Ringkamp, M.; Hartke, T.V.; Murinson, B.B.; Campbell, J.N.; Griffin, J.W.; Meyer, R.A. Early onset of spontaneous activity in uninjured C-fiber nociceptors after injury to neighboring nerve fibers. *J. Neurosci.* **2001**, *21*, RC140. [CrossRef]
15. Caterina, M.J.; Julius, D. The vanilloid receptor: A molecular gateway to the pain pathway. *Ann. Rev. Neurosci.* **2001**, *24*, 487–517. [CrossRef] [PubMed]
16. Serra, J.; Sola, R.; Quiles, C.; Casanova-Molla, J.; Pascual, V.; Bostock, H.; Valls-Sole, J. C-nociceptors sensitized to cold in a patient with small-fiber neuropathy and cold allodynia. *Pain* **2009**, *147*, 46–53. [CrossRef]
17. Wasner, G.; Schattschneider, J.; Binder, A.; Baron, R. Topical menthol—A human model for cold pain by activation and sensitization of C nociceptors. *Brain* **2004**, *127*, 1159–1171. [CrossRef]
18. Chaparro, L.E.; Wiffen, P.J.; Moore, R.A.; Gilron, I. Combination pharmacotherapy for the treatment of neuropathic pain in adults. *Cochrane Database Syst. Rev.* **2012**, *2020*, CD008943. [CrossRef]
19. Smith, E.S.J. Advances in understanding nociception and neuropathic pain. *J. Neurol.* **2018**, *265*, 231–238. [CrossRef]
20. De Sire, A.; Ammendolia, A.; Lippi, L.; Farì, G.; Cisari, C.; Invernizzi, M. Percutaneous Electrical Nerve Stimulation (PENS) as a Rehabilitation Approach for Reducing Mixed Chronic Pain in Patients with Musculoskeletal Disorders. *Appl. Sci.* **2021**, *11*, 4257. [CrossRef]
21. De Sire, A.; Lippi, L.; Curci, C.; Calafiore, D.; Cisari, C.; Ammendolia, A.; Invernizzi, M. Effectiveness of Combined Treatment Using Physical Exercise and Ultrasound-Guided Radiofrequency Ablation of Genicular Nerves in Patients with Knee Osteoarthritis. *Appl. Sci.* **2021**, *11*, 4338. [CrossRef]
22. Harden, N.; Cohen, M. Unmet Needs in the Management of Neuropathic Pain. *J. Pain Symptom Manag.* **2003**, *25*, S12–S17. [CrossRef]
23. Intiso, D.; Basciani, M.; Santamato, A.; Intiso, M.; Di Rienzo, F. Botulinum Toxin Type A for the Treatment of Neuropathic Pain in Neuro-Rehabilitation. *Toxins* **2015**, *7*, 2454–2480. [CrossRef] [PubMed]
24. Fishman, L.M.; Anderson, C.; Rosner, B. BOTOX and physical therapy in the treatment of piriformis syndrome. *Am. J. Phys. Med. Rehabil.* **2002**, *81*, 936–942. [CrossRef] [PubMed]
25. Baricich, A.; Picelli, A.; Carda, S.; Smania, N.; Cisari, C.; Santamato, A.; de Sire, A.; Invernizzi, M. Electrical stimulation of antagonist muscles after botulinum toxin type A for post-stroke spastic equinus foot. A randomized single-blind pilot study. *Ann. Phys. Rehabil. Med.* **2019**, *62*, 214–219. [CrossRef]
26. Attal, N.; de Andrade, D.C.; Adam, F.; Ranoux, D.; Teixeira, M.J.; Galhardoni, R.; Raicher, I.; Uceyler, N.; Sommer, C.; Bouhassira, D. Safety and efficacy of repeated injections of botulinum toxin A in peripheral neuropathic pain (BOTNEP): A randomised, double-blind, placebo-controlled trial. *Lancet Neurol.* **2016**, *15*, 555–565. [CrossRef]
27. Foster, L.; Clapp, L.; Erickson, M.; Jabbari, B. Botulinum toxin A and chronic low back pain: A randomized, double-blind study. *Neurology* **2001**, *56*, 1290–1293. [CrossRef]
28. Park, H.J.; Lee, Y.; Lee, J.; Park, C.; Moon, D.E. The effects of botulinum toxin A on mechanical and cold allodynia in a rat model of neuropathic pain. *Can. J. Anaesth.* **2006**, *53*, 470–477. [CrossRef]
29. Finnerup, N.B.; Attal, N.; Haroutounian, S.; McNicol, E.; Baron, R.; Dworkin, R.H.; Gilron, I.; Haanpaa, M.; Hansson, P.; Jensen, T.S.; et al. Pharmacotherapy for neuropathic pain in adults: A systematic review and meta-analysis. *Lancet Neurol.* **2015**, *14*, 162–173. [CrossRef]

30. Moisset, X.; Bouhassira, D.; Avez Couturier, J.; Alchaar, H.; Conradi, S.; Delmotte, M.H.; Lanteri-Minet, M.; Lefaucheur, J.P.; Mick, G.; Piano, V.; et al. Pharmacological and non-pharmacological treatments for neuropathic pain: Systematic review and French recommendations. *Rev. Neurol.* **2020**, *176*, 325–352. [CrossRef]
31. Morra, M.E.; Elgebaly, A.; Elmaraezy, A.; Khalil, A.M.; Altibi, A.M.; Vu, T.L.; Mostafa, M.R.; Huy, N.T.; Hirayama, K. Therapeutic efficacy and safety of Botulinum Toxin A Therapy in Trigeminal Neuralgia: A systematic review and meta-analysis of randomized controlled trials. *J. Headache Pain* **2016**, *17*, 63. [CrossRef] [PubMed]
32. Do, T.M.; Unis, G.D.; Kattar, N.; Ananth, A.; McCoul, E.D. Neuromodulators for Atypical Facial Pain and Neuralgias: A Systematic Review and Meta-Analysis. *Laryngoscope* **2021**, *131*, 1235–1253. [CrossRef] [PubMed]
33. Li, X.L.; Zeng, X.; Zeng, S.; He, H.P.; Zeng, Z.; Peng, L.L.; Chen, L.G. Botulinum toxin A treatment for post-herpetic neuralgia: A systematic review and meta-analysis. *Exp. Med.* **2020**, *19*, 1058–1064. [CrossRef] [PubMed]
34. Battista, S.; Buzzatti, L.; Gandolfi, M.; Finocchi, C.; Falsiroli Maistrello, L.; Viceconti, A.; Giardulli, B.; Testa, M. The Use of Botulinum Toxin A as an Adjunctive Therapy in the Management of Chronic Musculoskeletal Pain: A Systematic Review with Meta-Analysis. *Toxins* **2021**, *13*, 640. [CrossRef]
35. Dawson, A.; Dawson, J.; Ernberg, M. The effect of botulinum toxin A on patients with persistent idiopathic dentoalveolar pain—A systematic review. *J. Oral Rehabil.* **2020**, *47*, 1184–1191. [CrossRef]
36. Apalla, Z.; Sotiriou, E.; Lallas, A.; Lazaridou, E.; Ioannides, D. Botulinum toxin A in postherpetic neuralgia: A parallel, randomized, double-blind, single-dose, placebo-controlled trial. *Clin. J. Pain* **2013**, *29*, 857–864. [CrossRef]
37. Chun, A.; Levy, I.; Yang, A.; Delgado, A.; Tsai, C.Y.; Leung, E.; Taylor, K.; Kolakowsky-Hayner, S.; Huang, V.; Escalon, M.; et al. Treatment of at-level spinal cord injury pain with botulinum toxin A. *Spinal Cord Ser. Cases* **2019**, *5*, 77. [CrossRef]
38. Finlayson, H.C.; O'Connor, R.J.; Brasher, P.M.A.; Travlos, A. Botulinum toxin injection for management of thoracic outlet syndrome: A double-blind, randomized, controlled trial. *Pain* **2011**, *152*, 2023–2028. [CrossRef]
39. Ghasemi, M.; Ansari, M.; Basiri, K.; Shaigannejad, V. The effects of intradermal botulinum toxin type a injections on pain symptoms of patients with diabetic neuropathy. *J. Res. Med. Sci.* **2014**, *19*, 106–111.
40. Han, Z.A.; Song, D.H.; Oh, H.M.; Chung, M.E. Botulinum toxin type A for neuropathic pain in patients with spinal cord injury. *Ann. Neurol.* **2016**, *79*, 569–578. [CrossRef]
41. Ranoux, D.; Attal, N.; Morain, F.; Bouhassira, D. Botulinum toxin type A induces direct analgesic effects in chronic neuropathic pain. *Ann. Neurol.* **2008**, *64*, 274–283. [CrossRef] [PubMed]
42. Salehi, H.; Moussaei, M.; Kamiab, Z.; Vakilian, A. The effects of botulinum toxin type A injection on pain symptoms, quality of life, and sleep quality of patients with diabetic neuropathy: A randomized double-blind clinical trial. *Iran. J. Neurol.* **2019**, *18*, 99–107. [CrossRef]
43. Taheri, M.; Sedaghat, M.; Solhpour, A.; Rostami, P.; Lima, B.S. The Effect of Intradermal Botulinum Toxin a injections on painful diabetic polyneuropathy. *Diabetes Metab. Syndr. Clin. Res. Rev.* **2020**, *14*, 1823–1828. [CrossRef] [PubMed]
44. Xiao, L.; Mackey, S.; Hui, H.; Xong, D.; Zhang, Q.; Zhang, D. Subcutaneous injection of botulinum toxin a is beneficial in postherpetic neuralgia. *Pain Med.* **2010**, *11*, 1827–1833. [CrossRef] [PubMed]
45. Yuan, R.Y.; Sheu, J.J.; Yu, J.M.; Chen, W.T.; Tseng, I.J.; Chang, H.H.; Hu, C.J. Botulinum toxin for diabetic neuropathic pain: A randomized double-blind crossover trial. *Neurology* **2009**, *72*, 1473–1478. [CrossRef] [PubMed]
46. Breuer, B.; Sperber, K.; Wallenstein, S.; Kiprovski, K.; Calapa, A.; Snow, B.; Pappagallo, M. Clinically significant placebo analgesic response in a pilot trial of botulinum B in patients with hand pain and carpal tunnel syndrome. *Pain Med.* **2006**, *7*, 16–24. [CrossRef] [PubMed]
47. Jadad, A.R.; Moore, R.A.; Carroll, D.; Jenkinson, C.; Reynolds, D.J.; Gavaghan, D.J.; McQuay, H.J. Assessing the quality of reports of randomized clinical trials: Is blinding necessary? *Control. Clin. Trials* **1996**, *17*, 1–12. [CrossRef]
48. Sterne, J.A.C.; Savovic, J.; Page, M.J.; Elbers, R.G.; Blencowe, N.S.; Boutron, I.; Cates, C.J.; Cheng, H.Y.; Corbett, M.S.; Eldridge, S.M.; et al. RoB 2: A revised tool for assessing risk of bias in randomised trials. *BMJ* **2019**, *366*, l4898. [CrossRef]
49. Egeo, G.; Fofi, L.; Barbanti, P. Botulinum Neurotoxin for the Treatment of Neuropathic Pain. *Front. Neurol.* **2020**, *11*, 716. [CrossRef]
50. Oh, H.M.; Chung, M.E. Botulinum Toxin for Neuropathic Pain: A Review of the Literature. *Toxins* **2015**, *7*, 3127–3154. [CrossRef]
51. Park, J.; Chung, M.E. Botulinum Toxin for Central Neuropathic Pain. *Toxins* **2018**, *10*, 224. [CrossRef] [PubMed]
52. Park, J.; Park, H.J. Botulinum Toxin for the Treatment of Neuropathic Pain. *Toxins* **2017**, *9*, 260. [CrossRef]
53. Wei, J.; Zhu, X.; Yang, G.; Shen, J.; Xie, P.; Zuo, X.; Xia, L.; Han, Q.; Zhao, Y. The efficacy and safety of botulinum toxin type A in treatment of trigeminal neuralgia and peripheral neuropathic pain: A meta-analysis of randomized controlled trials. *Brain Behav.* **2019**, *9*, e01409. [CrossRef] [PubMed]
54. Lakra, C.; Cohen, H. A clinical review of the use of Botulinum Toxin type A in managing central neuropathic pain in patients with spinal cord injury. *J. Spinal Cord Med.* **2020**, *2*, 1–5. [CrossRef]
55. Ahmed, S.; Subramaniam, S.; Sidhu, K.; Khattab, S.; Singh, D.; Babineau, J.; Kumbhare, D.A. Effect of Local Anesthetic Versus Botulinum Toxin-A Injections for Myofascial Pain Disorders: A Systematic Review and Meta-Analysis. *Clin. J. Pain* **2019**, *35*, 353–367. [CrossRef] [PubMed]
56. Forstenpointner, J.; Rice, A.S.C.; Finnerup, N.B.; Baron, R. Up-date on Clinical Management of Postherpetic Neuralgia and Mechanism-Based Treatment: New Options in Therapy. *J. Infect. Dis.* **2018**, *218*, S120–S126. [CrossRef] [PubMed]

57. Hary, V.; Schitter, S.; Martinez, V. Efficacy and safety of botulinum A toxin for the treatment of chronic peripheral neuropathic pain: A systematic review of randomized controlled trials and meta-analysis. *Eur. J. Pain* **2022**, *26*, 980–990. [CrossRef] [PubMed]
58. Van Boekel, R.L.M.; Vissers, K.C.P.; van der Sande, R.; Bronkhorst, E.; Lerou, J.G.C.; Steegers, M.A.H. Moving beyond pain scores: Multidimensional pain assessment is essential for adequate pain management after surgery. *PLoS ONE* **2017**, *12*, e0177345. [CrossRef]
59. Teasell, R.; Foley, N.; Pereira, S.; Sequeira, K.; Miller, T. Evidence to practice: Botulinum toxin in the treatment of spasticity post stroke. *Top. Stroke Rehabil.* **2012**, *19*, 115–121. [CrossRef]
60. Baricich, A.; Picelli, A.; Santamato, A.; Carda, S.; de Sire, A.; Smania, N.; Cisari, C.; Invernizzi, M. Safety Profile of High-Dose Botulinum Toxin Type A in Post-Stroke Spasticity Treatment. *Clin. Drug Investig.* **2018**, *38*, 991–1000. [CrossRef]
61. Hernandez Herrero, D.; Miangolarra Page, J.C. Descriptive analysis of the annual cost of treating spasticity with different types of botulinum toxin A. *Neurologia* 2019, *in press*. [CrossRef] [PubMed]
62. Smith, B.H.; Hebert, H.L.; Veluchamy, A. Neuropathic pain in the community: Prevalence, impact, and risk factors. *Pain* **2020**, *161* (Suppl. 1), S127–S137. [CrossRef] [PubMed]
63. Wang, C.; Zhang, Q.; Wang, R.; Xu, L. Botulinum Toxin Type A for Diabetic Peripheral Neuropathy Pain: A Systematic Review and Meta-Analysis. *J. Pain Res.* **2021**, *14*, 3855–3863. [CrossRef] [PubMed]
64. Fitzmaurice, B.C.; Rayen, A.T.A. Treatments for neuropathic pain: Up-to-date evidence and recommendations. *BJA Educ.* **2018**, *18*, 277–283. [CrossRef]
65. Rhon, D.I.; Fritz, J.M.; Greenlee, T.A.; Dry, K.E.; Mayhew, R.J.; Laugesen, M.C.; Dragusin, E.; Teyhen, D.S. Move to health-a holistic approach to the management of chronic low back pain: An intervention and implementation protocol developed for a pragmatic clinical trial. *J. Transl. Med.* **2021**, *19*, 357. [CrossRef] [PubMed]
66. Sdrulla, A.; Chen, G. Minimally invasive procedures for neuropathic pain. *Pain Manag.* **2016**, *2*, 6:103-9. [CrossRef]
67. Szok, D.; Tajti, J.; Nyari, A.; Vecsei, L. Therapeutic Approaches for Peripheral and Central Neuropathic Pain. *Behav. Neurol.* **2019**, *2019*, 8685954. [CrossRef]
68. Varshney, V.; Osborn, J.; Chaturvedi, R.; Shah, V.; Chakravarthy, K. Advances in the interventional management of neuropathic pain. *Ann. Transl. Med.* **2021**, *9*, 187. [CrossRef]
69. De Sire, A.; Moggio, L.; Demeco, A.; Fortunato, F.; Spano, R.; Aiello, V.; Marotta, N.; Ammendolia, A. Efficacy of rehabilitative techniques in reducing hemiplegic shoulder pain in stroke: Systematic review and meta-analysis. *Ann. Phys. Rehabil. Med.* **2021**, *65*, 101602. [CrossRef]
70. Cavalli, E.; Mammana, S.; Nicoletti, F.; Bramanti, P.; Mazzon, E. The neuropathic pain: An overview of the current treatment and future therapeutic approaches. *Int. J. Immunopathol. Pharm.* **2019**, *33*, 2058738419838383. [CrossRef]
71. Petrosino, S.; Schiano Moriello, A. Palmitoylethanolamide: A Nutritional Approach to Keep Neuroinflammation within Physiological Boundaries-A Systematic Review. *Int. J. Mol. Sci.* **2020**, *21*, 9526. [CrossRef] [PubMed]
72. Rowin, J. Integrative neuromuscular medicine: Neuropathy and neuropathic pain: Consider the alternatives. *Muscle Nerve* **2019**, *60*, 124–136. [CrossRef] [PubMed]
73. Leitzelar, B.N.; Koltyn, K.F. Exercise and Neuropathic Pain: A General Overview of Preclinical and Clinical Research. *Sports Med. Open* **2021**, *7*, 21. [CrossRef] [PubMed]
74. Andersen Hammond, E.; Pitz, M.; Shay, B. Neuropathic Pain in Taxane-Induced Peripheral Neuropathy: Evidence for Exercise in Treatment. *Neurorehabil. Neural. Repair.* **2019**, *33*, 792–799. [CrossRef] [PubMed]
75. Mokhtari, T.; Ren, Q.; Li, N.; Wang, F.; Bi, Y.; Hu, L. Transcutaneous Electrical Nerve Stimulation in Relieving Neuropathic Pain: Basic Mechanisms and Clinical Applications. *Curr. Pain Headache Rep.* **2020**, *24*, 14. [CrossRef]
76. Domoto, R.; Sekiguchi, F.; Tsubota, M.; Kawabata, A. Macrophage as a Peripheral Pain Regulator. *Cells* **2021**, *10*, 1881. [CrossRef]
77. Matak, I.; Bolcskei, K.; Bach-Rojecky, L.; Helyes, Z. Mechanisms of Botulinum Toxin Type A Action on Pain. *Toxins* **2019**, *11*, 459. [CrossRef]
78. Lai, J.; Hunter, J.C.; Porreca, F. The role of voltage-gated sodium channels in neuropathic pain. *Curr. Opin. Neurobiol.* **2003**, *13*, 291–297. [CrossRef]
79. Black, J.A.; Nikolajsen, L.; Kroner, K.; Jensen, T.S.; Waxman, S.G. Multiple sodium channel isoforms and mitogen-activated protein kinases are present in painful human neuromas. *Ann. Neurol.* **2008**, *64*, 644–653. [CrossRef]
80. Siqueira, S.R.; Alves, B.; Malpartida, H.M.; Teixeira, M.J.; Siqueira, J.T. Abnormal expression of voltage-gated sodium channels Nav1.7, Nav1.3 and Nav1.8 in trigeminal neuralgia. *Neuroscience* **2009**, *164*, 573–577. [CrossRef]
81. Kim, D.W.; Lee, S.K.; Ahnn, J. Botulinum Toxin as a Pain Killer: Players and Actions in Antinociception. *Toxins* **2015**, *7*, 2435–2453. [CrossRef] [PubMed]
82. Liampas, A.; Rekatsina, M.; Vadalouca, A.; Paladini, A.; Varrassi, G.; Zis, P. Non-Pharmacological Management of Painful Peripheral Neuropathies: A Systematic Review. *Adv. Ther.* **2020**, *37*, 4096–4106. [CrossRef] [PubMed]
83. Couto, D.S.; Goulart, G.; Luciano, L.L.; Cardoso, E.J.R. Quality of life in neuropathic pain: A literature review. *Res. Soc. Dev.* **2021**, *10*, e416101422250. [CrossRef]
84. Binder, A.; Baron, R. The Pharmacological Therapy of Chronic Neuropathic Pain. *Dtsch. Arztebl. Int.* **2016**, *113*, 616–625. [CrossRef] [PubMed]
85. Higgins, J.P.; Thomas, J.; Chandler, J.; Cumpston, M.; Li, T.; Page, M.J.; Welch, V.A. (Eds.) *Cochrane Handbook for Systematic Reviews of Interventions Version 6.2*; John Wiley & Sons: Hoboken, NJ, USA, 2021.

86. Page, M.J.; McKenzie, J.E.; Bossuyt, P.M.; Boutron, I.; Hoffmann, T.C.; Mulrow, C.D.; Shamseer, L.; Tetzlaff, J.M.; Akl, E.A.; Brennan, S.E.; et al. The PRISMA 2020 statement: An updated guideline for reporting systematic reviews. *BMJ* **2021**, *372*, n71. [CrossRef] [PubMed]
87. Huang, X.; Lin, J.; Demner-Fushman, D. Evaluation of PICO as a knowledge representation for clinical questions. *AMIA Annu. Symp. Proc.* **2006**, *2006*, 359–363.

toxins

Review

Novel Anatomical Guidelines on Botulinum Neurotoxin Injection for Wrinkles in the Nose Region

Kyu-Ho Yi [1,2], Ji-Hyun Lee [2], Hye-Won Hu [2] and Hee-Jin Kim [2,*]

1 COVID-19 Division, Wonju Public Health Center, Wonju-si 26417, Korea; kyuho90@daum.net
2 Division in Anatomy and Developmental Biology, Department of Oral Biology, Human Identification Research Institute, BK21 PLUS Project, Yonsei University College of Dentistry, 50-1 Yonsei-ro, Seoul 03722, Korea; jh_anatomy@yuhs.ac (J.-H.L.); wonhuh@yuhs.ac (H.-W.H.)
* Correspondence: hjk776@yuhs.ac; Tel.: +82-2-2228-3047

Abstract: Botulinum neurotoxin injection surrounding the nose area is frequently used in aesthetic settings. However, there is a shortage of thorough anatomical understanding that makes it difficult to treat wrinkles in the nose area. In this study, the anatomical aspects concerning the injection of botulinum neurotoxin into the nasalis, procerus, and levator labii superioris alaeque muscles are assessed. In addition, the present knowledge on localizing the botulinum neurotoxin injection point from a newer anatomy study is assessed. It was observed that, for the line-associated muscles in the nose region, the injection point may be more precisely defined. The optimal injection sites are the nasalis, procerus, and levator labii superioris alaeque muscles, and the injection technique is advised. We advise the best possible injection sites in association with anatomical standards for commonly injected muscles to increase efficiency in the nose region by removing the wrinkles. Similarly, these suggestions support a more precise procedure.

Keywords: nasalis muscle; procerus muscle; levator labii superioris alaeque muscles; botulinum neurotoxin; bunny line; horizontal radix line; facial wrinkle; injection point

Key Contribution: The research proposes a guide for effective botulinum neurotoxin injection for wrinkles in the nose region.

Citation: Yi, K.-H.; Lee, J.-H.; Hu, H.-W.; Kim, H.-J. Novel Anatomical Guidelines on Botulinum Neurotoxin Injection for Wrinkles in the Nose Region. *Toxins* **2022**, *14*, 342. https://doi.org/10.3390/toxins14050342

Received: 18 April 2022
Accepted: 13 May 2022
Published: 15 May 2022

1. Introduction

Botulinum neurotoxin (BoNT) prevents neural connections by stimulating the release of acetylcholine at the motor endplates, obstructing the muscle from contracting [1,2]. In aesthetic clinics, BoNT is commonly used primarily to eliminate wrinkles in the nose region by weakening the muscles involved in facial expression, such as the procerus, nasalis, and levator labii superioris alaeque nasi muscles. The primary aesthetic concerns in the nose region are the bunny lines, horizontal radix line, and nasal side wall scrunch wrinkles for many individuals (Figure 1).

As wrinkle removal using BoNT is being performed more often, the adverse effects, such as paralysis of the nearby muscles, diplopia, ptosis, and samurai eyebrows have been reported [3]. When treating wrinkles with BoNT in the nasal region, significant problems, such as diplopia may result from unintended paralysis of the rectus inferior or medialis [4,5]. To prevent these side effects, the injection should be administered at an anatomically accurate location of the targeting muscle, and the initial treatment should be at a reduced dosage.

Another factor that should be considered is that large doses and repetitive injections of BoNT create antibodies, leading to inadequate treatment outcomes [6–9]. According to previous research, antibody formation differs with types of botulinum neurotoxin [10,11].

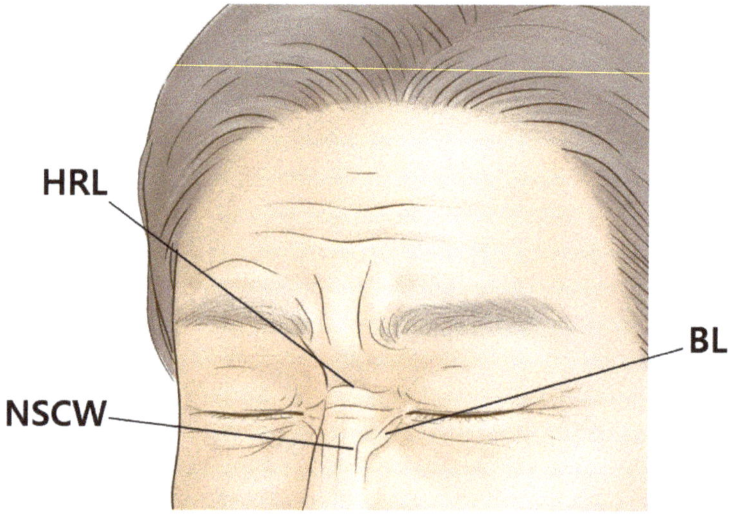

Figure 1. The wrinkles of the nose region are the bunny lines (BL), horizontal radix line (HRL), and nasal side wall scrunch wrinkles (NSCW).

Numerous studies on BoNT injection points in muscles have previously been published on external anatomical standards (Figure 2) [12–24]. We searched for articles using the following keywords: "botulinum neurotoxin in nose region" and "side effect of botulinum neurotoxin injection in nose wrinkle" on Pubmed and Scopus. A total of 16 articles and two textbooks were found; 10 articles were excluded owing to the irrelevance to this studies. The objective of this study is to propose a safe and efficient BoNT injection point and suggest injective techniques for wrinkles in the nasal region.

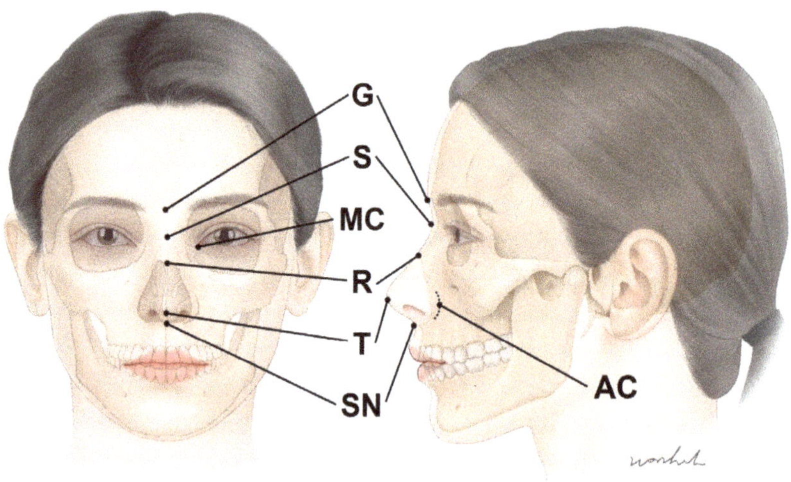

Figure 2. The external anatomical landmarks of the nose regions. G—glabella; S—sellion; R—rhinion; T—nose tip; SN—subnasale; MC—medial canthus; AC—alar crease.

2. The Anatomy of the Muscles in the Nasal Region

The schematic and dissected images of the muscles in the nasal region are presented below (Figures 3 and 4).

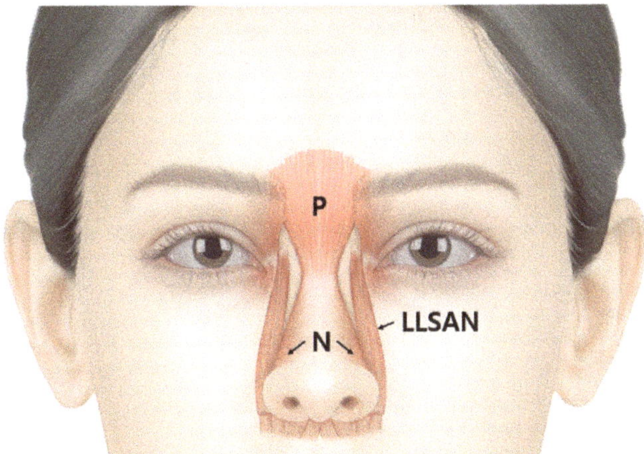

Figure 3. Schematic image of the procerus (P), nasalis (N), and levator labii superioris alaeque nasi (LLSAN).

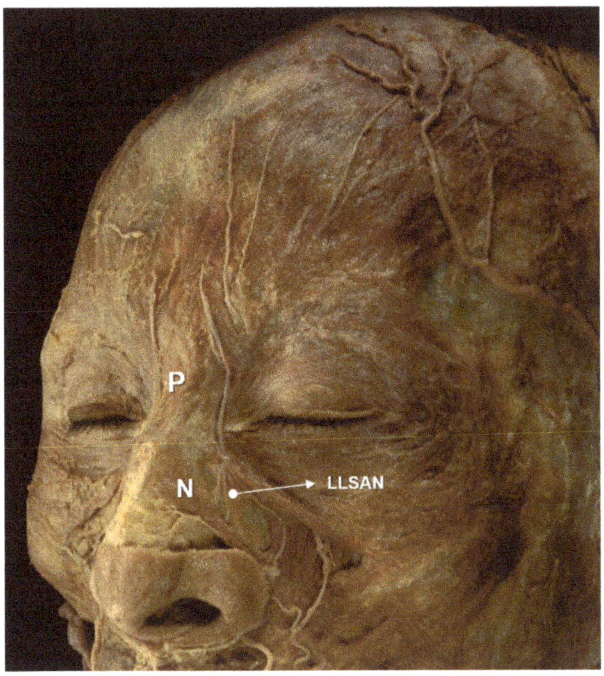

Figure 4. The dissected image of the procerus (P), nasalis (N), and levator labii superioris alaeque nasi (LLSAN).

2.1. Nasalis Muscle

The nasalis muscle is composed of transverse and alar parts [25]. The transverse part of the nasalis muscle is a morphologically triangular structure originating from the maxillary canine fossa which inserts into the lateral cartilage of the nose [26]. The alar part of the nasalis is a square-like muscle that originates from the maxillary lateral incisor and inserts into the lower alar cartilage [26]. These two parts of the nasalis muscle both contribute to the narrowing of the nostrils. However, the transverse muscle contracts the nasal aperture while the alar muscle widens the nostrils.

2.2. Procerus Muscle

The procerus muscle originates deep from the lateral cartilage of the nose and nasal bone, inserting superficially into the skin at the glabella and radix [27]. In the glabella, the muscle fibers of the procerus muscle combine with the frontalis muscle [28]. The procerus muscle acts by pulling down the medial portion of the eyebrow while creating a transverse wrinkle between the glabella and the sellion.

2.3. Levator Labii Superioris Alaeque Nasi Muscle

The levator labii superioris alaeque nasi muscle is a long running muscle originating in the maxillary frontal process, and involves the nasal ala and upper lip [29]. The levator labii superioris alaeque nasi muscle can be divided into deep and superficial bellies [30]. The deep belly runs deep to the levator labii superioris muscle, whereas the superficial belly runs superficial to the levator labii superioris muscle [29].

3. Injection Techniques

3.1. Horizontal Radix Line

The horizontal radix lines are mainly caused by the procerus muscle; thereby, targeting the procerus muscle is the critical injection point. A dose of 2 U should be injected into the nasal dorsum. An accurate point should be located in the middle of the glabella and sellion. The glabella is the midline bony prominence between the frontal bone and supraciliary arches. In addition, the glabella presents the most anterior part of the forehead (Figure 2). The sellion is located at the midline of the base of the nasal root. It is the most posteriorly located landmark of the frontonasal contour (Figure 2) [31].

3.2. Nasal Side Wall Scrunch Wrinkles (Vertical Lines)

Scrunch wrinkles on the nasal side wall are affected by the transverse part of the nasalis muscle. A dose of 2 U should be injected into the superior ala of the nose on both sides. The injection should be conducted in the middle of the rhinion and the medial end of the supra-alar crease (Figure 5).

3.3. Bunny Line (Oblique Nose Furrows)

The bunny lines are oblique wrinkles on both sides of the nose dorsum at a 45° angle. The lines are caused primarily by the levator labii superioris alaeque nasi muscle and secondarily by the medial muscular band of the orbicularis oculi muscle. A dose of 2 U should be injected into the upper part of the levator labii superioris alaeque nasi muscle on each side.

The injection should be conducted at the crossing point of the horizontal line at the level of the rhinion and the vertical line at the level of the medial canthus (Figure 5).

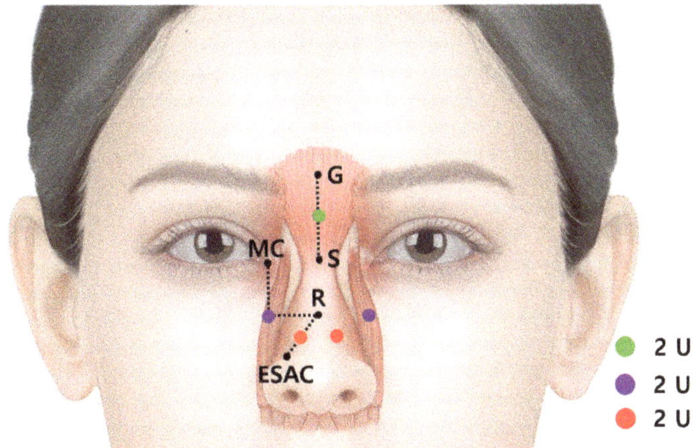

Figure 5. The injection point for the horizontal radix line (green dot) is in middle of the glabella (G) and sellion (S); for the nasal side scrunch wrinkles (red dots), it is in the middle of the rhinion (R) and the medial end of the supra-alar crease (ESAC); and for the bunny lines (purple dots) it is at the crossing point of the horizontal line at the level of the rhinion and the vertical line at the level of the medial canthus; 2 Units of botulinum neurotoxins should be injected per point.

4. Discussion

The nasal region has complex anatomical structures that may lead to adverse effects, such as BoNT rebalancing. The wrinkles in the nasal area can be exaggerated after BoNT injection because of this phenomenon. Therefore, it is important to identify and differentiate wrinkles in the nasal region. In addition, BoNT in the nasal region can cause major problems, such as diplopia resulting from the unintended blocking of the rectus inferior or medialis [4,5,32,33]. Although the incidence of BoNT causing diplopia is uncommon, it can be critical to some individuals [3,5]. Chen et al. reported a patient with BoNT in the lateral canthal region that caused lateral rectus paresis [4].

Side effects of BoNT injection in the upper nose region of the facial muscles, such as ptosis and samurai eyebrows, have been reported [31,32]. Although sensitivity to BoNT differs among individuals, there is no effective treatment for ptosis, which persists for several months [31,32]. Accurate BoNT injection points from an anatomical point of view have been suggested in various studies concerning BoNT injection in specific muscles [16–21,33–37]. According to the meta-analysis conducted by Camargo et al., most studies had duration of treatment of 5 months [38]. Notably, botulinum toxin effects take about two weeks to fully develop and last three to four months.

Precise injection guidelines can directly relate to fewer BoNT injections. When increased doses and repeated BoNT injections are administered, antibodies can be produced, leading to inadequate treatment outcomes [6–9]. Therefore, an extensive and detailed anatomical understanding of the muscles is crucial to achieve maximum results with the lowest possible amount of BoNT. If the desired outcomes are not attained, an additional retouching treatment may follow. Likewise, during an injection procedure, manual blocking of the inner boundary of the orbital rim should be carried out [39]. The injection should be performed gently and slowly to prevent BoNT from laterally diffusing to the eyelids [40]. The limitation of this study is that the review is an anatomy-based proposal for nasal-region wrinkles. These precise injection methods would be time consuming to be applied in clinics. Moreover, the suggested doses are not universal to all types of BoNT and may applied in increased or decreased doses [41].

In summary, the suggested injection point for the horizontal radix line is at the middle of the glabella and sellion; at the middle of the rhinion and the medial end of the supra-alar crease for the nasal side crunch scrunch wrinkles; and at the crossing point of the horizontal line at the level of the rhinion and the vertical line at the level of the medial canthus for the bunny line. An amount of 2 U of botulinum neurotoxin should injected per point.

This study carried out a broad analysis of published research on the anatomy of muscles in the nasal region to provide anatomical guidelines for BoNT indications.

Author Contributions: Conceptualization, K.-H.Y.; writing—Original Draft Preparation, K.-H.Y.; writing—Review and Editing, J.-H.L. and H.-J.K.; visualization, H.-W.H.; and supervision, H.-W.H. and H.-J.K. All authors have read and agreed to the published version of the manuscript.

Funding: This work was supported by the National Research Foundation of Korea (NRF) grant funded by the Korean government (MSIP) (NRF-2020R1A2B5B0100223811).

Institutional Review Board Statement: Not applicable.

Informed Consent Statement: Not applicable.

Data Availability Statement: Not applicable.

Conflicts of Interest: The authors declare no conflict of interest.

References

1. Dessy, L.A.; Mazzocchi, M.; Rubino, C.; Mazzarello, V.; Spissu, N.; Scuderi, N. An Objective Assessment of Botulinum Toxin A Effect on Superficial Skin Texture. *Ann. Plast. Surg.* **2007**, *58*, 469–473. [CrossRef] [PubMed]
2. Childers, M.K. Targeting the Neuromuscular Junction in Skeletal Muscles. *Am. J. Phys. Med. Rehabil.* **2004**, *83*, S38–S44. [CrossRef] [PubMed]
3. Wutthiphan, S.; Kowal, L.; O'Day, J.; Jones, S.; Price, J. Diplopia following subcutaneous injections of botulinum A toxin for facial spasms. *J. Pediatr. Ophthalmol. Strabismus* **1997**, *34*, 229–234. [CrossRef] [PubMed]
4. Chen, C.S.; Miller, N.R. Botulinum toxin injection causing lateral rectus palsy. *Br. J. Ophthalmol.* **2007**, *91*, 843. [CrossRef] [PubMed]
5. Aristodemou, P.; Watt, L.; Baldwin, C.; Hugkulstone, C. Diplopia Associated with the Cosmetic Use of Botulinum Toxin A for Facial Rejuvenation. *Ophthalmic Plast. Reconstr. Surg.* **2006**, *22*, 134–136. [CrossRef] [PubMed]
6. Hsu, T.S.J.; Dover, J.S.; Arndt, K.A. Effect of Volume and Concentration on the Diffusion of Botulinum Exotoxin A. *Arch. Dermatol.* **2004**, *140*, 1351–1354. [CrossRef]
7. Kinnett, D. Botulinum toxin A injections in children: Technique and dosing issues. *Am. J. Phys. Med. Rehabil.* **2004**, *83*, S59–S64. [CrossRef]
8. Lepage, D.; Parratte, B.; Tatu, L.; Vuiller, F.; Monnier, G. Extra- and intramuscular nerve supply of the muscles of the anterior antebrachial compartment: Applications for selective neurotomy and for botulinum toxin injection. *Surg. Radiol. Anat.* **2005**, *27*, 420–430. [CrossRef]
9. Pingel, J.; Nielsen, M.S.; Lauridsen, T.; Rix, K.R.; Bech, M.; Alkjær, T.; Andersen, I.T.; Nielsen, J.B.; Feidenhansl, R. Injection of high dose botulinum-toxin A leads to impaired skeletal muscle function and damage of the fibrilar and non-fibrilar structures. *Sci. Rep.* **2017**, *7*, 14746. [CrossRef]
10. Bellows, S.; Jankovic, J. Immunogenicity Associated with Botulinum Toxin Treatment. *Toxins* **2019**, *11*, 491. [CrossRef]
11. Campanati, A.; Giuliodori, K.; Martina, E.; Giuliano, A.; Ganzetti, G.; Offidani, A. Onabotulinumtoxin type A (Botox((R))) versus Incobotulinumtoxin type A (Xeomin((R))) in the treatment of focal idiopathic palmar hyperhidrosis: Results of a com-parative double-blind clinical trial. *J. Neural. Transm.* **2014**, *121*, 21–26. [CrossRef] [PubMed]
12. Yi, K.-H.; Cong, L.; Bae, J.-H.; Park, E.-S.; Rha, D.-W.; Kim, H.-J. Neuromuscular structure of the tibialis anterior muscle for functional electrical stimulation. *Surg. Radiol. Anat.* **2017**, *39*, 77–83. [CrossRef] [PubMed]
13. Yi, K.-H.; Rha, D.-W.; Lee, S.C.; Cong, L.; Lee, H.-J.; Lee, Y.-W.; Kim, H.-J.; Hu, K.-S. Intramuscular nerve distribution pattern of ankle invertor muscles in human cadaver using Sihler stain. *Muscle Nerve* **2016**, *53*, 742–747. [CrossRef] [PubMed]
14. Rha, D.-W.; Yi, K.-H.; Park, E.S.; Park, C.; Kim, H.-J. Intramuscular nerve distribution of the hamstring muscles: Application to treating spasticity. *Clin. Anat.* **2016**, *29*, 746–751. [CrossRef]
15. Yi, K.; Choi, Y.; Cong, L.; Lee, K.; Hu, K.; Kim, H. Effective botulinum toxin injection guide for treatment of cervical dystonia. *Clin. Anat.* **2020**, *33*, 192–198. [CrossRef]
16. Yi, K.; Lee, H.; Lee, J.; Lee, K.; Kim, H. Effective botulinum neurotoxin injection in treating iliopsoas spasticity. *Clin. Anat.* **2021**, *34*, 431–436. [CrossRef]
17. Yi, K.; Lee, H.; Choi, Y.; Lee, K.; Lee, J.; Kim, H. Anatomical guide for botulinum neurotoxin injection: Application to cosmetic shoulder contouring, pain syndromes, and cervical dystonia. *Clin. Anat.* **2021**, *34*, 822–828. [CrossRef]
18. Yi, K.; Lee, K.; Lee, J.; Hu, H.; Lee, K.; Seo, K.K.; Kim, H. Guidelines for botulinum neurotoxin injections in piriformis syndrome. *Clin. Anat.* **2021**, *34*, 1028–1034. [CrossRef]
19. Yi, K.-H.; Lee, H.-J.; Lee, J.-H.; Seo, K.; Kim, H.-J. Application of Botulinum Neurotoxin Injections in TRAM Flap for Breast Reconstruction: Intramuscular Neural Arborization of the Rectus Abdominis Muscle. *Toxins* **2021**, *13*, 269. [CrossRef]

20. Yi, K.-H.; Lee, J.-H.; Lee, D.K.; Hu, H.-W.; Seo, K.K.; Kim, H.-J. Anatomical locations of the motor endplates of sartorius muscle for botulinum toxin injections in treatment of muscle spasticity. *Surg. Radiol. Anat.* **2021**, *43*, 2025–2030. [CrossRef]
21. Yi, K.H.; Lee, J.H.; Kim, H.J. Intramuscular Neural Distribution of the Serratus Anterior Muscle: Regarding Botulinum Neu-rotoxin Injection for Treating Myofascial Pain Syndrome. *Toxins* **2022**, *14*, 271. [CrossRef] [PubMed]
22. Yi, K.-H.; Lee, J.-H.; Hu, H.-W.; Kim, H.-J. Anatomical Proposal for Botulinum Neurotoxin Injection for Glabellar Frown Lines. *Toxins* **2022**, *14*, 268. [CrossRef] [PubMed]
23. Yi, K.-H.; Lee, H.-J.; Seo, K.K.; Kim, H.-J. Intramuscular Neural Arborization of the Latissimus Dorsi Muscle: Application of Botulinum Neurotoxin Injection in Flap Reconstruction. *Toxins* **2022**, *14*, 107. [CrossRef] [PubMed]
24. Jeong, J.; Terence, G.; Kim, J. Understanding the Anatomy of the Transverse Nasalis Aponeurotic Fibers and Its Importance in Asian Rhinoplasty. *Ann. Plast. Surg.* **2018**, *81*, 516–522. [CrossRef]
25. Hur, M.-S.; Hu, K.-S.; Youn, K.-H.; Song, W.-C.; Abe, S.; Kim, H.-J. New anatomical profile of the nasal musculature: Dilator naris vestibularis, dilator naris anterior, and alar part of the nasalis. *Clin. Anat.* **2011**, *24*, 162–167. [CrossRef] [PubMed]
26. Starkman, S.J.; Sherris, D.A. Association of Corrugator Supercilii and Procerus Myectomy With Endoscopic Browlift Outcomes. *JAMA Fac. Plast. Surg.* **2019**, *21*, 375–380. [CrossRef] [PubMed]
27. Hwang, K.; Jin, S.; Park, J.H.; Chung, I.H. Innervation of the Procerus Muscle. *J. Craniofacial Surg.* **2006**, *17*, 484–486. [CrossRef]
28. Hur, M.S.; Hu, K.S.; Park, J.T.; Youn, K.H.; Kim, H.J. New anatomical insight of the levator labii superioris alaeque nasi and the transverse part of the nasalis. *Surg. Radiol. Anat.* **2010**, *32*, 753–756. [CrossRef]
29. Hur, M.-S. Anatomical relationships of the procerus with the nasal ala and the nasal muscles: Transverse part of the nasalis and levator labii superioris alaeque nasi. *Surg. Radiol. Anat.* **2017**, *39*, 865–869. [CrossRef]
30. Katina, S.; McNeil, K.; Ayoub, A.; Guilfoyle, B.; Khambay, B.; Siebert, P.; Sukno, F.; Rojas, M.; Vittert, L.; Waddington, J.; et al. The definitions of three-dimensional landmarks on the human face: An interdisciplinary view. *J. Anat.* **2016**, *228*, 355–365. [CrossRef]
31. Kothari, M.; Shukri, N.; Quayyum, A. Transient superior oblique paresis after injection of Botulinum Toxin A for facial reju-venation. *Indian J. Ophthalmol.* **2012**, *60*, 77–78. [CrossRef] [PubMed]
32. Sheen-Ophir, S.; Almog, Y. Diplopia following subcutaneous injections of botulinum toxin for cosmetic or medical use. *Harefuah* **2013**, *152*, 98–100. [PubMed]
33. Nestor, M.S.; Han, H.; Gade, A.; Fischer, D.; Saban, Y.; Polselli, R. Botulinum toxin-induced blepharoptosis: Anatomy, etiology, prevention, and therapeutic options. *J. Cosmet. Dermatol.* **2021**, *20*, 3133–3146. [CrossRef] [PubMed]
34. Kwon, K.-H.; Shin, K.S.; Yeon, S.H.; Kwon, D.G. Application of botulinum toxin in maxillofacial field: Part II. Wrinkle, intraoral ulcer, and cranio-maxillofacial pain. *Maxillofac. Plast. Reconstr. Surg.* **2019**, *41*, 42. [CrossRef] [PubMed]
35. Galadari, H.; Galadari, I.; Smit, R.; Prygova, I.; Redaelli, A. Use of AbobotulinumtoxinA for Cosmetic Treatments in the Neck, and Middle and Lower Areas of the Face: A Systematic Review. *Toxins* **2021**, *13*, 169. [CrossRef]
36. Yi, K.-H.; Lee, K.-L.; Lee, J.-H.; An, M.-H.; Kim, H.-S.; Kim, H.-J. Elucidating intramuscular neural distribution of the quadratus lumborum muscle to propose an optimal trigger point injection for myofascial pain syndrome. 2022; *in press*. [CrossRef]
37. Yi, K.H.; Lee, H.J.; Seo, K.K.; Kim, H.J. Botulinum neurotoxin injection guidelines regarding flap surgeries in breast recon-struction. *J. Plast. Reconstr. Aesthetic Surg.* **2022**, *75*, 503–505. [CrossRef]
38. Yi, K.-H.; Lee, J.-H.; Kim, H.-M.; Kim, H.-J. The botulinum neurotoxin for pain control after breast reconstruction: Neural distribution of the pectoralis major muscle. *Reg. Anesth. Pain Med.* **2022**, *47*, 322–326. [CrossRef]
39. Camargo, C.P.; Xia, J.; Costa, C.S.; Gemperli, R.; Tatini, M.D.; Bulsara, M.K.; Riera, R. Botulinum toxin type A for facial wrinkles. *Cochrane Database Syst. Rev.* **2021**, *7*, CD011301.
40. Seo, K.K. *Botulinum Toxin for Asians*; Springer: Berlin/Heidelberg, Germany, 2017.
41. Campanati, A.; Martina, E.; Giuliodori, K.; Consales, V.; Bobyr, I.; Offidani, A. Botulinum Toxin Off-Label Use in Dermatology: A Review. *Skin Appendage Disord.* **2017**, *3*, 39–56. [CrossRef]

toxins MDPI

Article

Can Botulinum Toxin-A Contribute to Reconstructing the Physiological Homeostasis of the Masticatory Complex in Short-Faced Patients during Occlusal Therapy? A Prospective Pilot Study

Xin Li [1], Xiaoyan Feng [1], Juan Li [1], Xinyu Bao [1], Jinghong Xu [2] and Jun Lin [1,*]

[1] Department of Stomatology, The First Affiliated Hospital, College of Medicine, Zhejiang University, Hangzhou 310003, China; 22018628@zju.edu.cn (X.L.); 21718013@zju.edu.cn (X.F.); lijuan5650@zju.edu.cn (J.L.); 22118751@zju.edu.cn (X.B.)
[2] Department of Plastic Surgery, The First Affiliated Hospital, College of Medicine, Zhejiang University, Hangzhou 310003, China; 1304017@zju.edu.cn
* Correspondence: linjun2@zju.edu.cn; Tel.: +86-0571-8723-6338

Abstract: The physiological homeostasis of the masticatory complex in short-faced patients is too robust to be disintegrated and reconstructed due to the powerful masseter muscle. This study innovatively introduced the botulinum toxin-A (BTX-A) into the field of dental occlusal treatment, providing a novel and minimally invasive therapy perspective for the two major clinical problems in these patients (low treatment efficiency and high rates of complications). In total, 10 adult patients with skeletal low angle seeking occlusal treatment (age: 27.0 ± 6.1 years; 4 males and 6 females) were administered 30–50 U of BTX-A in each masseter muscle and evaluated before and 3 months after injection based on cone-beam computed tomography (CBCT). We found a significant reduction in the thickness of the masseter muscle (MMT) ($p < 0.0001$). With regards to occlusion, we found a significant increase in the height of the maxillary second molar (U7-PP) ($p < 0.05$) with significantly flattened occlusal curves (the curve of Spee [COS] ($p < 0.01$), and the curve of Wilson [COW] ($p < 0.05$)). Furthermore, the variations in the temporomandibular joint exhibited a significant reduction in the anterior joint space (AJS) ($p < 0.05$) and superior joint space (SJS) ($p < 0.05$). In addition, the correlation analysis of the masticatory complex provided the basis for the following multiple regression equation: MMT = 10.08 − 0.11 COW + 2.73 AJS. The findings from our pilot study indicate that BTX-A, as a new adjuvant treatment attempt of occlusal therapy for short-faced patients, can provide a more favorable muscular environment for subsequent occlusal therapy through the adjustment of the biting force and may contribute to the reconstruction of healthier homeostasis of the masticatory complex. However, further research is required to establish the reliability and validity of these findings.

Keywords: botulinum toxin-A; masticatory complex; cone-beam computed tomography; masseter muscle; short-faced patients

Key Contribution: This study preliminarily explores the feasibility of clinical application of BTX-A to reconstruct the physiological homeostasis of the masticatory complex by researching the variation of the masticatory complex in short-faced patients, providing an innovative and minimally invasive treatment perspective for existing major clinical problems in the future.

Citation: Li, X.; Feng, X.; Li, J.; Bao, X.; Xu, J.; Lin, J. Can Botulinum Toxin-A Contribute to Reconstructing the Physiological Homeostasis of the Masticatory Complex in Short-Faced Patients during Occlusal Therapy? A Prospective Pilot Study. *Toxins* **2022**, *14*, 374. https://doi.org/10.3390/toxins14060374

Received: 25 April 2022
Accepted: 26 May 2022
Published: 28 May 2022

1. Introduction

Since the birth of dentistry, the discussion on the target of occlusal treatment has been continuously progressing [1–3]. In the early stage, dentists only focused on the teeth and occlusal relationship, which was generally static and non-functional [4–6]. Until the prevalence of functional occlusion in recent years [7,8], the concept of complete dentistry was put forward [9], with the studies on dynamic occlusion and masticatory function valued by interdisciplinary oral researchers [1–3,10–18], especially in the fields of orthodontics [1–3], occlusal reconstruction [10,11], dental implant [12,13], periodontal disease [14,15], dental caries treatment [16], temporomandibular joint disorders (TMD) [17], bruxism [18], etc. Dentists have begun to realize the integrality of the masticatory system (MS) (i.e., the stomatognathic system (SS)), which is generally composed of teeth and occlusion, the temporomandibular joint (TMJ), neuromuscular factors, jaw, periodontal tissue, etc. [19]. Therefore, occlusal therapy is regarded as the process of dynamic reconstruction of each component, whose ultimate goal is to maintain the health and stability of the whole masticatory system [1–3,9,19].

The significant role of muscle factors in dental and maxillofacial development has been expounded by a lot of scholars [20,21]. It is proved that the occlusion and jaw are the consequence of the balance among multiple functional systems involving the masticatory system, respiratory system, etc., which also serves as the theoretical basis to prevent and correct jaw deformity and malocclusion in adolescence [22]. In addition, the function of TMJ is also closely related to the masticatory muscles, whose long-term discoordination and hyperactivity results in structural variation of TMJ [23], which was observed even in adult patients with a stable tendency [24]. Therefore, we conducted this study on the masticatory complex composed of the three (muscle-occlusion-joint complex) (Figure 1A).

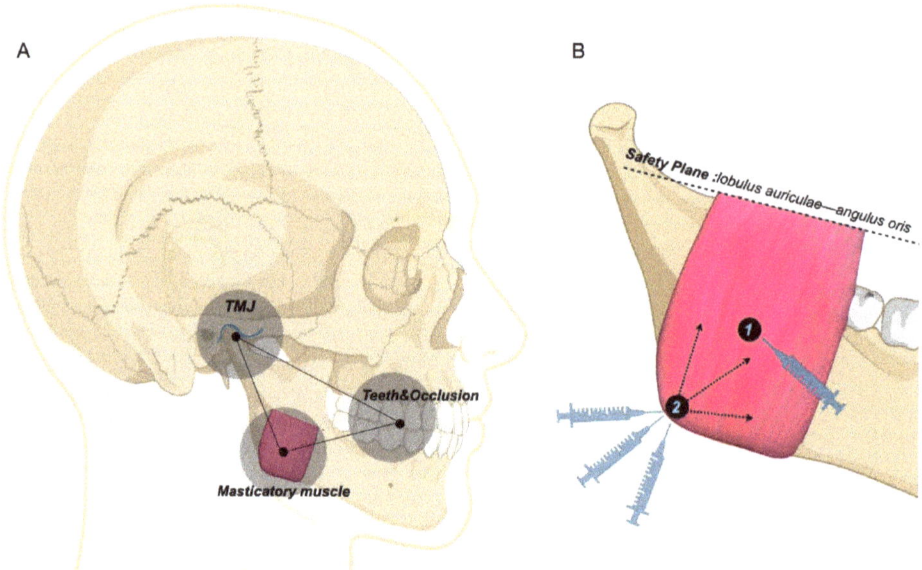

Figure 1. Schematic diagram of the masticatory complex and BTX-A injection method. (**A**) It demonstrates the triangle homeostasis of teeth and occlusion, temporomandibular joint (TMJ), and masticatory muscle. The three interrelate and have mutual influence under normal physiological conditions to guarantee the normal operation of the stomatognathic system, with any alteration beyond the body's adaptability in one of the links resulting in occlusal diseases. (**B**) Each patient was administrated 30–50 U of BTX-A (Botox®, 100 U/vial, Allergan, Irvine, CA, USA) in each masseter muscle (30 G, 25 mm needle length), with a maximum bilateral difference of 5 U. Injection should be conducted below the safety plane connecting the lobulus auriculae and the angulus oris. Palpation was used to select the most significant location of masseter muscle swelling as the first point for injection, with the dose controlled at 15–30 U. The second point was the mandibular angle region, with injection upward from the mandibular edge in a fan-shaped manner. In general, 3 channels with less than 10 U each can cover the entire injection area.

The masseter muscle is the most powerful masticatory muscle in the maxillofacial region, which ensures the stomatognathic system is fully stressed [21,22,25]. Specifically, variation in the masseter muscle can tremendously affect the function of the masticatory system, where weakness in the masseter muscle elicits the mandible to rotate clockwise, resulting in the opening of the mandible angle and the formation of high angle with a long-faced pattern [26]. In contrast, in short-faced patients with skeletal low angle, the masseter muscle is powerful and thick, which causes a square-shaped jaw accompanied by deep overbite and reduced vertical height [25–29]. As an extremely strong muscle-occlusion-joint triangular balance is too robust to be broken and reconstructed, it is hard for short-faced people to reestablish a new physiological homeostasis of the masticatory complex, indicating that occlusal treatment for these patients is awfully tough [29,30]. For example, it is generally rather difficult to increase the vertical height or open bite in occlusal therapy due to the powerful muscle-occlusion balance, and the curative effect is unstable with a high recurrence rate [31,32]. Moreover, these patients usually have a stout jaw and high bone density [21], which poses a challenge for successful tooth movement in orthodontic treatment, especially in cases of tooth extraction. In addition, because of the frequent impact of locking occlusion, posteriorly positioned mandible and condyle [33,34], steep occlusal curve and occlusal plane [35,36], patients are prone to posterior occlusion interference, which easily elicits bruxism or tooth wear, facial pain, muscle dysfunction, and TMD [17,18].

Therefore, the establishment of coordinated muscle function is the crux to effectively modify the occlusal relationship with long-term stability [10,37]. For dentists, two major technical difficulties impede the establishment of harmonious muscle function. One is the limitation of conventional dental treatment, for example, despite the multiple intervention measures with good clinical achievements for the occlusion, jaw, and TMJ, dentists seem to be powerless to adjust the neuromuscular factors. Although the occlusal splints can relax masticatory muscles while treating TMD [38,39], it is achieved indirectly by isolating the occlusal contact. Therefore, a minimally invasive, simple, and effective approach to directly intervene in neuromuscular factors is required, and botulinum toxin-A (BTX-A) can meet these treatment needs. BTX-A works by inhibiting the release of acetylcholine at the nerve–muscle junction [40] and has been applied in the auxiliary treatment of some diseases in the oral and maxillofacial region [41]. At present, its effectiveness in the treatment of bruxism [42,43], facial paralysis [44], TMD [45,46], neuropathic pain [42], sialorrhea [47], orofacial dystonia [48], and other diseases has been verified by high-level evidence-based medical evidence, providing a new direction and method for alternative treatment of oral diseases. However, the studies above [41–48] mostly focused on the field of oral and maxillofacial surgery, with insufficient attention given to occlusal therapy. Currently, only some animal experiments have been carried out [49–53] but almost no clinical studies.

Another issue is the difficulty in obtaining muscle diagnosis using dental imaging. Traditional two-dimensional dental imaging data can only present the hard tissue components with the possibility of "distortion" due to the influence of overlap and magnification. With the application of digitalization, cone-beam computed tomography (CBCT) and its three-dimensional reconstruction technology [54], which can reproduce both the soft and hard tissue in equal proportions [55–57], have been widely utilized in dental clinical work. CBCT can greatly meet the requirements of convenient diagnosis and treatment in dental clinics, with a much lower price, cost of time, and radiation, overcoming the disadvantages of traditional 3D imaging approaches such as CT [39,58] and MRI [59,60].

Based on CBCT images, this study preliminarily explored the feasibility of clinical application of BTX-A to reconstruct the physiological homeostasis of the masticatory complex in short-faced patients by researching the variation in the masticatory complex, aiming to provide an innovative and minimally invasive treatment perspective for the two major clinical problems, involving the low treatment efficiency (e.g., slow tooth movement, difficult bite opening, easy recurrence, etc.) and various potential complications (e.g., TMD, myofascial pain, bruxism, periodontal problems, etc.). In addition, the coordination mechanisms of the muscle-occlusion-joint complex are also discussed here by providing the regression equation for the first time, extending clinicians' cognition of dynamic reconstruction of the stomatognathic system in occlusal therapy.

2. Patients and Methods

2.1. Subjects

Patients requiring orthodontic treatment in the Department of Stomatology, The First Affiliated Hospital of Zhejiang University Medical College from May 2021 to January 2022 were selected. In total, 10 eligible patients with an average age of 27.0 ± 6.1 years (4 males and 6 females) were selected according to the following criteria. Inclusion and exclusion criteria: (1) Short-faced patients with skeletal low angle (FMA< 24°) and benign masseter hypertrophy; (2) healthy adults, aged 18–35 years; (3) eruption of all permanent teeth (except the third molar); (4) a basically symmetrical maxillofacial region without significant deviation; (5) no malocclusion such as open bite or crossbite, and no serious dental and periodontal diseases; (6) no previous history record of maxillofacial surgery (such as orthognathic surgery, plastic surgery, etc.), orthodontic treatment, or prosthetic treatment; and (7) no masticatory system dysfunction, temporomandibular joint disorder syndrome, etc. This study was approved by the Clinical Research Ethics Committee of the First Affiliated Hospital, College of Medicine, Zhejiang University (protocol code: (2021) IIT (171) and date of approval: 10 March 2021). Before the study, all subjects were informed of the purpose of this study and gave informed consent.

2.2. BTX-A Injection Method

Each patient was treated with a single injection of BTX-A (Botox®, 100 U/vial, Allergan, Irvine, CA, USA), with 2 mL of NS added, diluting the concentration to 5 U/0.1mL. The patient was placed in the supine position and injected with 30–50 U/side into the masseter muscle (30 G, 25 mm needle length), with a maximum bilateral difference of 5 U. An expert plastic surgeon determined the specific injection dose of BTX-A based on his clinical expertise, and the injection was administered below the safety plane linking the lobulus auriculae and the angulus oris. Palpation was used to select the most significant location of masseter muscle swelling as the first point for injection, with the dose controlled at 15–30 U. The mandibular angle region was selected as the second point, with injection performed upward from the mandibular edge in a fan-shaped manner. In general, three channels with less than 10 U each can cover the entire injection area (Figure 1B). When injecting BTX-A, adrenaline (1:1000) should be prepared, followed by routine nursing care and short-term observation.

2.3. CBCT Measurement Items and Methods

CBCT (NewTom VGi, Verona, Italy) was taken in all patients 1 week before BTX-A injection and 3 months after injection. During the shooting, the patient was required to maintain the maximum intercuspal position, with consistent scanning parameters (tube voltage 110 kV, tube current 3.5 mA, exposure time 3.6 s, and definition 0.3 mm). The scanning ranged from the superior orbital edge to the lower mandibular edge. All CBCT data were saved in DICOM format and imported into Dolphin Imaging 11.95 software (Chatsworth, Los Angeles, CA, USA) for 3D reconstruction and analysis (Figure 2). In this study, the three components of the masticatory complex were measured and analyzed, and the data were measured twice by the same staff with an interval of one week, and the average value was taken. If the data difference between the two measurements exceeded 0.5 mm, one more measurement was carried out again.

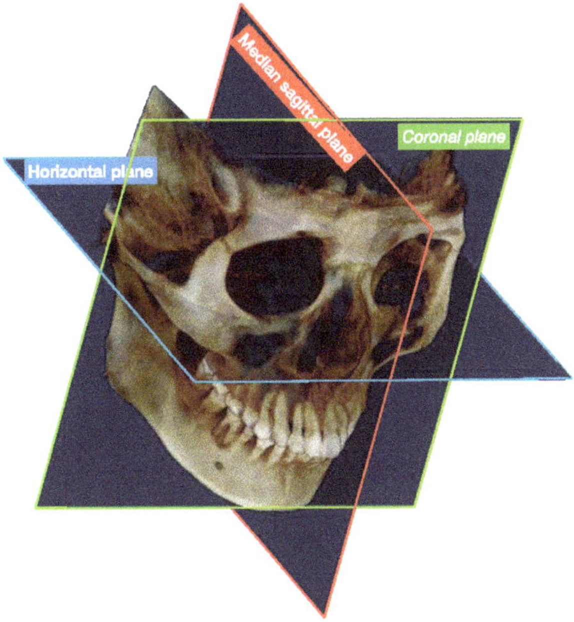

Figure 2. Schematic diagram of 3D reconstruction of CBCT. It displays the outcomes of the 3D reconstruction images using Dolphin Imaging 11.95 software (Chatsworth, Los Angeles, CA, USA) along with three calibration planes, which can be slid or rotated to obtain subsequent measurement planes for the measurement of various indexes.

2.3.1. Masseter Muscle Index

When measuring the masseter muscle thickness (MMT), the head position in the 3D reconstruction was required to be repositioned in three dimensions using line calibration, and the CBCT image was divided into several 2 mm slices parallel to the mandibular plane (MP) (Figure 3A). The MP was defined as the plane from the Gnathion point (Gn) to the Gonion point (Go). The MMT of both sides was measured on the axial slice passing through the mandibular lingual structure (a sharp and thin bony piece in front of the mandibular foramen) (Figure 3B), as previous studies [61] have suggested that the maximum section of the masseter muscles of most individuals is generally at the level of the mandibular foramen.

2.3.2. Dental and Occlusal Indexes

There were 14 items in total, and the indexes in this part were divided into static occlusal indexes (6 posterior tooth height items, 4 transverse width items) and dynamic functional indexes (2 occlusal plane items and 2 occlusal curve items).

The posterior tooth height (U5-PP, U6-PP, U7-PP, L5-MP, L6-MP, L7-MP) [62] and the occlusal plane angle (AOP-FH, POP-FH) [36] were measured on the lateral cephalogram from CBCT by projecting the 3D reconstruction image into the midsagittal plane from right to left (Figure 4A,B). U5, U6, and U7 represent the buccal cusp of the maxillary second premolar, the midpoint of the maxillary first molar at the occlusal surface, and the midpoint of the maxillary second molar at the occlusal surface, respectively. L5, L6, and L7 represent the points of the mandibular posterior teeth. The palatal plane (PP) was determined using a line drawn from the anterior nasal spine point (ANS) to the posterior nasal spine point (PNS). U5-PP, U6-PP, and U7-PP were defined as the vertical distance from the U5, U6, and U7 points to PP, respectively, while L5-MP, L6-MP, and L7-MP were defined as the vertical distance from the L5, L6, and L7 points to MP, respectively (Figure 4C).

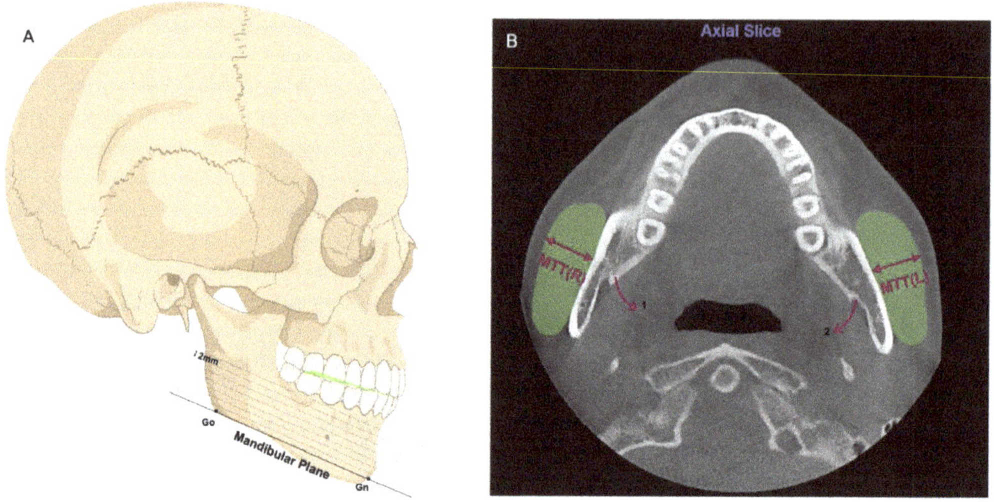

Figure 3. Measurement of the masseter muscle thickness (MTT). (**A**) The head position in the 3D reconstruction was required to be repositioned, ensuring the CBCT image was divided parallel to the mandibular plane (MP). The distance between each slice was 2mm; (**B**) The MMT of both sides was measured on the axial slice passing through the mandibular lingual structure (arrow 1), which is a sharp and thin bony piece in front of the mandibular foramen (arrow 2).

The Frankfort horizontal plane (FH) was determined using a line drawn from the orbitale point (Or) to the anatomical porion point (Po), which was generally parallel to the ground level. The posterior occlusal plane (POP) was determined by the line between U5 and U7, and the anterior occlusal plane (AOP) was determined by the line between U5 and the incisal edge point of the maxillary central incisor (U1). AOP-FH was defined as the angle between FH and AOP while POP-FH was defined as the angle between FH and POP. The location of the anatomical reference landmarks and specific measurement methods can be seen in the schematic diagram of the lateral cephalogram (Figure 4C).

After the measurement above, the head position was re-adjusted in three dimensions, passing through the FH plane in the bilateral sagittal position, and maintaining the head position in the subsequent measurement. Regarding the transverse width (NF, HP, BAC, LAC) [63] and the curve of Wilson (COW), they were all measured on the coronal slice passing through the maximum section of the maxillary first molars (Figure 5A). COW itself is defined as a convex downward curve formed by connecting the buccal and lingual cusps of the homonymous maxillary molars on both sides. In this study, we measured the curvature of COW [64] by drawing two lines connecting the bilateral central fovea and root bifurcation, respectively, and the sum of the included angles between each line and the line perpendicular to FH was defined as the curvature of COW. The curve of Spee (COS) itself is defined as a concave upward curve formed by the incisal edge of the mandibular central incisors passing backwards through the cusp of the mandibular canines, the buccal cusps of the mandibular premolars, and the mesial and distal buccal cusps of the mandibular molars. The measurement of COS was carried out in 3D reconstruction images using 3D points and calibration. Different from the COW measured by the angle, the depth of COS [64] was more clinically meaningful. A plane was determined at the incisal edge of the mandibular central incisors and the distal buccal cusp of the bilateral mandibular second molars (Figures 3A and 4A show this plane in green). The depth of COS was considered as the maximum distance from the lowest point of the bilateral COS to this plane.

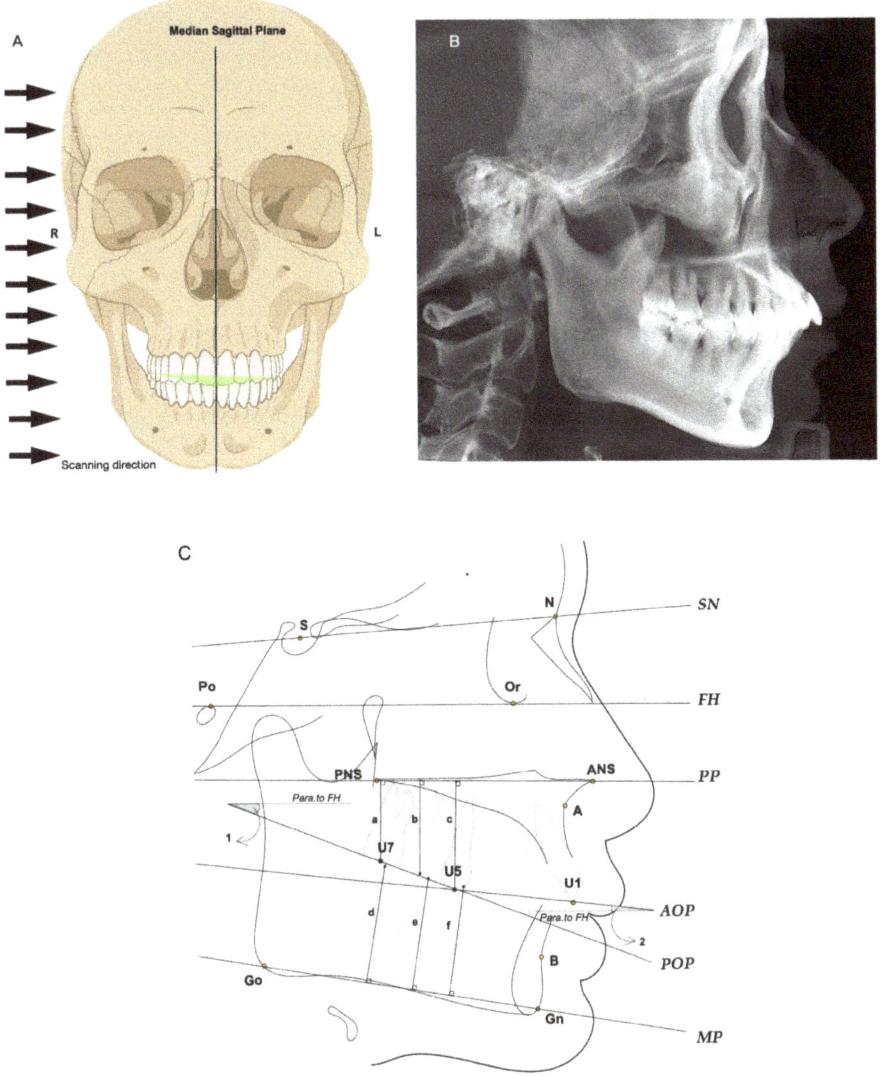

Figure 4. Measurement of the posterior tooth height (U5-PP, U6-PP, U7-PP, L5-MP, L6-MP, L7-MP) and the occlusal plane angle (AOP-FH, POP-FH). (**A**) The method of projecting the 3D reconstruction image into the midsagittal plane from right to left; (**B**) The lateral cephalogram formed after projection; (**C**) Tracing diagram of the lateral cephalogram. It shows the anatomical reference landmarks and measurement planes used for cephalometric measurements. Landmarks are defined as follows: sella point (S); nasion point (N); anatomical porion point (Po); orbitale point (Or); anterior nasal spine point (ANS); posterior nasal spine point (PNS); subspinale point (**A**); supramental point (**B**); gnathion point (Gn); gonion point (Go); the buccal cusp of the maxillary second premolar (U5); the midpoint of the maxillary first molar at the occlusal surface (U6, not shown); the midpoint of the maxillary second molar at the occlusal surface (U7); the buccal cusp of the mandibular second premolar (L5, not shown); the midpoint of the mandibular first molar at the occlusal surface (L6, not shown); the midpoint of the mandibular second molar at the occlusal surface (L7, not shown); the incisal edge point of the maxillary central incisor (U1). In addition, a-f represent U7-PP, U6-PP, U5-PP, L7-MP, L6-MP, and L5-MP, respectively, while angle 1 and angle 2 represent POP-FH and AOP-FH, respectively.

2.3.3. Temporomandibular Joint Indexes

There were 7 items in total, and all measurements were completed on the same measurement plane. The measurement plane (condylar sagittal plane) was determined by moving layer by layer in the horizontal and coronal views until the inner and outer diameter of the condyle was the largest in the two views. Then, in the sagittal view, scanning from right to left step by step was performed until the slice passed through the highest point of the glenoid fossa (Figure 5B).

The measurements of the joint indexes [65] were conducted as follows:

1. Draw four parallel lines of FH from the superior to the inferior, tangent to the glenoid fossa, condyle, the lowest end of the articular eminence, and the sigmoid notch of the mandible, respectively, named L1, L2, L3, and L4. The vertical distance between L2 and L4 was defined as the condyle height (Ht.Co) and the vertical distance between L1 and L3 was defined as the fossa height (Ht.Fo). L3 intersected the posterior wall of the fossa, and the horizontal distance between the intersection point and the tangent point was defined as the fossa width (Wd.Fo).
2. Taking the tangent point of L1 (superior fossa (SF) point) as the starting point, draw three new tangent lines from the anterior to the posterior, which were tangent to the posterior slope of the articular eminence, the anterior edge of the condyle, and the posterior edge of the condyle, respectively, named T1, T2, and T3. The angle between T1 and FH was defined as the articular eminence inclination (AEI).
3. Joint spaces were measured using the Kamelchuk method [66]. The distance between the tangent point of L1 and L2 was measured as the superior joint space (SJS). The perpendicular distance from T2 tangent point to the posterior slope of articular eminence was defined as anterior joint space (AJS), and the perpendicular distance from T3 tangent point to the posterior wall of fossa was defined as posterior joint space (PJS).

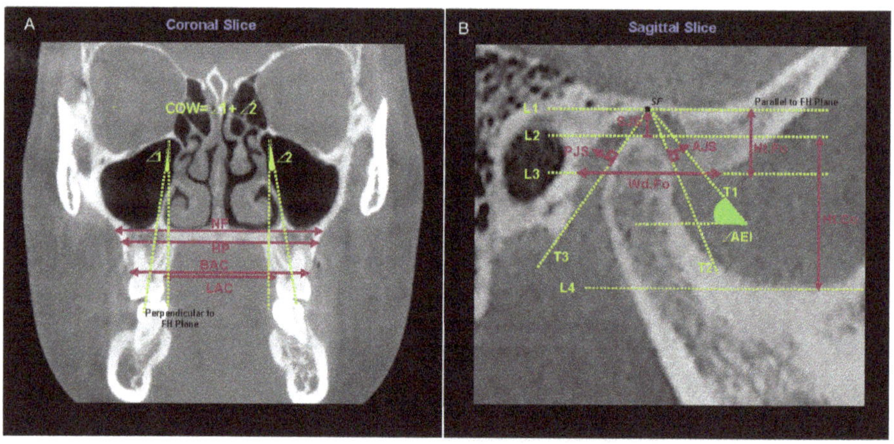

Figure 5. Measurement of the transverse width (NF, HP, BAC, LAC), the curve of Wilson (COW), and TMJ indexes. (**A**) NF: maxillary basal bone width at tangent to the nasal floor; HP: maxillary basal bone width at tangent to the hard palate; BAC: maxillary alveolar bone width between the bilateral buccal alveolar ridge; LAC: Maxillary alveolar bone width between the bilateral lingual alveolar ridge. The curvature of COW was defined as the sum of the bilateral angles between the line connecting the central fovea and root bifurcation and the line perpendicular to the FH plane. (**B**) Seven tangent lines were drawn as mentioned in the text. Ht.Co: vertical distance between L2 and L4; Ht.Fo: vertical distance between L1 and L3; Wd.Fo: horizontal distance between the intersection point as mentioned and the tangent point of L3; AEI: the angle between the T1 and FH of the TMJ index; joint spaces were measured using the Kamelchuk method.

2.4. Statistical Analysis

SPSS 26 software (IBM Corp., Armonk, NY, USA) was applied. All the items were measurement data, expressed as (mean ± standard deviation). Kolmogorov–Smirnov was utilized to examine whether the analysis data were in accordance with a normal distribution. The 10 patients with the short-faced pattern were compared before and after injection. A paired t test was performed for those with a normal distribution and Wilcoxon paired test was performed for those with a non-normal distribution. Meanwhile, the symmetry of the bilateral indexes was also checked using a paired t test. In addition, we also carried out correlation analysis for each measurement index before injection. Pearson correlation analysis was conducted for the measurement items conforming to a normal distribution, and Spearman rank correlation analysis was conducted for those that did not conform to a normal distribution. The bilateral test level was set at $\alpha = 0.05$ and $p < 0.05$ was considered statistically significant.

3. Results

3.1. Changes in the CBCT Masticatory Complex

3.1.1. Changes in the Masseter Muscle Thickness

There was no significant difference in the bilateral MMT before or after BTX-A injection ($p > 0.05$), indicating symmetrical bilateral masseter muscle in all patients. Therefore, the average value of the bilateral MMT was utilized in this study. The average MMT of 10 patients with low angle before and 3 months after BTX-A injection was 16.14 ± 3.44 mm and 13.90 ± 3.14 mm, respectively, significantly reduced by 2.25 ± 0.73 mm ($p < 0.0001$) (Figure 6).

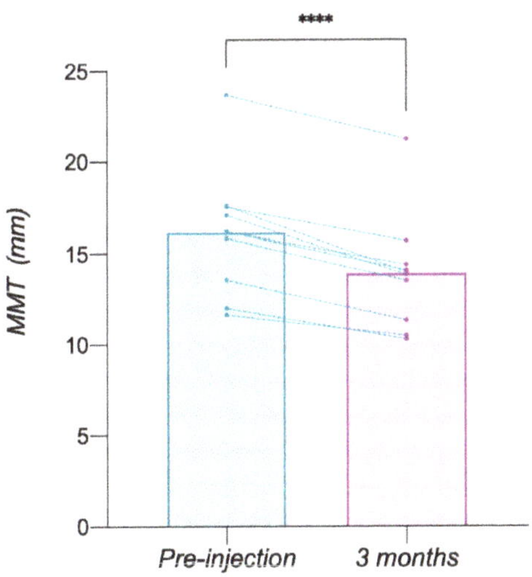

Figure 6. Changes in the masseter muscle thickness. It illustrates a significant reduction in MMT 3 months after BTX-A injection (****: $p < 0.0001$).

3.1.2. Changes in Dental Occlusion

In terms of the variation in the posterior tooth height, U7-PP increased from 20.51 ± 2.56 mm before BTX-A injection to 21.52 ± 2.92 mm after injection, with a significant difference of 1.01 ± 1.27 mm ($p < 0.05$) (Figure 7A). These results suggested that BTX-A caused compensatory elongation of the maxillary second molars by about 1mm, and elongation of the maxillary posterior teeth far from the masseter muscle fibers and all mandibular posterior teeth was not evident, as U5-PP, U6-PP, L5-MP, L6-MP, and L7-MP showed no significant change. There were also no significant differences in NF, HP, BAC, and LAC before and 3 months after Botox injection (Figure 7B), indicating the unavailability of BTX-A in the reconstruction of both the maxillary basal bone width and maxillary alveolar bone width.

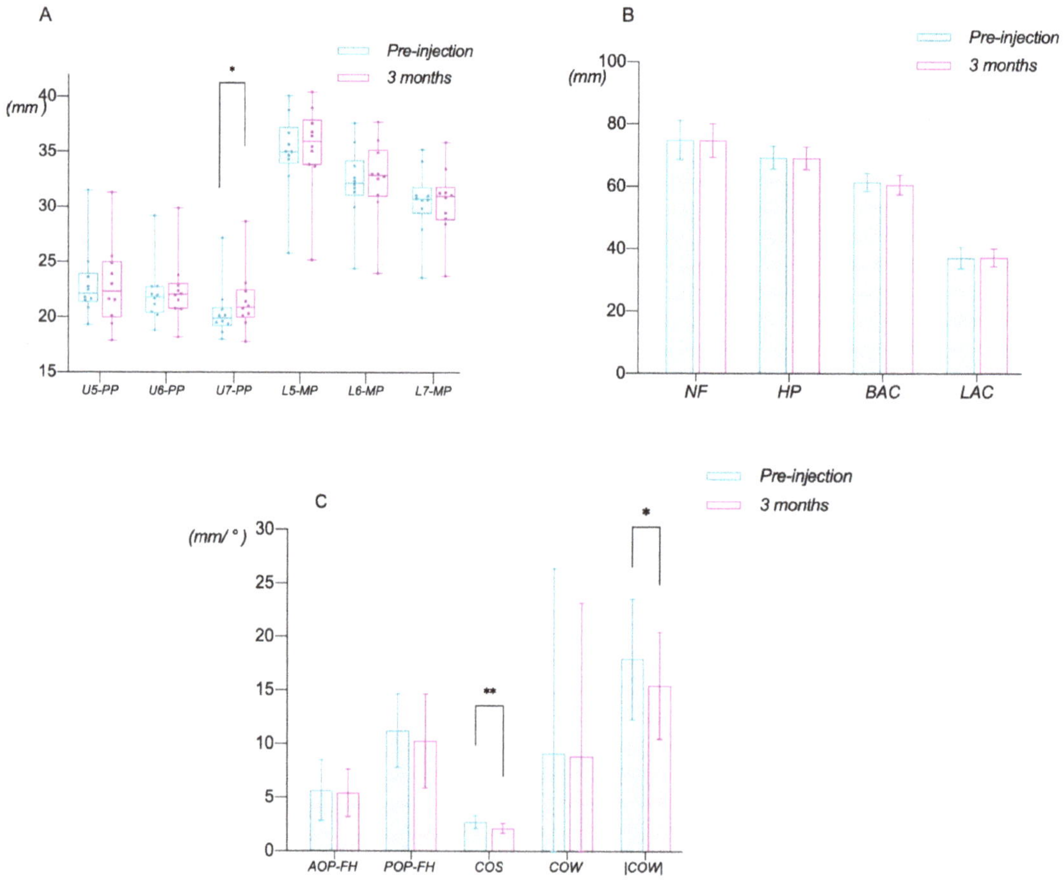

Figure 7. Changes in dental occlusion. (**A**) Variations in the posterior tooth height. It illustrates that U7-PP of 10 patients significantly increased 3 months after BTX-A injection while U5-PP, U6-PP, L5-MP, L6-MP, and L7-MP showed no obvious changes. (**B**) Variations in the transverse width. It illustrates no significant differences in both the basal bone (NF, HP) and alveolar bone (BAC, LAC) before and 3 months after BTX-A injection. (**C**) Variations in the functional occlusion. It illustrates that neither AOP-FH nor POP-FH showed a significant change while the depth of COS showed a significant decrease. As for the curvature of COW, its value itself does not change significantly, but a significant flattening was found when the change in the absolute value of COW before and 3 months after BTX-A injection was compared (*: $p < 0.05$; **: $p < 0.01$).

Neither AOP-FH nor POP-FH showed significant changes, despite a nearly 1° flattened value of the POP. However, the changes in the occlusal curve were striking, as the depth of COS decreased from 2.76 ± 0.60 mm to 2.17 ± 0.46 mm after BTX-A injection, significantly decreasing by 0.59 ± 0.55 mm ($p < 0.01$), suggesting an astonishing leveling effect of COS. As for the curvature of COW, its value itself did not change significantly. However, considering the existence of a negative value, when comparing the changes in the absolute value of COW before and after injection, we also found a significant flattening ($p < 0.05$), decreasing from 17.97 ± 5.66 ° to 15.49 ± 4.97 °. This indicates the occurrence of an upright effect of 2.48 ± 2.87 ° in the inclination angle of the maxillary posterior teeth (Figure 7C).

3.1.3. Changes in TMJ

The TMJ underwent the same bilateral measurement as that of the masseter muscle. As each index of the bilateral joint exhibited no significant difference before and after injection ($p > 0.05$), the average value was adopted. In terms of the joint structure, there were no significant changes in Ht.Co, Ht.Fo, Wd.Fo, or AEI after BTX-A injection in 10 patients with a short-faced pattern (Figure 8A), indicating that BTX-A injection in this study did not promote remodeling of the TMJ structure, regardless of the condylar height, the fossa height and width, and the posterior slope of the articular eminence. In terms of the joint space, AJS and SJS were significantly reduced ($p < 0.05$) by 0.13 ± 0.16 mm and 0.21 ± 0.24 mm, respectively, suggesting a conceivable anterior-upper rotational shift of the condyle position, while PJS showed an insignificant increase of about 0.10 mm (Figure 8B).

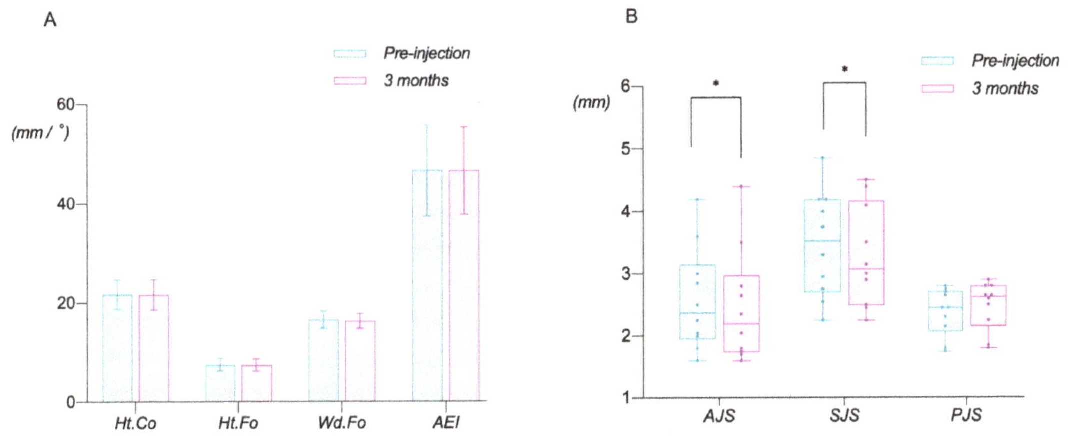

Figure 8. Changes in TMJ. (**A**) Variations in the joint structure. It illustrates no significant changes in Ht.Co, Ht.Fo, Wd.Fo, or AEI 3 months after BTX-A injection. (**B**) Variations in the joint space. It illustrates that AJS and SJS were significantly reduced 3 months after BTX-A injection while PJS showed an insignificant increase (*: $p < 0.05$).

3.2. *Correlation Analysis of the Components of the Masticatory Complex in Short-Faced Patients*

Pairwise correlation analysis was conducted between all variables before BTX-A injection, and a correlation heatmap was generated (Figure 9). We found that MMT exhibited a significantly negative correlation with COW ($p < 0.05$) while a significantly positive correlation with AJS and SJS ($p < 0.01$, $p < 0.05$). After synthetically considering the absolute value of the correlation coefficient r and the clinical significance, COW was taken as the occlusal index and AJS as the TMJ index in the final regression equation.

Eventually, we performed multiple linear regression analysis with pre-injection MMT as the dependent variable while COW and AJS as independent variables, and obtained the regression model: MMT = 10.08 − 0.11 COW + 2.73 AJS. Here, all variables and constants possessed robust statistical significance in the regression equation ($p < 0.01$), and the whole regression model exhibited extremely high significance ($p = 0.0003$). The high value of the adjusted R^2 indicated the capability of the regression model to explain the variation of MMT of about 87.1% (Table 1).

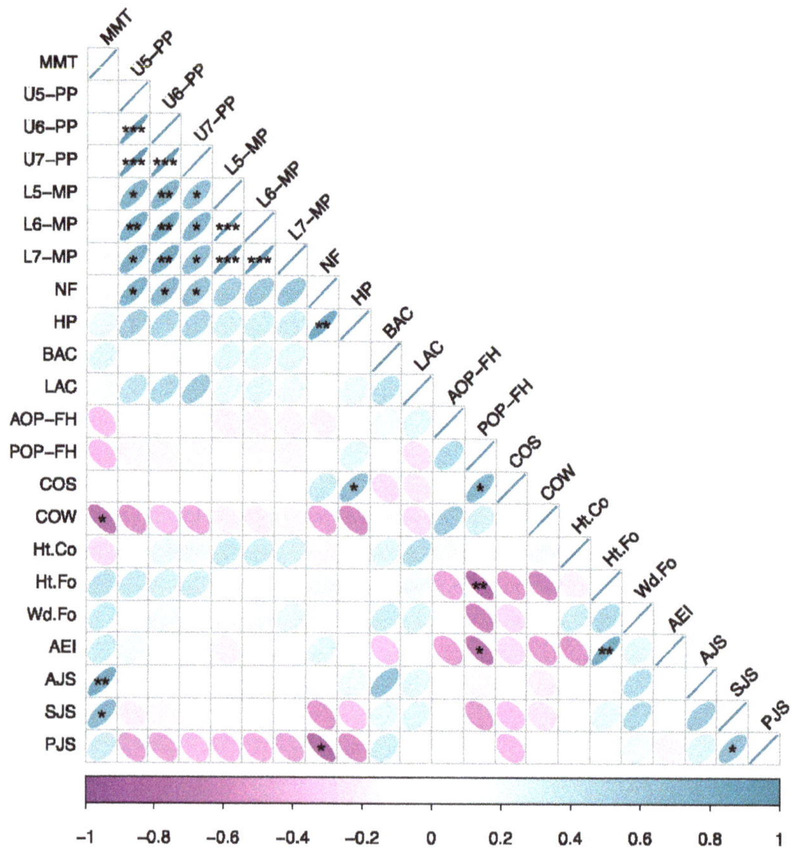

Figure 9. Correlation heatmap of the masticatory complex. Positive correlation is represented by cyan ellipses, whereas negative correlation is represented by purple ellipses, with a deeper hue indicating a stronger correlation. Specifically, the darker the cyan, the closer the r is to 1, and the darker the purple, the closer the r is to −1. Similarly, when the correlation varies, the size of the ellipse changes. The closer the r is to 1 (cyan) or −1 (purple), the closer the ellipse is to a line, whereas the closer the r is to 0, the closer the ellipse is to a perfect circle. Correlations with significant differences are marked in the figure. (*: $p < 0.05$; **: $p < 0.01$; ***: $p < 0.001$).

Table 1. Regression analysis of MMT.

	Standardization Coefficient β	Significance	Adjusted R^2	*p* Value
(Intercept)	10.08	0.0002		
COW	−0.11	0.003	0.871	0.0003
AJS	2.73	0.001		

MMT = 10.08 − 0.11 COW + 2.73 AJS.

4. Discussion

4.1. Effects of BTX-A on Homeostasis Reconstruction of the Masticatory Complex

4.1.1. BTX-A and Masticatory Muscle

Patients with a short-faced pattern generally experience the problem of benign masseter hypertrophy due to the mutual shaping of the muscle and bone in the process of growth and development [21,22,25]. BTX-A is regarded as the most effective approach to treat benign masseter hypertrophy [41,67]. In this study, BTX-A significantly promoted a reduction in the thickness of the masseter muscle in 10 patients, consistent with the previous results obtained from B-ultrasound assessment by Diracoglu et al. [68], CT assessment by Hong et al. [39], and 3D laser scanning assessment by Lee et al. [69]. The results above demonstrate the effectiveness of BTX-A in masticatory muscle adjustment, which also provide the premise for subsequent study, and verify the effective diagnostic value of CBCT for the masseter muscle [60].

Zhang et al. [70] injected BTX-A into the masseter muscle of patients with bruxism and found that the biting force in the maximum intercuspal position significantly decreased, which directly proved the efficiency of the BTX-A treatment in reducing the biting force. A similar property was indirectly verified through electromyography by Lee et al. [71]. Through 3D CT reconstruction, Hong et al. [39] detected changes in the cortical bone thickness at the insertion sites of the masseter muscle and temporalis muscle after BTX-A injection. The imaging evidence also indirectly reflected a reduction in the muscle functional load. Although the inhibitory effect of BTX-A on neuromuscular junction transmission is not permanent, with a period of about 3 months from the peripheral nerve germination to functional recovery of the neuromuscular complex, it has also been reported that repeated injections of BTX-A help maintain reduced biting force [72].

4.1.2. BTX-A and Teeth and Occlusion

In untreated patients, the functional contraction direction of the masseter muscle should be at a stable angle to the functional occlusal plane of the posterior teeth, generating three-dimensional forces on the occlusal surface of the posterior teeth, which are composed of vertical, anterior, and lateral force [73]. Occlusal treatment (such as orthodontic treatment) is a dynamic process in which the original muscle-tooth balance is constantly destructed, and the new balance is rebuilt as the tooth position is modulated. For skeletal high angle patients with a long face, their weak masticatory muscles and low strength provide conditions for an efficient and smooth treatment process as generally soft and light force can counter the old balance [21,37]. On the contrary, for low angle patients with a short-faced pattern, due to the strong masticatory muscles and robust biting forces in the three-dimensional direction, the processes involved during treatment against the vertical force to open bite [29,30], against the anterior force to prevent molar mesial drift and COS deepening [74,75], or against the lateral force to control the maxillary width and buccolingual molar inclination (COW) are significantly difficult [76,77]. Even if these are achieved, high rates of recurrence and periodontal risks were reported. Therefore, during the early stage of occlusal treatment, the muscle strength and biting force should be adjusted as soon as possible by injecting BTX-A, which can directly affect the load of the occlusal surface of teeth, and then manipulate the changes in the three-dimensional dental eruption state to adapt to the remodeling of balance in a more efficient and healthy way [29].

Most studies on tooth changes caused by BTX injection have focused on animal experiments [49–53]. Multiple previous studies have proved that BTX injection can induce a double decline in the shape and function of masticatory muscles, and that the reduced biting force will lead to manipulation of the tooth height. Tsai et al. [51] injected BTX unilaterally into rat masseter muscle, showing an excessive eruption of the upper and lower molars, in addition to a decrease in the size and weight of the masseter muscle. Choi et al. [49] injected BTX into the unilateral masseter muscle and temporal muscle of developing rats to weaken the muscle force, and discovered that compensatory alveolar eruption caused deviation in the occlusal plane. Navarrete et al. [53] found that BTX injection administered to rabbits elicited excessive tooth eruption, and argued that it would affect facial morphology changes, such as increasing the vertical dimension or tendency of a long-faced pattern. Considering the differences in the anatomy of the muscles, jaw, and drug metabolism between humans and different animals, this evidence is unsuitable for direct extrapolation to humans but confirms the feasibility of injecting BTX into the masticatory muscles to induce paralysis as a means of adjusting occlusion. Currently, in clinical studies, only some examples of

BTX-A in maintaining occlusal stability after orthognathic surgery have been reported [78]; the efficacy of BTX-A in non-surgical occlusal therapy has not been studied. Dai et al. [79] found that the biting force was almost distributed on the molars during static masticatory movement, in which the force on the incisors was almost zero. Therefore, this study mainly focused on the posterior teeth. The U7-PP of 10 patients significantly increased 3 months after BTX-A injection, indicating compensatory elongation of the maxillary second molar while the height of other posterior teeth showed no obvious changes. This may indicate that the lower bone mineral density of the maxilla, and the more abundant nerves and vessels in the maxilla make it more vulnerable to adjustment and remodeling by external forces than the mandible. In addition, the physical distance between the masseter fibers and teeth will also directly affect the eruption state of the teeth. In terms of the transverse width, NF, HP, BAC, and LAC exhibited no significant changes before and after injection, indicating that the basal and alveolar bone widths of adult patients were relatively stable and unaffected by changes in the masseter muscle strength. This seems to be inconsistent with Kiliaridis et al.'s study [80], possibly because the latter samples were involved in the juvenile development period and gender differences were differentiated.

In addition to static occlusal indexes, modern concepts of functional occlusion pay more attention to balance and stability during dynamic mastication. The effect of BTX-A on functional occlusion in animals or humans remains to be studied, and this study is an innovative exploration. In normal chewing and grinding movement, the mandible moves laterally or forward, the posterior teeth should be separated, and the anterior teeth contact to form the canine guide (lateral movement) or incisor guide (forward movement), where the anterior teeth can protect the posterior teeth during functional movements [10,81]. However, short-faced people often have a steep posterior occlusal plane or deep occlusal curves, meaning a high possibility of occlusal interference [35], which may cause tooth abrasion, periodontal disease, and TMD. In this study, the process of dynamic occlusion was reflected by two occlusal planes and two occlusion curves. Under extremely ideal conditions, COS and COW should be represented as an arc located on the same sphere (Monson sphere) [64]. Here, both COS and COW were significantly flattened 3 months after Botox injection. Apart from the influence of the orthodontic force itself, the weakened occlusal force broke the original muscle-tooth balance and promoted the three-dimensional movement of teeth in a more efficient way. Studies [82] have demonstrated the close relation between the occlusal curve and masticatory function. Generally, individuals who have flatter curves may obtain a better food mixing capability, higher masticatory efficiency, less occlusal interference, and lower incidence of periodontal disease and TMD. As for the occlusal planes, neither POP nor BOP showed significant changes, which may be related to the fact that the interference of the sagittal skeletal type was not excluded in this study, with a variety of sagittal skeletal types. Professor Sato [36] believed that the increased POP-FH was the source of the occurrence of the Class II pattern; the crux for successful treatment in such patients was to flatten the POP [83]. It has been found that POP does tend to flatten after BTX-A injection, which may be related to the orthodontic force itself or the compensatory elongation of the maxillary second molars caused by masseter paralysis. The variation characters in the occlusal curves and occlusal planes reflect the dynamic reconstruction process of the stomatognathic system, and BTX-A injection may effectively accelerate the rate of homeostasis reconstruction in low angle patients with a lower incidence of complications [32].

4.1.3. BTX-A and TMJ

According to the international classification of DC/TMD, TMD can be divided into two categories [84,85]. The first category is pain diseases, which generally involve joint pain itself and pain secondary to masticatory muscle hyperfunction [86–88]. BTX-A has a good therapeutic effect on pain diseases as it can inhibit the release of local pain neuropeptides, such as substance P, calcitonin gene-related peptide, glutamate, and transient receptor potential V1, which has also attracted multiple clinical studies [86–88]. Due to the positive therapeutic effect of this kind of disease, this study did not continue its exploration.

The second type is joint organic disease. TMJ remodeling often occurs after mechanical load changes in the stomatognathic system, mainly including the process of morphological adaptation (such as changes in the condyle bone) and spatial adaptation (such as changes in the position of the condyle and joint disc) [89]. The former is often applied as a model of paralysis in animal experiments, where BTX-A induces bone changes over a short period of time. Rafferty et al. [90] found bone loss of the mandibular condyle in rabbits after an injection of BTX-A into the unilateral masseter muscle, which was also found in the masseter muscle attachment area in rats by Tsai et al. [52] and Kun-Darbois et al. [91]. However, due to the discrepancy between animal and human bone biomechanics, the effect of BTX-A injection on the condylar bone in the human body remains controversial. Based on 3D CT reconstruction, Hong et al. [39] found that reduced muscle functional load affected the cortical bone mass of the condyle, which was more significant in postmenopausal women. While Raphael et al. [92] seemed to reach the opposite result, finding no clinically significant changes in the density or volume of the TMJ complex in women with TMD who had received at least two (and often three) cycles of BTX-A treatment in the past year. Since the experiments above were mostly based on TMD patients who had repeated injections of botulinum, this study aimed to explore whether a single botulinum toxin injection would adversely affect the condylar bone in the normal population, which may run counter to our original intention. Fortunately, no bone changes in TMJ were found in the 10 patients 3 months after the single BTX-A injection, which is consistent with the findings of Lee et al. [55], regardless of whether the condyle, glenoid fossa, or articular eminence was examined. During the follow-up of the 10 patients, no adverse indications such as TMJ clicks and pain were reported. On the contrary, two patients said that their mild bruxism symptoms were relieved, with acid swelling of the masticatory muscles in the morning disappearing, which actually resulted from the weakened biting force caused by masseter muscle atrophy. The weakened biting force reduced the load on the TMJ area, which was conducive to the decompressed and balanced repair environment of the TMJ.

Moreover, this experiment innovatively involved the joint space indexes to verify the dual response of TMJ to the mechanical load. In addition to morphology adaptation, there is also spatial relocation [23]. The masticatory muscle function is well known to be critical for proper TMJ development during ontogeny [93]. In patients with skeletal low angle, the joint structure is usually thick and broad due to strong muscle force while the condyle position is often posterior and inferior due to excessive counterclockwise closure of the mandible. An abnormal condyle position and abnormal occlusal relationship are risk factors for TMD [94]. The 10 patients in this study showed a reduction in AJS and SJS after BTX-A injection, which is conducive to the stabilization of the condyle in the central relation (CR) position. As the father of modern occlusions, Dr. Roth [94] believed that a stable stomatognathic system should be equipped with stable joint position and coordinated occlusal relationship, as Dr. Dawson summed up that when the upper and lower teeth are in the maximum intercuspal position, the condyle should be at the front and top of the glenoid fossa [3]. Therefore, BTX-A can coordinate the condyle position with the final target position of the occlusal treatment, contributing to the long-term stability of the curative effect and low rates of TMD.

4.2. Preliminary Study on the Homeostasis Mechanisms of the Masticatory Complex

To explore the homeostasis mechanisms of the masticatory complex in short-faced patients, we analyzed the pairwise correlation between all variables before injection. The MMT was defined as a novel entry point for the homeostasis balance of the occlusal treatment to explore how the changes in occlusion and TMJ under the state of physiological balance manipulate and adapt to the changes in the masticatory muscle.

When selecting variables for regression analysis, not only the absolute value of correlation coefficient r should be considered but also the clinical factors and possible collinearity problems. For example, in addition to COW's robust correlation, there also exists a certain correlation between COW and COS depending on the clinical conditions, and the correlation heatmap showed a significant positive correlation between COS and POP-FH ($p < 0.05$). Therefore, the final inclusion of COW was consistent with both the experimental results and the clinical significance. As for why AJS instead of SJS was chosen, in addition to the greater r of AJS, we also considered the correlation of SJS and PJS depicted in the heatmap ($p < 0.05$). Considering the correlation of PJS and AJS in the clinic, AJS and SJS may show collinearity in a clinical sense, so eventually only AJS was included as the only TMJ index. Consequently, the whole regression model was highly significant ($p = 0.0003$) and had an explanation capability of 87.1% of the clinical outcomes.

Here, the homeostasis of the masticatory complex in patients with a short-faced pattern was intuitively and quantitatively described firstly by means of a multiple linear regression model. As mentioned above, mastication is one of the main oral functions, and its efficient performance requires a healthy and stable stomatognathic system. It generally consists of occlusion, TMJ, and masticatory muscles, which are defined as a functional complex in this study. They are interrelated and have mutual influence, forming a stable balance triangle to ensure the normal operation of the stomatognathic system under normal physiological conditions, where change in any one of the links will lead to an imbalance of the triangular relationship. When the complex exceeds the adaptability of the body, corresponding diseases will arise from the weakest link [10,17,19], such as malocclusion, TMD, myofascial pain, etc. These diseases are collectively referred to as occlusal diseases, the treatment of which is so-called occlusal treatment, where a holistic treatment concept is apparently required.

4.3. Limitations

Due to the innovative significance of this study, as a brand-new exploration of BTX-A in occlusal therapy, the sample size included was limited. Moreover, similar to the new application of BTX in other clinical fields, it lacked the evidence support of a long-term follow-up, such as the injection dose, injection frequency, and side effects.

4.4. Future Perspective

More research, particularly well-designed randomized controlled trials with high evidence levels, is needed in the future to verify the reliability and long-term stability of the above findings by expanding the sample size, lengthening the follow-up time, and increasing the number of BTX-A injections. Simultaneously, an individualized BTX-A dosage, injection method, administration frequency, and cycle should be researched further in order to provide a novel adjuvant treatment approach for short-faced individuals within a safe range of occlusal therapy.

5. Conclusions

BTX-A can provide a more favorable muscular environment for short-faced patients through adjustment of the biting force, contributing to the reconstruction of healthier homeostasis of the masticatory complex. However, further research is required to validate these findings.

Author Contributions: Conceptualization, X.L. and X.F.; methodology, X.L. and J.L. (Juan Li); software, X.L.; validation, X.F., X.B. and J.X.; formal analysis, X.L.; investigation, X.L.; resources, J.L. (Jun Lin) and J.X.; data curation, X.L.; writing—original draft preparation, X.L.; writing—review and editing, X.F.; visualization, X.L.; supervision, J.L. (Jun Lin); project administration, J.L. (Jun Lin); funding acquisition, J.L (Jun Lin). All authors have read and agreed to the published version of the manuscript.

Funding: This research was funded by the National Natural Science Foundation of China (No. 81970978) and Zhejiang Provincial Natural Science Foundation of China (No. LY22H140006).

Institutional Review Board Statement: The study was conducted in accordance with the Declaration of Helsinki, and approved by the Institutional Review Board of The First Affiliated Hospital,

College of Medicine, Zhejiang University (protocol code: (2021) IIT (171) and date of approval: 10 March 2021).

Informed Consent Statement: Informed consent was obtained from all subjects involved in the study.

Data Availability Statement: The data presented in this study are available on request from the corresponding author due to restrictions e.g. privacy or ethical.

Conflicts of Interest: The authors declare no conflict of interest. The funders had no role in the design of the study; in the collection, analyses, or interpretation of data; in the writing of the manuscript, or in the decision to publish the results.

Abbreviations

BTX-A	botulinum toxin-A	FH	frankfort horizontal plane
MS	masticatory system	AOP	anterior occlusal plane
SS	stomatognathic system	POP	posterior occlusal plane
TMJ	temporomandibular joint	COS	curve of Spee
TMD	temporomandibular joint disorders	COW	curve of Wilson
CBCT	cone-beam computed tomography	AEI	articular eminence inclination
MMT	masseter muscle thickness	AJS	anterior joint space
MP	mandibular plane	SJS	superior joint space
PP	palatal plane	PJS	posterior joint space

References

1. Okeson, J.P. Evolution of occlusion and temporomandibular disorder in orthodontics: Past, present, and future. *Am. J. Orthod. Dentofac. Orthop.* **2015**, *147*, S216–S223. [CrossRef] [PubMed]
2. Kandasamy, S.; Greene, C.S. The evolution of temporomandibular disorders: A shift from experience to evidence. *J. Oral Pathol. Med.* **2020**, *49*, 461–469. [CrossRef] [PubMed]
3. Zonnenberg, A.J.J.; Türp, J.C.; Greene, C.S. Centric relation critically revisited—What are the clinical implications? *J. Oral Rehabil.* **2021**, *48*, 1050–1055. [CrossRef] [PubMed]
4. Angle, E.H. Classification of malocclusion. *Dent. Cosmos.* **1899**, *41*, 246–264.
5. Summers, C.J. The occlusal index: A system for identifying and scoring occlusal disorders. *Am. J. Orthod.* **1971**, *59*, 552–567. [CrossRef]
6. Andrews, L.F. The six keys to normal occlusion. *Am. J. Orthod.* **1972**, *62*, 296–309. [CrossRef]
7. Roth, R.H. The maintenance system and occlusal dynamics. *Dent. Clin. N. Am.* **1976**, *20*, 761–788. [CrossRef]
8. Roth, R.H. Functional occlusion for the orthodontist. *J. Clin. Orthod.* **1981**, *15*, 32–51.
9. Dawson, P.E. *Functional Occlusion: From TMJ to Smile Design*, 1st ed.; Elsevier Health Sciences: St. Louis, MO, USA, 2006; pp. 4–5.
10. Goodacre, C.J.; Roberts, W.E.; Goldstein, G.; Wiens, J.P. Does the Stomatognathic System Adapt to Changes in Occlusion? Best Evidence Consensus Statement. *J. Prosthodont.* **2021**, *30*, 5–11. [CrossRef]
11. Nakamura, S.S.; Donatelli, D.; Rosenberg, E.S. Posterior bite collapse: Guidelines for treatment based on form and function. *Int. J. Periodontics Restor. Dent.* **2022**, *42*, 351–359. [CrossRef]
12. D'Amico, C.; Bocchieri, S.; Sambataro, S.; Surace, G.; Stumpo, C.; Fiorillo, L. Occlusal load considerations in implant-supported fixed restorations. *Prosthesis* **2020**, *2*, 252–265. [CrossRef]
13. Sheridan, R.A.; Decker, A.M.; Plonka, A.B.; Wang, H.L. The role of occlusion in implant therapy: A comprehensive updated review. *Implant. Dent.* **2016**, *25*, 829–838. [CrossRef] [PubMed]
14. Dommisch, H.; Walter, C.; Difloe-Geisert, J.C.; Gintaute, A.; Jepsen, S.; Zitzmann, N.U. Efficacy of tooth splinting and occlusal adjustment in patients with periodontitis exhibiting masticatory dysfunction: A systematic review. *Clin. Periodontol.* 2021; *ahead of print*.
15. Passanezi, E.; Sant'Ana, A.C.P. Role of occlusion in periodontal disease. *Periodontol 2000.* **2019**, *79*, 129–150. [CrossRef] [PubMed]
16. Gund, M.P.; Wrbas, K.-T.; Hannig, M.; Rupf, S. Apical periodontitis after intense bruxism. *BMC Oral Health* **2022**, *22*, 91. [CrossRef] [PubMed]
17. Liu, X.; Zhang, C.; Liu, Q.; Zhou, K.; Yin, N.; Zhang, H.; Shi, M.; Liu, X.; Wang, M. Dental malocclusion stimulates neuromuscular circuits associated with temporomandibular disorders. *Eur. J. Oral Sci.* **2018**, *126*, 466–475. [CrossRef]
18. Ribeiro-Lages, M.B.; Martins, M.L.; Magno, M.B.; Masterson Ferreira, D.; Tavares-Silva, C.M.; Fonseca-Gonçalves, A.; Serra-Negra, J.M.; Maia, L.C. Is there association between dental malocclusion and bruxism? A systematic review and meta-analysis. *J. Oral Rehabil.* **2020**, *47*, 1304–1318. [CrossRef]
19. Buvinic, S.; Balanta-Melo, J.; Kupczik, K.; Vásquez, W.; Beato, C.; Toro-Ibacache, V. Muscle-Bone Crosstalk in the Masticatory System: From Biomechanical to Molecular Interactions. *Front. Endocrinol.* **2021**, *11*, 606947. [CrossRef]
20. Pepicelli, A.; Woods, M.; Briggs, C. The mandibular muscles and their importance in orthodontics: A contemporary review. *Am. J. Orthod. Dentofac. Orthop.* **2005**, *128*, 774–780. [CrossRef]

21. Saadeh, M.; Fayyad-Kazan, H.; Haddad, R.; Ayoub, F. Facial soft tissue thickness differences among different vertical facial patterns. *Forensic. Sci. Int.* **2020**, *317*, 110468. [CrossRef]
22. Yamada, T.; Sugiyama, G.; Mori, Y. Masticatory muscle function affects the pathological conditions of dentofacial deformities. *Jpn. Dent. Sci. Rev.* **2020**, *56*, 56–61. [CrossRef]
23. Kattiney de Oliveira, L.; Fernandes Neto, A.J.; Moraes Mundim Prado, I.; Guimarães Henriques, J.C.; Beom Kim, K.; de Araújo Almeida, G. Evaluation of the condylar position in younger and older adults with or without temporomandibular symptoms by using cone beam computed tomography. *J. Prosthet. Dent.* **2022**, *127*, 445–452. [CrossRef] [PubMed]
24. Chamberland, S. Progressive idiopathic condylar resorption: Three case reports. *Am. J. Orthod. Dentofac. Orthop.* **2019**, *156*, 531–544. [CrossRef] [PubMed]
25. Almukhtar, R.M.; Fabi, S.G. The masseter muscle and its role in facial contouring, aging, and quality of life: A literature review. *Plast. Reconstr. Surg.* **2019**, *143*, 39e–48e. [CrossRef] [PubMed]
26. Biondi, K.; Lorusso, P.; Fastuca, R.; Mangano, A.; Zecca, P.A.; Bosco, M.; Caprioglio, A.; Levrini, L. Evaluation of masseter muscle in different vertical skeletal patterns in growing patients. *Eur. J. Paediatr. Dent.* **2016**, *17*, 47–52.
27. Takeuchi-Sato, T.; Arima, T.; Mew, M.; Svensson, P. Relationships between craniofacial morphology and masticatory muscle activity during isometric contraction at different interocclusal distances. *Arch. Oral. Biol.* **2019**, *98*, 52–60. [CrossRef]
28. Lione, R.; Franchi, L.; Noviello, A.; Bollero, P.; Fanucci, E.; Cozza, P. Three-dimensional evaluation of masseter muscle in different vertical facial patterns: A cross-sectional study in growing children. *Ultrason Imaging* **2013**, *35*, 307–317. [CrossRef]
29. Piancino, M.G.; Tortarolo, A.; Di Benedetto, L.; Crincoli, V.; Falla, D. Chewing patterns and muscular activation in deep bite malocclusion. *J. Clin. Med.* **2022**, *11*, 1702. [CrossRef]
30. Arponen, H.; Suominen, A.; Svedström-Oristo, A.L. Longitudinal analysis of the quality of orthodontic treatment outcome and stability of occlusal traits. *Acta Odontol. Scand.* **2022**, *80*, 234–240. [CrossRef]
31. Pollard, D.; Akyalcin, S.; Wiltshire, W.A.; Rody, W.J. Relapse of orthodontically corrected deepbites in accordance with growth pattern. *Am. J. Orthod. Dentofac. Orthop.* **2012**, *141*, 477–483. [CrossRef]
32. Rozzi, M.; Mucedero, M.; Pezzuto, C.; Lione, R.; Cozza, P. Long-term stability of curve of Spee levelled with continuous archwires in subjects with different vertical patterns: A retrospective study. *Eur. J. Orthod.* **2019**, *41*, 286–293. [CrossRef]
33. Santander, P.; Quast, A.; Olbrisch, C.; Rose, M.; Moser, N.; Schliephake, H.; Meyer-Marcotty, P. Comprehensive 3D analysis of condylar morphology in adults with different skeletal patterns–a cross-sectional study. *Head Face Med.* **2020**, *16*, 33. [CrossRef]
34. Barbosa, L.A.G.; Araujo, E.; Behrents, R.G.; Buschang, P.H. Longitudinal cephalometric growth of untreated subjects with Class II Division 2 malocclusion. *Am. J. Orthod. Dentofac. Orthop.* **2017**, *151*, 914–920. [CrossRef] [PubMed]
35. Paes-Souza, S.D.A.; Marañón-Vásquez, G.A.; Galisteu-Luiz, K.; Ferreira, D.M.T.P.; Maia, L.C.; Nojima, M.D.C.G. Is there variation in the depth of the curve of Spee in individuals with different dentoskeletal patterns? A systematic review with meta-analysis. *Eur. J. Orthod.* **2022**, *2*, cjab078. [CrossRef] [PubMed]
36. Tanaka, E.M.; Sato, S. Longitudinal alteration of the occlusal plane and development of different dentoskeletal frames during growth. *Am. J. Orthod. Dentofac. Orthop.* **2008**, *134*, 602.e1–602.e11. [CrossRef]
37. Charalampidou, M.; Kjellberg, H.; Georgiakaki, I.; Kiliaridis, S. Masseter muscle thickness and mechanical advantage in relation to vertical craniofacial morphology in children. *Acta Odontol. Scand.* **2008**, *66*, 23–30. [CrossRef]
38. Oliveira, S.S.I.; Pannuti, C.M.; Paranhos, K.S.; Tanganeli, J.P.C.; Laganá, D.C.; Sesma, N.; Duarte, M.; Frigerio, M.L.M.A.; Cho, S.C. Effect of occlusal splint and therapeutic exercises on postural balance of patients with signs and symptoms of temporomandibular disorder. *Clin. Exp. Dent. Res.* **2019**, *5*, 109–115. [CrossRef]
39. Hong, S.W.; Kang, J.H. Decreased mandibular cortical bone quality after botulinum toxin injections in masticatory muscles in female adults. *Sci. Rep.* **2020**, *10*, 3623. [CrossRef]
40. Sellin, L.C. The action of botulinum toxin at the neuromuscular junction. *Med. Biol.* **1981**, *59*, 11–20.
41. Serrera-Figallo, M.A.; Ruiz-De-León-Hernández, G.; Torres-Lagares, D.; Castro-Araya, A.; Torres-Ferrerosa, O.; Hernández-Pacheco, E.; Gutierrez-Perez, J.L. Use of botulinum toxin in orofacial clinical practice. *Toxins* **2020**, *12*, 112. [CrossRef]
42. Kumar, A.; Spivakovsky, S. Bruxism-is botulinum toxin an effective treatment? *Evid. Based Dent.* **2018**, *19*, 59. [CrossRef]
43. Sendra, L.A.; Montez, C.; Vianna, K.C.; Barboza, E.P. Clinical outcomes of botulinum toxin type A injections in the management of primary bruxism in adults: A systematic review. *J. Prosthet. Dent.* **2020**, *126*, 33–40. [CrossRef] [PubMed]
44. Cooper, L.; Lui, M.; Nduka, C. Botulinum toxin treatment for facial palsy: A systematic review. *J. Plast. Reconstr. Aesthetic Surg.* **2017**, *70*, 833–841. [CrossRef] [PubMed]
45. Sipahi Calis, A.; Colakoglu, Z.; Gunbay, S. The use of botulinum toxin-a in the treatment of muscular temporomandibular joint disorders. *J. Stomatol. Oral Maxillofac. Surg.* **2019**, *120*, 322–325. [CrossRef] [PubMed]
46. Kim, H.-S.; Yun, P.-Y.; Kim, Y.-K. A clinical evaluation of botulinum toxin-A injections in the temporomandibular disorder treatment. *Maxillofac. Plast. Reconstr. Surg.* **2016**, *38*, 5. [CrossRef]
47. Berweck, S.; Bonikowski, M.; Kim, H.; Althaus, M.; Flatau-Baqué, B.; Mueller, D.; Banach, M.D. Placebo-Controlled Clinical Trial of IncobotulinumtoxinA for Sialorrhea in Children. *Neurology* **2021**, *97*, e1425–e1436. [CrossRef]
48. Bakke, M.; Baram, S.; Dalager, T.; Biernat, H.B.; Møller, E. Oromandibular dystonia, mental distress and oro-facial dysfunction—A follow-up 8-10 years after start of treatment with botulinum toxin. *J. Oral Rehabil.* **2019**, *46*, 441–449. [CrossRef]

49. Choi, J.W.; Kim, H.J.; Moon, J.W.; Kang, S.H.; Tak, H.J.; Lee, S.H. Compensatory dentoalveolar supraeruption and occlusal plane cant after botulinum-induced hypotrophy of masticatory closing muscles in juvenile rats. *Arch. Oral Biol.* **2019**, *101*, 34–42. [CrossRef]
50. Yoshimi, T.; Koga, Y.; Nakamura, A.; Fujishita, A.; Kohara, H.; Moriuchi, E.; Yoshimi, K.; Tsai, C.Y.; Yoshida, N. Mechanism of motor coordination of masseter and temporalis muscles for increased masticatory efficiency in mice. *J. Oral Rehabil.* **2017**, *44*, 363–374. [CrossRef]
51. Tsai, C.Y.; Chiu, W.C.; Liao, Y.H.; Tsai, C.M. Effects on craniofacial growth and development of unilateral botulinum neurotoxin injection into the masseter muscle. *Am. J. Orthod. Dentofac. Orthop.* **2009**, *135*, 142.e1–142.e6. [CrossRef]
52. Liu, J.; Liu, S.Y.; Zhao, Y.J.; Gu, X.; Li, Q.; Jin, Z.L.; Chen, Y.J. Effects of occlusion on mandibular morphology and architecture in rats. *J. Surg. Res.* **2016**, *200*, 533–543. [CrossRef]
53. Navarrete, A.L.; Rafferty, K.L.; Liu, Z.J.; Ye, W.; Greenlee, G.M.; Herring, S.W. Botulinum neurotoxin type A in the masseter muscle: Effects on incisor eruption in rabbits. *Am. J. Orthod. Dentofac. Orthop.* **2013**, *143*, 499–506. [CrossRef] [PubMed]
54. Kim, I.H.; Singer, S.R.; Mupparapu, M. Review of cone beam computed tomography guidelines in North America. *Quintessence Int.* **2019**, *50*, 136–145.
55. Lee, H.J.; Kim, S.J.; Lee, K.J.; Yu, H.S.; Baik, H.S. Repeated injections of botulinum toxin into the masseter muscle induce bony changes in human adults: A longitudinal study. *Korean J. Orthod.* **2017**, *47*, 222–228. [CrossRef] [PubMed]
56. Demehri, S.; Muhit, A.; Zbijewski, W.; Stayman, J.W.; Yorkston, J.; Packard, N.; Senn, R.; Yang, D.; Foos, D.; Thawait, G.K.; et al. Assessment of image quality in soft tissue and bone visualization tasks for a dedicated extremity cone-beam CT system. *Eur. Radiol.* **2015**, *25*, 1742–1751. [CrossRef] [PubMed]
57. Becht, M.P.; Mah, J.; Martin, C.; Razmus, T.; Gunel, E.; Ngan, P. Évaluation De La Morphologie Du Muscle Masséter Dans Différents Types De Malocclusions En Utilisant La Tomographie Volumétrique À Faisceau Conique. *Int. Orthod.* **2014**, *12*, 32–48. [CrossRef]
58. Iyer, A.; Thor, M.; Onochie, I.; Hesse, J.; Zakeri, K.; Locastro, E.; Jiang, J.; Veeraraghavan, H.; Elguindi, S.; Lee, N.Y.; et al. Prospectively-validated deep learning model for segmenting swallowing and chewing structures in CT. *Phys. Med. Biol.* **2022**, *67*, 024001. [CrossRef]
59. Hu, Z.J.; He, J.; Zhao, F.D.; Fang, X.Q.; Zhou, L.N.; Fan, S.W. An assessment of the intra- and inter-reliability of the lumbar paraspinal muscle parameters using CT scan and magnetic resonance imaging. *Spine* **2011**, *36*, 868–874. [CrossRef]
60. Pan, Y.; Wang, Y.; Li, G.; Chen, S.; Xu, T. Validity and reliability of masseter muscles segmentation from the transverse sections of Cone-Beam CT scans compared with MRI scans. *Int. J. Comput. Assist. Radiol. Surg.* **2022**, *17*, 751–759. [CrossRef]
61. Xu, J.A.; Yuasa, K.; Yoshiura, K.; Kanda, S. Quantitative analysis of masticatory muscles using computed tomography. *Dentomaxillofac. Radiol.* **1994**, *23*, 154–158. [CrossRef]
62. Esenlik, E.; Sabuncuoglu, F.A. Alveolar and symphysis regions of patients with skeletal class ii division 1 anomalies with different vertical growth patterns. *Eur. J. Dent.* **2012**, *6*, 123–132. [CrossRef]
63. Ning, R.; Guo, J.; Li, Q.; Martin, D. Maxillary width and hard palate thickness in men and women with different vertical and sagittal skeletal patterns. *Am. J. Orthod. Dentofac. Orthop.* **2021**, *159*, 564–573. [CrossRef] [PubMed]
64. Al-Qawasmi, R.; Coe, C. Genetic influence on the curves of occlusion in children seeking orthodontic treatment. *Int. Orthod.* **2021**, *19*, 82–87. [CrossRef] [PubMed]
65. Yoon, S.Y.; Song, J.M.; Kim, Y.D.; Chung, I.K.; Shin, S.H. Clinical changes of TMD and condyle stability after two jaw surgery with and without preceding TMD treatments in class III patients. *Maxillofac. Plast. Reconstr. Surg.* **2015**, *37*, 9. [CrossRef]
66. Kamelchuk, L.S.; Grace, M.G.A.; Major, P.W. Post-imaging temporomandibular joint space analysis. *Cranio* **1996**, *14*, 23–29. [CrossRef] [PubMed]
67. Cheng, J.; Hsu, S.H.; McGee, J.S. Botulinum toxin injections for masseter reduction in east asians. *Dermatol. Surg.* **2019**, *45*, 566–572. [CrossRef]
68. Diracoglu, D.; Sahbaz, T.; Alptekin, K.; Dogan, N. Effects of ultrasound-assisted botulinum neurotoxin-A injection in patients with bruxism and masseter hypertrophy. *Turk. J. Phys. Med. Rehabil.* **2021**, *67*, 351–356. [CrossRef]
69. Lee, H.H.; Kim, S.T.; Lee, K.J.; Baik, H.S. Effect of a second injection of botulinum toxin on lower facial contouring, as evaluated using 3-dimensional laser scanning. *Dermatol. Surg.* **2015**, *41*, 439–444. [CrossRef]
70. Zhang, L.D.; Liu, Q.; Zou, D.R.; Yu, L. feng. Occlusal force characteristics of masseteric muscles after intramuscular injection of botulinum toxin A(BTX–A)for treatment of temporomandibular disorder. *Br. J. Oral Maxillofac. Surg.* **2016**, *54*, 736–740. [CrossRef]
71. Kim, J.H.; Shin, J.H.; Kim, S.T.; Kim, C.Y. Effects of two different units of botulinum toxin type a evaluated by computed tomography and electromyographic measurements of human masseter muscle. *Plast. Reconstr. Surg.* **2007**, *119*, 711–717. [CrossRef]
72. Kim, K.S.; Byun, Y.S.; Kim, Y.J.; Kim, S.T. Muscle weakness after repeated injection of botulinum toxin type A evaluated according to bite force measurement of human masseter muscle. *Dermatol. Surg.* **2009**, *35*, 1902–1907. [CrossRef]
73. Wong, A.; Woods, M.G.; Stella, D. Three-dimensional computed tomographic assessment of mandibular muscles in growing subjects with different vertical facial patterns. *Aust. Orthod. J.* **2016**, *32*, 2–17. [CrossRef] [PubMed]
74. Acar, A.; Alcan, T.; Erverdi, N. Evaluation of the relationship between the anterior component of occlusal force and postretention crowding. *Am. J. Orthod. Dentofac. Orthop.* **2002**, *122*, 366–370. [CrossRef] [PubMed]
75. Teng, F.; Du, F.Y.; Chen, H.Z.; Jiang, R.P.; Xu, T.M. Three-dimensional analysis of the physiologic drift of adjacent teeth following maxillary first premolar extractions. *Sci. Rep.* **2019**, *9*, 14549. [CrossRef] [PubMed]

76. Alkhatib, R.; Chung, C.H. Buccolingual inclination of first molars in untreated adults: A CBCT study. *Angle Orthod.* **2017**, *87*, 598–602. [CrossRef]
77. Golshah, A.; Rezaei, N.; Heshmati, S. Buccolingual Inclination of Canine and First and Second Molar Teeth and the Curve of Wilson in Different Sagittal Skeletal Patterns of Adults Using Cone-Beam Computed Tomography. *Int. J. Dent.* **2020**, *2020*, 8893778. [CrossRef]
78. Seok, H.; Kim, S.G. Correction of malocclusion by botulinum neurotoxin injection into masticatory muscles. *Toxins* **2018**, *10*, 27. [CrossRef]
79. Dai, F.; Wang, L.; Chen, G.; Chen, S.; Xu, T. Three-dimensional modeling of an individualized functional masticatory system and bite force analysis with an orthodontic bite plate. *Int. J. Comput. Assist. Radiol. Surg.* **2016**, *11*, 217–229. [CrossRef]
80. Kiliaridis, S.; Georgiakaki, I.; Katsaros, C. Masseter muscle thickness and maxillary dental arch width. *Eur. J. Orthod.* **2003**, *25*, 259–263. [CrossRef]
81. Koyano, K.; Tsukiyama, Y.; Kuwatsuru, R. Rehabilitation of occlusion–science or art? *J. Oral Rehabil.* **2012**, *39*, 513–521. [CrossRef]
82. Fueki, K.; Yoshida, E.; Okano, K.; Igarashi, Y. Association between occlusal curvature and masticatory movements with different test foods in human young adults with permanent dentitions. *Arch. Oral Biol.* **2013**, *58*, 674–680. [CrossRef]
83. Yu, X.; Lu, D.; Tian, S.; Yuan, X.; Cheng, Q. Influence of cranial base angle and posterior occlusal plane inclination on sagittal dentoskeletal types. *Chin. J. Tissue Eng. Res.* **2020**, *24*, 1659–1665.
84. Santhosh Kumar, M.P. The emerging role of botulinum toxin in the treatment of orofacial disorders: Literature update. *Asian J. Pharm. Clin. Res.* **2017**, *10*, 21–29. [CrossRef]
85. Schiffman, E.; Ohrbach, R.; Truelove, E.; Look, J.; Anderson, G.; Goulet, J.-P.; List, T.; Svensson, P.; Gonzalez, Y.; Lobbezoo, F.; et al. Diagnostic criteria for temporomandibular disorders (DC/TMD) for clinical and research applications: Recommendations of the International RDC/TMD Consortium Network* and Orofacial Pain Special Interest Group†. *J. Oral Facial Pain Headache* **2014**, *28*, 6–27. [CrossRef]
86. Oksanen, E.; Männistö, V.; Kormi, E.; Vallioniemi, H.; Suojanen, J. Temporomandibular disorder patients benefit from intramuscular botulinum toxin type A injections. *J. Craniofac. Surg.* 2021; *ahead of print*.
87. Pihut, M.; Ferendiuk, E.; Szewczyk, M.; Kasprzyk, K.; Wieckiewicz, M. The efficiency of botulinum toxin type A for the treatment of masseter muscle pain in patients with temporomandibular joint dysfunction and tension-type headache. *J. Headache Pain* **2016**, *17*, 29. [CrossRef] [PubMed]
88. Al-Wayli, H. Treatment of chronic pain associated with nocturnal bruxism with botulinum toxin. A prospective and randomized clinical study. *J. Clin. Exp. Dent.* **2017**, *9*, e112–e117. [CrossRef]
89. Shi, Z.; Lv, J.; Xiaoyu, L.; Zheng, L.W.; Yang, X.W. Condylar degradation from decreased occlusal loading following masticatory muscle atrophy. *Biomed. Res. Int.* **2018**, *2018*, 6947612. [CrossRef]
90. Rafferty, K.L.; Liu, Z.J.; Ye, W.; Navarrete, A.L.; Nguyen, T.T.; Salamati, A.; Herring, S.W. Botulinum toxin in masticatory muscles: Short- and long-term effects on muscle, bone, and craniofacial function in adult rabbits. *Bone* **2012**, *50*, 651–662. [CrossRef] [PubMed]
91. Kün-Darbois, J.D.; Libouban, H.; Chappard, D. Botulinum toxin in masticatory muscles of the adult rat induces bone loss at the condyle and alveolar regions of the mandible associated with a bone proliferation at a muscle enthesis. *Bone* **2015**, *77*, 75–82. [CrossRef]
92. Raphael, K.G.; Janal, M.N.; Tadinada, A.; Santiago, V.; Sirois, D.A.; Lurie, A.G. Effect of multiple injections of botulinum toxin into painful masticatory muscles on bone density in the temporomandibular complex. *J. Oral Rehabil.* **2020**, *47*, 1319–1329. [CrossRef]
93. Dickinson, E.; Fitton, L.C.; Kupczik, K. Ontogenetic changes to muscle architectural properties within the jaw-adductor musculature of Macaca fascicularis. *Am. J. Phys. Anthropol.* **2018**, *167*, 291–310. [CrossRef]
94. Jiménez-Silva, A.; Tobar-Reyes, J.; Vivanco-Coke, S.; Pastén-Castro, E.; Palomino-Montenegro, H. Centric relation–intercuspal position discrepancy and its relationship with temporomandibular disorders. A systematic review. *Acta Odontol. Scand.* **2017**, *75*, 463–474. [CrossRef] [PubMed]

toxins

MDPI

Review

How Does Botulinum Toxin Inhibit Itch?

Parisa Gazerani [1,2]

[1] Department of Life Sciences and Health, Faculty of Health Sciences, Oslo Metropolitan University, 0130 Oslo, Norway; parisaga@oslomet.no or gazerani@hst.aau.dk
[2] Department of Health Science and Technology, Faculty of Medicine, Aalborg University, 9220 Aalborg East, Denmark

Abstract: Two decades after reports of the anti-pruritic effects of botulinum neurotoxins (BoNTs), there is still no approved product for the anti-itch indication of BoNTs, and most clinical case reports still focus on the off-label use of BoNTs for various itchy conditions. Few randomized clinical trials have been conducted with controversial results, and the beneficial effects of BoNTs against itch are mainly based on case studies and case series. These studies are valuable in presenting the potential application of BoNTs in chronic pruritic conditions, but due to the nature of these studies, they are categorized as providing lower levels of evidence or lower grades of recommendation. To obtain approval for the anti-pruritic indication of BoNTs, higher levels of evidence are required, which can be achieved through conducting large-scale and well-designed studies with proper control groups and established careful and reliable primary and secondary outcomes. In addition to clinical evidence, presenting the mechanism-based antipruritic action of BoNTs can potentially strengthen, accelerate, and facilitate the current efforts towards further investments in accelerating the field towards the potential approval of BoNTs for itchy conditions. This review, therefore, aimed to provide the state-of-the-art mechanisms underlying the anti-itch effect of BoNTs from basic studies that resemble various clinical conditions with itch as a hallmark. Evidence of the neuronal, glial, and immune modulatory actions of BoNTs in reducing the transmission of itch are presented, and future potential directions are outlined.

Keywords: botulinum neurotoxin; itch; anti-pruritic; mechanism of action

Key Contribution: This focused review provides the current understanding of mechanisms underlying the antipruritic effects of BoNTs.

Citation: Gazerani, P. How Does Botulinum Toxin Inhibit Itch?. *Toxins* **2022**, *14*, 701. https://doi.org/10.3390/toxins14100701

Received: 2 September 2022
Accepted: 10 October 2022
Published: 12 October 2022

1. Introduction

1.1. Botulism, Clostridium Botulinum, and Botulinum Toxin

An illness characterized by muscle paralysis following the consumption of spoiled sausage was first termed botulism by Muller [1]. Botulus means lunch meat, salami, sausage, and similar in Latin [2], and it most likely dates back to earlier historical concerns about foodborne toxicity in the Byzantine era [3]. A comprehensive description of a somewhat similar illness by Justinus Kerner [1], is in line with what is currently considered in the diagnosis of botulism, which is marked by muscle weakness or paralysis, swallowing difficulty, and signs of autonomic dysfunction, such as dry mouth [4]. Krener was the first to propose that the illness is potentially caused by a biological toxin [5]. Emile Van Ermengem, in 1895, provided the first description of the organism that could cause botulism, an anaerobic Gram-positive, rod-shaped bacterium, which he named bacillus botulinum [6]. The name was changed in 1924 by Ida Bengstrom to *Clostridium botulinum*, based on the spindle-like shape of the bacteria [6], where kloster in Greek means spindle. The purification and development of botulinum toxin, the toxin from clostridium botulinum, was eventually dated back to the time of World War II, when it could potentially be used as a biological weapon [7,8]. Carl Lamanna and Richard Duff were the first to extract and crystalize botulinum toxin [9]. Later in 1946, Edward Shantz and Erik Johnson could purify larger amounts and also further refine the toxin for clinical research [10]. Burgen and his colleagues, in 1949, explained how the toxin can produce muscle paralysis and presented the effect of the toxin on the neuromuscular junction and blockade of acetylcholine (Ach) release [11]. In 1964, when Daniel Drachman demonstrated muscle weakness in the hind limb of a chicken following toxin administration, which was dose-dependent, Allen Scott and his colleagues became interested in using it for strabismus [7,12]. The first paper on this subject was published in 1980, presenting a clinical trial's results of 67 patients, where a selected injection of BoNT into eye muscles could correct strabismus [12]. Evidence of the benefits of the BoNT injection, as shown in blepharospasm and hemifacial spasm, finally led to the approval of botulinum toxin-A by the FDA for conditions of strabismus, blepharospasm, and hemifacial spasm in 1989 [13,14]. Since then, toxin product development and testing for various medical conditions has followed, which has resulted in FDA approvals [15]. There are also various off-label uses of BoNT in various medical fields [16,17]. Along the way, efforts by basic scientists have resulted in presenting the molecular structures of botulinum toxins and their mechanisms of action for various clinical conditions [18–20].

Several toxin serotypes are produced by *Clostridium botulinum* that, from the immunological point of view, are considered distinct and are named A, B, C1, C2, D, E, F, and G [21]. All serotypes are neurotoxins, except for C2. Human botulism is known to be caused by serotypes A, B, E, and rarely F, while botulism in fish, birds, and non-human mammals is mainly caused by serotypes C1 and D [22]. Botulinum neurotoxin serotype A (BoNTA) is the most toxic substance known to man with an estimated intravenous lethal dose (LD50) of 1–2 nanogram per kilogram (ng/kg) in humans [23]. Now, however, the lethal toxin has been developed as a safe medicinal product for clinical pharmacotherapy in a large number of medical conditions in humans [24,25], and it is also used in the field of veterinary medicine [26].

1.2. Botulinum Toxin Products and Uses

Several products are marketed around the globe with different product characteristics [27–30]. The chronological footprint tracking of the US-marketed botulinum neurotoxins approved by the FDA [31,32] shows OnabotulinumtoxinA (Botox) by Allergan Inc., approved for several conditions such as blepharospasm, hemifacial spasm, strabismus, cervical dystonia, glabellar lines, axillary hyperhidrosis, chronic migraine, upper limb spasticity, neurogenic bladder, lateral canthal lines, overactive bladder, lower limb spasticity in adults, forehead lines, and upper limb spasticity in children. IncobotulinumtoxinA (Xeomin) by Merz Pharma was approved by the FDA for a number of conditions such as cervical dystonia, blepharospasm, glabellar lines, upper limb spasticity in adults, and sialorrhea. Other approved products are AbobotulinumtoxinA (Dysport) by Ipsen Biopharmaceuticals, RimabotulinumtoxinB (Myobloc/Neurobloc) by US World Med-Solstice, and PrabotulinumtoxinA (Jeuveau) by Evolus Inc.

There are several conditions for which BoNT is not yet approved for clinical use, but off-label use has been presented, with beneficial effects, for example, in dermatology [33–35], primary headaches other than migraine [36], depression [37], and neuropathic pain [38,39]. One large area in which the off-label use of BoNT has been practiced for the last two decades is dermatology, for various conditions that appear with or without itch [40–43]. This focused review is dedicated to conditions accompanied by itch. Please consider that the use of BoNT in the field of cosmetology [44] is also active, but it is not the focus of this review. Interested readers are referred to a recent publication in this domain [45]. In the following, clinical evidence from the literature is presented first to indicate the current status for the use of BoNT in clinical conditions with itch. Thereafter, evidence is presented from basic science to outline the mechanistic basis for the antipruritic effects of BoNT. The purpose of this review is to accelerate the work in the field and motivate progression towards presenting the mechanism-based antipruritic action of BoNTs, which can potentially strengthen, accelerate, and facilitate their approval for itchy conditions.

2. Clinical Evidence for the Use of BoNT in Clinical Conditions with Itch

Itch is a sensory modality that is also called pruritus [46]. It often appears as an uncomfortable and unpleasant sensation and usually provokes a strong desire to scratch [46,47]. Dermatological conditions are often accompanied by itching [48]. Chronic itch [49], in particular, reduces the overall quality of life in affected patients, such as patients with atopic dermatitis (AD) [50,51]. Several types of itch have been defined, e.g., pruriceptive itch, neurogenic itch, neuropathic itch, and psychogenic itch [52,53]. This classification has been mainly based on the underlying mechanisms together with clinical manifestations [53]. Itch accompanies some systemic disorders, for example, some chronic kidney and liver diseases [54]. Diabetic neuropathy and shingles are among neurological disorders with itching as one of their debilitating symptoms [55–57]. Although probably not fully accurate, the first report of the antipruritic effect of Botulinum Toxin Type A (BoNTA) appeared in 2002, which presented the beneficial effects of BoNTA for lichen simplex in an open-label pilot study [58]. In the same year, the antipruritic effect of BoNTA was reported in hand dyshidrotic dermatitis [59]. Following these initial reports of the antipruritic effect of BoNTA, further investigations have continued [43,60], and successful treatments are being reported in the literature for many other dermatological conditions, such as Hailey–Hailey disease [61] and psoriasis [62].

A limited number of review articles are available, for example, the 2017 review on botulinum toxin off-label use in dermatology [63]. Another review [64] from the same year is also available, with a focus on novel indications of BoNT in dermatology. The most recent systematic review came out in 2021 [60]. This review included 167 studies from 1994 to 2020, and based on the evidence, diseases for which BoNTs could have therapeutic potentials were listed. These also included conditions with itch (e.g., post-herpetic neuralgia, notalgia paresthetica, Hailey–Hailey disease, and psoriasis) [60]. Most of these reports are based on case studies or case series that are considered to have an evidence level of three or four; hence, the grade of recommendation for use will be C and D [65]. On the other hand, there are cases of randomized control trials where results show no statistically significant effect of BoNT on itch, for example, the level of itch on the visual analog scale (VAS) or hyperpigmentation in notalgia parasthetica [66]. It is not clear as to what reason similar results of observational or case series are not repeated when studies are conducted as RCTs. This might be related to a true lack of effect, or other influential factors such as criteria for inclusion of participants (diversity, responsiveness, or lack of response) or selection of outcome measures (sensitivity, specificity) that might mask a statistically significant antipruritic effect. A review from 2018 [67], with a focus on localized chronic itch, evaluated available studies on the effect of BoNTA following intradermal administration. The authors found 25 studies conducted between 1996 and 2016 [67], where 11 articles were identified as relevant for further evaluation. This review [67] followed PRISMA and rated the studies based on the grade of recommendation. The toolkit provided by the Oxford Centre for Evidence-based Medicine Levels of Evidence was used in this review [67] to grade each study. The findings [67] showed that prospective observational studies and case reports mostly resulted in complete relief, strong improvement, and long-lasting effects (Table 1).

Table 1. Summary of studies examining the effect of botulinum toxin type A in chronic pruritus. CR, case report; POS, prospective observational study. ∗ Determined based on the Oxford Centre for Evidence-based Medicine levels of evidence. Reused with permission (license number: 5376990524166) from Elsevier and Copyright Clearance Center.

Grade of Recommendation ∗	Study Design	Diagnosis	N	Regimen	Degree of Pruritus Reduction	Reference
C	POS	Lichen simplex	4	100 U/mL BoNT/A spaced 2 cm apart	One treatment led to complete relief of symptoms lasting 4 months	[58]
C	POS	Partial-thickness to full-thickness burns	9	25 U/mL BoNT/A (spacing unknown)	One treatment led to a reduction in symptoms to <3/10 lasting an average of 6.3 months	[68]
C	POS	Inverse psoriasis	15	20 U/mL BoNT/A spaced 2.8 cm apart	One treatment led to a reduction in the mean visual analog score to 2.1/10 lasting 3 months	[41]
C	POS	Notalgia paresthetica	5	40 U/mL BoNT/A spaced 2 cm apart	In 3 patients: 1 treatment led to "symptomatic improvement" lasting 1 month; in 2 patients: 1 treatment led to "worsening of pruritus"	[69]
C	POS	Notalgia paresthetica, meralgia paresthetica, and neuropathic itch	6	0.27–0.47 U/mL BoTN/A (spacing unknown)	One treatment led to a 28% reduction in the mean visual analog score lasting ≥6 weeks	[70]
C	POS	Notalgia paresthetica	2	1.3 U/mL BoNT/A spaced 2 cm apart	First patient: 1 treatment led to complete relief of symptoms lasting ≥18 months; second patient: 2 treatments spaced 18 months apart led to complete relief of symptoms lasting an unknown duration	[42]
D	CR	Brachioradial pruritus	1	33.3 U/mL BoNT/A spaced 1.5 cm apart	One treatment led to complete relief of symptoms lasting 6 months	[71]
D	CR	Intractable pruritus after facial surgery	1	15 U BoNT/A (unknown dilution or spacing)	One treatment led to a "significant reduction" in symptoms lasting 2 months	[72]
D	CR	Axillary granular parakeratosis	1	50 U BoNT/A (unknown dilution or spacing)	One treatment led to a complete relief of symptoms lasting 4 months	[73]
D	CR	Inverse Psoriasis	1	100 U BoNT/A (unknown dilution or spacing)	One treatment led to a complete relief of symptoms lasting ≥1 month	[74]
D	CR	Hailey–Hailey disease	1	200 U BoNT/A (unknown dilution or spacing)	One treatment led to the complete relief of symptoms lasting 3 months	[75]

Collectively, systematic and non-systematic literature reviews, including all types of studies (e.g., case reports, case series, and RCTs), show that many applications of BoNT for dermatological conditions with and without itch are still off-label. Moreover, it seems that in the majority of cases, BoNTA might not be considered first-line therapy, but perhaps an option for patients with persistent or recurrent issues that remain unsolved or irresponsive to other treatments [43,76]. Toyama et al., in a recent review [76], presented a long list of available options for troublesome itch, where, among other options (e.g., medications used for depression and phosphodiesterase-4 inhibitors), BoNTA is also listed [76]. These authors also indicated that itch was relieved by BoNTA in patients with several conditions, such as chronic lichen simplex, inverse psoriasis, and post-burn itch [40,58,71]. As outcome measures, when the eczematic area and severity index (EASI) was used to determine the therapeutic effect, BoNTA reduced this score in chronic lichen simplex. Moreover, in inverse psoriasis, BoNTA reduced the PASI (psoriasis area and severity index) [40]. Pruritus and hyperhidrosis were both shown to be diminished in pediatric Fox–Fordyce disease [77]. BoNTA could also reduce itching related to keloid scars [78], in particular, when it was combined with triamcinolone [79], against atopic dermatitis [80] and post-burn itch [68].

BoNTs provide long-term therapeutic effects, and this characteristic is in favor of patient compliance [81]. However, it can be a costly treatment, and its comparable effects with other available agents might place BoNTs in the third or fourth line of therapy. It is too early to recommend the regular use of BoNTs at this stage, but it is not unlikely that we will witness the appearance of more reports on conditions in which stronger evidence exists for the beneficial effects of BoNTs. Further clinical trials must consider careful design and patient inclusion criteria, safe and effective dosing, and an optimal interval in repeated administration. In addition, a strategy for determining and implementing primary and secondary outcome measures would help in the more accurate identification of responders. The determination of influential factors, such as age, gender, and ethnic background, would also aid in targeting the right group of patients for the optimal effect. The formation of neutralizing antibodies [82] and the influence on safety and efficacy must also be taken into consideration for future clinical studies.

3. Mechanism(s) of Action of Botulinum Neurotoxins

As summarized above, BoNTA has been shown to reduce itch in several clinical dermatological conditions with itch as a common troublesome symptom [67,76]. Therefore, the next logical question would be to ask how this occurs [43,83]. Before addressing this question, two main clarifications are needed: (1) how botulinum toxin interacts with cells, receptors, and neurotransmitter release, and (2) how itch is generated and transmitted.

In the following section, the mechanism of action of BoNT in inhibiting the Ach release at the neuromuscular junction [84] is presented first. This is the proposed mechanism underlying muscle paralysis that can lead to respiratory failure and death in botulism [4,85]. The muscle relaxation effect of BoNT explains its medical use for several conditions, such as spasticity and strabismus, where abnormal or excessive muscle spasm is problematic [86,87]. Secondly, to understand the inhibitory action of BoNTA on itch, itch mechanisms are presented.

3.1. Mechanism of Action of Botulinum Neurotoxin in Blocking Neurotransmitter Release

Botulinum toxin consists of two chains, light and heavy chains, that are connected through a disulfide bond. The heavy chain, from its C-terminal region, binds to receptors at the presynaptic neurons [88]. A two-step receptor binding has been proposed: the first is binding to polysialoganglioside (PSG), which is followed by binding to SV2 (synaptic vesicle glycoprotein) [88]. This binding forms a complex that undergoes an endocytosis process. Within the cytoplasm, the disulfide bond is cleaved, permitting the light chain to act on the SNAP-25 (synaptosomal-associated protein-25) and cleave it. This is one of the protein components required for the vesicle fusion and exocytosis process. Interestingly, different serotypes cleave different proteins, which are Soluble N-ethylmaleimide-sensitive factor Attachment protein REceptors (SNAREs) [89]. Once vesicle fusion is prohibited, there is no vesicular release of transmitters from nerve terminals [18,19,90,91]. Figure 1 presents the proposed mechanism of action of BoNT in blocking the Ach release [87].

Identifying the blockade of Ach by BoNTA has sparked the interest of some researchers to look into other potential neurotransmitters that are blocked by BoNT. In one of the active areas, pain research, the analgesic action of BoNT was investigated extensively [92]. BoNTA shows analgesic efficacy in various pain conditions, such as musculoskeletal pain [93], dental medicine [94], and neuropathic pain [95], but it is only FDA approved for chronic migraine [96,97], among other headache and related disorders [98]. The proposed mechanism of action of BoNTA for migraine [99–101] has also been postulated by several researchers based on findings from basic and clinical research [102–104]. At a cellular–molecular level, some of the identified neurotransmitters and pathways targeted by BoNTA in pain are applicable in explaining the antipruritic effects of BoNTA [43]. This is perhaps because itch and pain interact and share some similarities while holding to the uniqueness of each modality [105–108]. Our studies from cells to animals and human experimental models of pain [83,109–114] revealed that BoNTA inhibits the release of several neurotransmitters involved in pain, including glutamate [113,114]. Other studies have also shown that BoNTA blocks the release of glutamate, CGRP, and substance P [115–118]. Readers are directed to a recent review by Matak et al. [92] for details on the mechanism of action of BoNT in pain.

A series of investigations by Burstein's group highlighted that BoNTA acts on the C-fibers to reduce pain and that TRPV1 and TRPA1 channels play a critical role [119]. In addition, this group has proposed that BoNTA is capable of altering inflammatory gene expression and immune cells in migraine prevention, where it can reduce pre-existing inflammation [120].

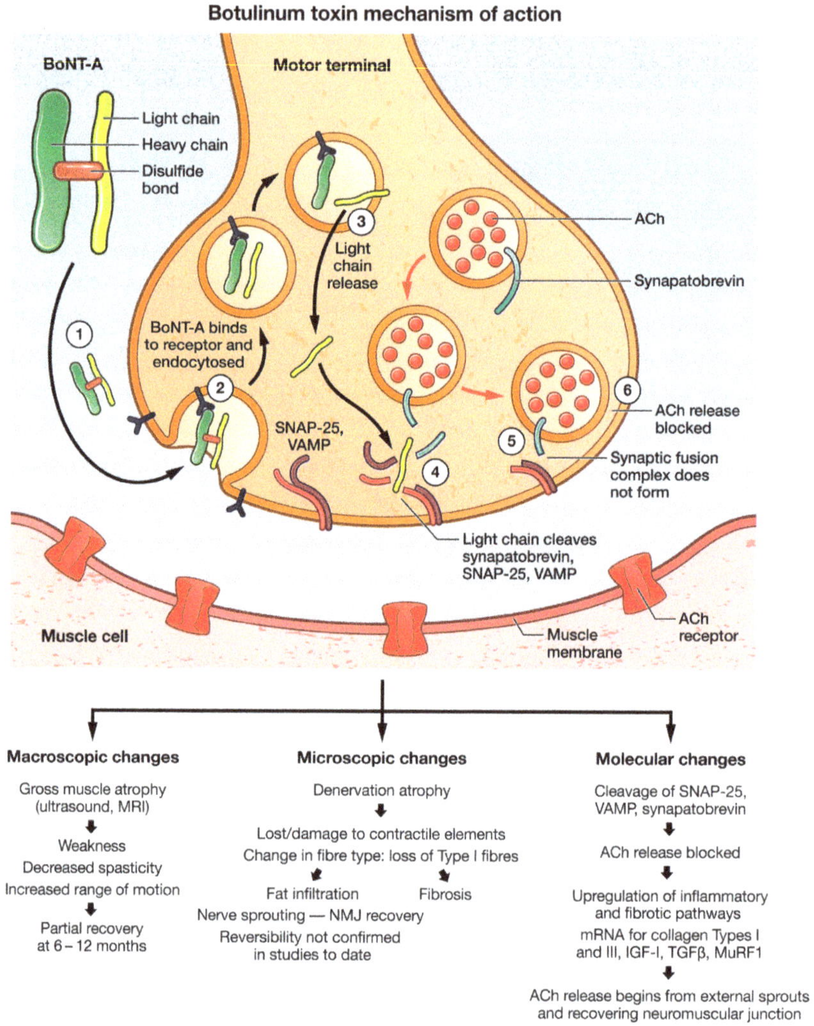

Figure 1. Botulinum toxin type A (BoNT-A) mechanism of action. The BoNT-A heavy chain is shown in green and the light chain in yellow, linked by a disulfide bond. Acetylcholine (Ach), the neurotransmitter that is blocked by BoNT-A, is shown as red dots within a circular vesicle in the nerve terminal. The effects of chemodenervation via injection of BoNT-A are summarized at macroscopic, microscopic and molecular levels. SNAP 25, soluble N-ethylmaleimide fusion protein/attachment protein; VAMP, vesicle-associated membrane protein. Reused from [87] for non-commercial/educational purposes under a Creative Commons license (Attribution-Noncommercial), Springer Nature.

Considering the evidence for the blockade of the vesicular release of substances by BoNTA presented above, it is not implausible to propose that BoNTA may block the release of substances that contribute to the sensation of itch [52]. To explain this, it is essential to present how itch is generated and transmitted. The section below presents what is known about pathways leading to an itching sensation.

3.2. Itch Mechanisms

Pruritus or itch is an uncomfortable sensation that generates a strong desire to scratch [121], which is seen in both humans and animals [122]. Itch is a unique sensory modality within the somatosensory system, and once it becomes chronic, it poses major distress and impairs the quality of life of the affected individuals [123]. In addition, people with chronic itch frequently suffer from self-harm when they are in the loop of uncontrollable itch–scratch cycles [124]. Itch is often associated with dermatological conditions, but it can also be a hallmark of systemic, neurological, and psychogenic conditions [123]. The mechanisms of itch have been extensively investigated in recent years, and as a result of the better understanding of itch pathways, several targeting sites and molecules have been identified and introduced to the field. Figure 2 presents a simplified sketch of the itch signaling pathway from the primary sensory neurons to the brain [125]. It is yet to be determined where and how exactly BoNT targets itch alongside this signaling pathway. Evidence is, however, being accumulated (Please refer to Section 4).

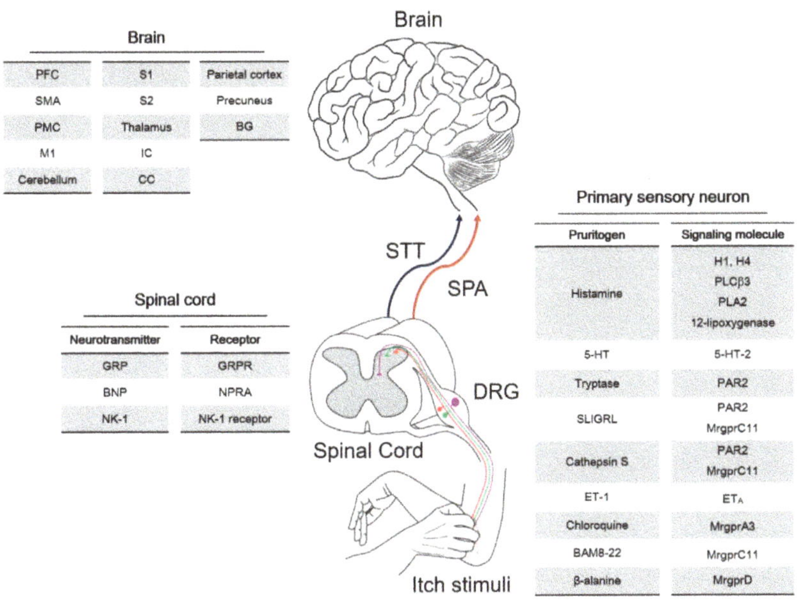

Figure 2. Itch signaling pathway: Schematic illustrating the transmission of itch from the primary sensory neurons to the brain. Itch stimuli (pruritogens) activate itch-sensing neurons in the dorsal root ganglion (DRG) that innervate the skin, which then stimulate second-order neurons in the spinal cord and multiple brain regions. Indicated in the tables are pruritogens, itch-selective molecules and receptors expressed in the primary sensory neurons and spinal cord, and brain regions activated by the cutaneous application of a pruritogen. STT, spinothalamic tract; SPA, spino-parabrachio-amygdaloid pathway; PFC, prefrontal cortex; SMA, supplementary motor area; PMC, premotor cortex; M1, primary motor cortex; S1, primary somatosensory cortex; S2, secondary somatosensory cortex; CC, cingulate cortex; IC, insular cortex; BG, basal ganglia; GRP, gastrin-releasing peptide; GRPR, gastrin-releasing peptide receptor; BNP, B-type natriuretic peptide; NPRA, natriuretic peptide receptor A; NK-1, neurokinin-1; H1, histamine H1 receptor; PLCβ3, phospholipase C β3; PLA2, phospholipase A2; 5-HT, 5-hydroxytryptamine (serotonin); 5-HT-2, 5-HT receptor subtype 2; PAR2, protease-activated receptor 2; Mrgpr, Mas-related G-protein-coupled receptor; ET-1, endothelin-1; ETA, endothelin-1 receptor A; BAM8-22, bovine adrenal medullary peptide 8–22. Reused from [125] under an open access terms of the Creative Commons Attribution Non-Commercial License, BMB Rep, Korean Society for Biochemistry and Molecular Biology.

Historically, two main categories of itch have been defined, histaminergic and nonhistaminergic itch [126], which are closely related but act independently from each other as two separate pathways. Chronic pruritus is proposed to involve the nonhistaminergic pathway [127]. In itch transmission, two families of receptors are found to contribute: G protein-coupled receptors (GPCRs) and transient receptor potential (TRP) channels [127,128]. Numerous molecules are found that activate these receptors, for example, at-the-periphery [129] histamine; serotonin; nerve growth factor; interleukins IL-4, 13, and 31, among many others. The itch sensation can be induced by the direct or indirect activation of these receptors and channels and activators of these are released from various cells, including T-cells, mast cells, and keratinocytes [129]. The number of players emphasizes that there is no singular cause of itch, and as a consequence, mechanisms underlying various chronic itch conditions differ [129]. Figure 3 depicts detailed peripheral mechanisms underlying some of the chronic itch conditions [129].

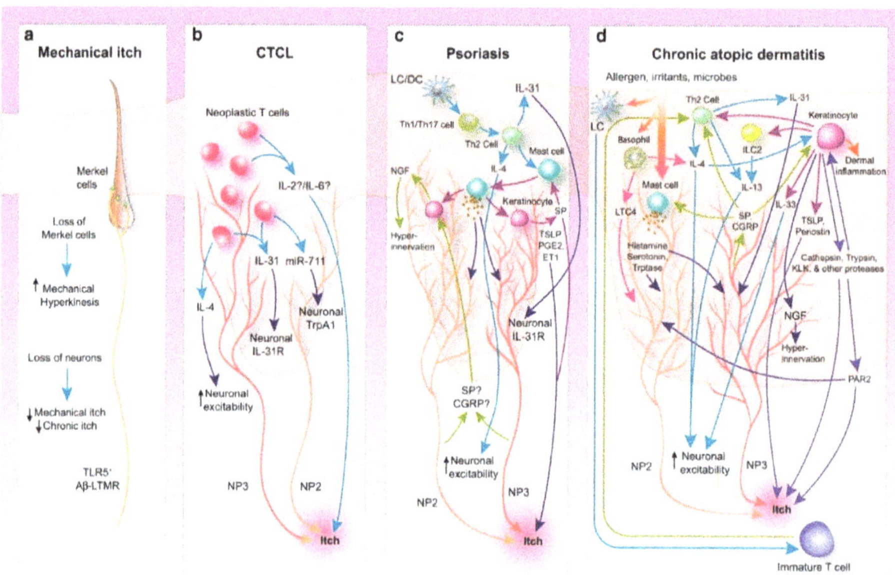

Figure 3. Mechanisms of chronic itch: (**a**) Age-related loss of touch-sensitive Merkel cells is associated with increased mechanical hyperkinesis. However, the loss of TLR5 Aβ-LTMRs that detect mechanical itch leads to the loss of mechanical itch sensitivity and the attenuation of chronic itch. (**b**) The current understanding of CTCL itch occurs through two main pathways. First, IL-31 released by neoplastic T cells can directly activate neuronal itch receptors, whereas IL-4, IL-2, and IL-6 enhance itch. (**c**) Psoriasis-associated itch mechanisms are not completely understood. Endogenous pruritogens, including IL-31, TSLP, and mast cell–derived PGE2 and ET1, have all been implicated, whereas NGF and type 2 cytokines may further potentiate itch by enhancing neuronal excitability. (**d**) AD-associated itch is primarily generated by type 2 cytokine-induced itch neuron excitability and reciprocal modulation by neurons. Endogenous pruritogens, released by immune cells and keratinocytes, include peptides, proteases, and mast cell-, basophil-, and ILC2-derived factors. AD, atopic dermatitis; CGRP, calcitonin gene-related peptide; CTCL, cutaneous T-cell lymphoma; DC, dendritic cell; ILC, innate lymphoid cell; KLK, kallikrein; LC, Langerhans cell; LTC4, leukotriene C4; NP, nonpeptidergic; SP, substance P; Th, T helper; TLR, Toll-like receptor. Illustration assistance provided by Ruvido Medical Illustration. Reused from [129] with permission (license number: 5379421293131) from Elsevier and Copyright Clearance Center.

Several reviews are available, and readers are encouraged to look deeper into itch pathogenesis and treatments [123,125,129–131].

Antipruritic effects of BoNT have been presented in clinical cases in the literature (see Section 2). Investigators have tried to identify how BoNT can reduce or stop itch. To understand the mechanisms underlying the antipruritic effects of BoNT, cell-based studies, laboratory animal investigations and human experimental models have been employed to provide the mechanism-based evidence that is presented below.

4. Mechanisms Underlying the Antipruritic Effect of BoNTs

As presented above, merging known mechanisms underlying BoTN effects at the cellular–molecular level, and mechanisms of itch generation and transmission can help understand the mechanism(s) underlying the antipruritic effect of BoNT. This knowledge, combined with evidence of the antipruritic effects of BoNTs at the clinic is beneficial and important because it can potentially strengthen, accelerate, and facilitate the current efforts towards further investments in pushing the field forward for the potential approval of BoNTs for itchy conditions. In addition, it can advance the scientific field in terms of gaining a better understanding or providing evidence of neuronal, glial, and immune-modulatory systems involved in itch and targeting them with BoNTs and compounds similar to BoNTs. Further information, such as dosing, interval, and safety information, can also be gained through such preclinical studies. Experimental models of itch [132] are helpful in this regard, as clinical conditions of itch accompany several other confounding factors [133] that cannot be eliminated while studying itch mechanisms. The concept of itch models is somewhat similar to the concept of modeling other medical conditions, e.g., pain. In experimental human models of pain [134] and central sensitization [135], healthy volunteers act temporarily as subjects for the provocation of pain, which is an ethical, controlled, and short-term condition. In this scenario, the application of a chemical algogen or other types of stimuli (thermal, ischemic, mechanical, or electric) produces pain and other measurable outcomes, such as pain sensitivity, measured by a visual analog scale (VAS) and vasomotor responses. In these experimental models, responsiveness to various analgesics [136], including BoNT, has been evaluated [110–112,137].

4.1. Human Surrogate Models of Itch—Antipruritic Effects of BoNT

Human surrogate models of itch have provided a platform for studying new antipruritic compounds, as well as the investigation of the mechanisms underlying antipruritic effects of already approved compounds for other conditions, for instance, BoNTs. Experimental itch [132,138,139] in humans is induced over a short period and is usually assessed psychophysically. Itch has been provoked by the application of electrical [140], mechanical [141], and chemical stimuli, the latter being through the application of, for example, histamine [142,143], cowhage [144,145], capsaicin [146], BAM8-22 [147], β-alanine [148], and serotonin [149].

In 2009, we tested if the administration of BoNTA subcutaneously can reduce itch in a human model of itch [83]. The study recruited 14 healthy male subjects, and the itch was artificially provoked by histamine, which was delivered to the volar forearm skin using a prick test [83]. An amount of 5 U of BoNTA (BOTOX®, Allergan) or a similar volume of saline (control) was administrated prior to the histamine provocation test [83]. Baseline assessments were conducted, and itch intensity and neurogenic inflammation produced by the histamine prick test were evaluated one day, three days, and a week after the administration of BoNTA or saline. The results of this study [83] showed that BoNTA was capable of diminishing itch intensity and reducing the area of itch perception compared with saline at all time points post-treatment. The itch resolution time was also shorter in BoNTA-treated areas, and the maximum effect was seen on day 7. The flare area (observable skin reaction in the form of redness) was smaller on the BoNTA-treated arm at all post-treatment time points. Histamine-induced elevated blood flow and skin temperature subsided following the application of BoNTA, with the largest effect seen on days 3 and 7. This study showed that BoNTA can inhibit histaminergic itch in humans [83].

Later, in 2017, a group of researchers [150] applied a non-histaminergic model of itch in healthy humans and investigated if BoNTA can also exert antipruritic effects. In this study, 35 (16 males and 19 females) healthy volunteers were enrolled, and experimental itch was provoked by the application of cowhage [145] (*Mucuna pruriens*). Arthur and Shelley, in 1955 [151], were the first to discover that cowhage induces itching and scratching. This is the effect of mucunain, which is the active component of cowhage and is chemically classified as a cysteine protease. This substance binds to proteinase-activated receptors 2 and 4, so-called PAR-2 and PAR-4, respectively [134]. This model is proposed to mimic chronic itch conditions in humans, which are often non-histaminergic and do not respond to antihistamines [152]. In this study, 10 U of BoNTA (BOTOX®, Allergan Inc., Irvin, CA, USA) administered intradermally could reduce the itch intensity at all time points compared with the saline that was used as the control treatment. Sensory tests in this study [150] included skin temperature sensitivity, pain, and itch following cowhage and post-treatment, and measurements were performed at baseline, week 1, month 1, and month 3 post-treatment. This study [150] provided evidence for the long-term effect of BoNTA against itch, lasting 3 months following a single application.

Taken together, these two studies, using both histaminergic and non-histaminergic itch models in humans, showed the antipruritic effects of BoNTA, where it reduced itch and related symptoms (neurogenic flare, skin temperature, skin blood flow, and paresthesia in the form of hyperknesis and alloknesis) in human skin as early as 1 week, and up to 3 months in the cowhage model [150].

This field is still open for further investigation by applying these or other surrogate models of itch, combined with other subjective (psychophysics and quantitative sensory testing) and objective measures, for example, the bioanalysis of biomarkers through microdialysis, skin micro biopsy, or imaging studies. We employed the human dermal microdialysis technique [153] and presented that BoNTA inhibits the release of neurotransmitters, e.g., glutamate, in human skin [114]. It is valuable to understand how BoNTs interact with various cell types, including immune cells and nerve endings. The modeling of some dermatological conditions that accompany itch is not ethical in healthy humans; therefore, it is proposed that at least selected measurement techniques be used in patients pre- and post-BoNT treatments for the identification of the mechanism of action. These conditions include, but are not limited to, post-herpetic neuralgia, notalgia paresthetica, Hailey–Hailey disease, and psoriasis.

4.2. Rodent Surrogate Models of Itch—Antipruritic Effects of BoNT

Animal studies [154] are of great importance to understand the underlying mechanisms of pathogenesis and the action of drugs in humans. These models, however, can only mimic limited aspects of pathogenesis and often accompany translational challenges from animals to humans due to multiple factors, including species differences [155]. Considering these limitations, several itch models [132] have been developed and tested in different animals, mainly rodents. These models resemble itch of both an acute and chronic nature and have been employed to understand how itch is transmitted and how it can be targeted at various points of transmission [156]. Followed by the discovery of Arthur and Shelley [151] about the pruritogenic properties of cowhage, in 1963, Joglekar and colleagues [157] applied cowhage ointment (5%) topically and reported the provocation of itching and scratching in dogs. Since then, several models have been introduced, for example, bombesin-induced itch in rats [158] and intrathecal morphine injection in monkeys [159]. An easy subcutaneous injection of pruritogen in experimental mice was introduced in 1995 by Kuraishi and his colleagues [160], in which they selected the necks of mice where scratching bouts using the hind legs could be quantified as the outcome. This model and similar ones helped in the identification of neurotransmitters and neuromodulators of itch, as well as itch signaling pathways. A long list of substances has been used to induce itching and scratching behavior, for example, serotonin [161], chloroquine [162], SLIGRL (Ser–Leu–Ile–Gly–Arg–Leu, a proteinase-activated receptor-2 agonist) [163], interleukin-31 (IL-31) [47], and phoenixin [164].

In addition, it is possible to develop animal models of diseases that accompany itch and other symptoms to study mechanisms and treatment options. Among many disease models, NC/Nga mice are, for example, used for studying AD-like skin lesions and atopic dermatitis [165]. The modeling of psoriasis [166] has also been achieved with the aid of a Toll-like receptor 7 agonist, imiquimod, which is applied to the backs of mice. Another model is the dry skin model [167], which is induced by the application of a 1:1 mixture of acetone and ether to the nape of mice's necks. In addition to in vivo models, and the behavioral outcomes of itch-evoked scratching [122], other measurements can be planned, such as in vitro investigations, to help in the identification of cellular and molecular aspects and changes following the application of various compounds to provoke or inhibit itch.

Although the antinociceptive effects of BoNTs have been extensively tested in animal models of pain [168–170], the antipruritic effects of BoNTs have only been investigated in a couple of itch provocation and disease model studies. These studies have proposed the potential antipruritic mechanisms of BoNTs to be the blockade of neurotransmitter release, the blockade of mast cell degranulation, the downregulation of TRPs, and the inhibition of neuroimmune key players, such as IL-17. In the following, these findings are presented in detail.

In the study by Ramachandran and colleagues [171], compound 48/80 or chloroquine was injected intradermally into mice to induce itch. The compound 48/80 is known to induce mast cell-dependent scratching, while chloroquine is known to be a mast cell-independent compound to provoke itching and scratching. The researchers applied both $BoNTA_1$ and $BoNTB_1$ (1.5 U, intradermal injection) to test their effects on days 2, 7, 14, and 21. In this study [171], saline was used as a control. They also investigated human and murine mast cells in culture and investigated the direct effect of BoNTs in vitro [171]. This study identified the interaction of BoNTs with mast cells, and findings demonstrated that both compound 48/80 and chloroquine provoked itching and scratching behavior, and BoNTs could reduce these outcomes. An explanation of the mechanisms underlying these effects, however, did not appear to be straightforward [171]. Pre-treatment with $BoNTA_1$ and $BoNTB_1$ inhibited compound 48/80-provoked mast cell degranulation in culture. This finding indicated that these toxins may directly affect mast cells and prevent the degranulation of these cells. Since BoNTs target SNAREs, the authors [171] investigated the target component of this complex by simulating the inhibition of SNAP-25 or VAMPs by BoNTs A_1 and B_1, respectively. The mRNA expression of SNAP-25 in mast cell cultures was very low in both cases of mice and human cell cultures. To confirm this finding, the authors [171] performed immune staining for SNAP-25 and VAMPs, and the cleavage of these isoforms became evident. The Western blot result, however, did not show SNAP-25 in mast cells, and there was no indication of $BoNTB_1$ on mast-cell-expressed VAMPs. This finding [171] raised speculation that BoNTs may inhibit vesicular release from mast cells by a mechanism other than the SNARE-related mechanism. The authors [171] speculated that TRP channels might be involved [172,173], based on the literature exhibiting that $BoNTA_1$ inhibits TRPV1 receptor function. According to the authors [171], BoNTs may also inhibit the depolarization-evoked calcium currents in mast cells. This can explain the inhibitory effect on the compound 48/80–evoked response, because compound 48/80 induces mast cell degranulation via calcium-dependent exocytosis [174,175]. Further investigation is required to determine if BoNTs inhibit the degranulation of mast cells in itch via direct or indirect mechanisms.

Another study [176] investigated the antipruritic effects of BoNTA in itch models of acute and chronic itch in mice. The authors applied compound 48/80, chloroquine, and a mixture of acetone–diethyl and ether–water to provoke itch [176]. Intradermal BoNTA could present a long-term inhibition of itch in both compound 48/80 and chloroquine models of acute itch. The effect was seen from day 1 to day 14. Itch induced by acetone–

diethyl ether–water was also reduced by BoNTA pretreatment up to day 14. To study the potential mechanisms, the authors [176] looked at the levels of receptor expression in mice dorsal root ganglia, where they found that BoNTA could reduce the expression of TRPV1 and TRPA1 in both acute itch models provoked by compound 48/80 and chloroquine. However, in the dry skin model of chronic itch, BoNTA only reduced the DRG upregulation of TRPA1 [176]. Collectively, based on these findings [176], the authors proposed that, at least in part, the downregulation of TRPA1 and TRPV1 in DRG can contribute to the antipruritic effects of BoNTA.

A recent study [177] employed one of the disease models, psoriasis, and investigated if BoNTB could interfere with the immune axis of IL-23/Th17, which is proposed to act as one of the primary modulators in psoriasis [178]. The study investigators [177] applied imiquimod to artificially model psoriasis-like dermatitis [179] in mice. Following pretreatment with BoNTB, they found the significant suppression of cytokine production in skin lesions. In addition, the cell counts for CD4+ T cells, CD11c+ dendritic cells, and IL-17 were reduced dramatically [177]. The authors also found that BoNTB reduced the expression of substances P and CGRP on PGP9.5+ nerve fibers, which are reported to be increased in the lesions of psoriatic skin [180,181]. Results from this study [177] emphasize the importance of the neuroimmune system in psoriasis and that BoNTB via inhibitory action on this system could reverse the condition of lesioned skin. In this study, however, itch or scratch behavior was not measured, and the effect of BoNTB was mainly based on the scoring of erythema, scale, the thickness of skin, and overall scale [177]. Therefore, it is speculated that BoNTB in this study might have also reduced itch and scratch behavior in mice.

The cleaving effect of BoNTs on SNARE has also been proposed as an ideal long-lasting solution to break a vicious cycle of immune–nerve communication in pathologic conditions such as AD [182]. BoNTA also blocks substance P and CGRP release [183], similar to what is observed for BoNTB [177]. BoNTA can also prevent the upregulation of TRPA1 and TRPV1, which occurs following the activation of Th_2 by cytokines [184] and breaks the vicious cycle of neuro-immune contribution in AD [182]. It is reasonable to consider that when SNAP-25 is cleaved by BoNTA, the release of natriuretic peptide (BNP) from pruriceptive neurons is inhibited [185], and this results in blocking itch transmission. BNP is known to potentiate TRPV3 expression on keratinocytes, and it is proposed that BoNTs, most likely B and D serotypes, can prevent this by cleaving VAMPs [186] (vesicle-associated membrane protein).

Although a large amount of research work remains to be done regarding the antipruritic effects of BoNTs, we can benefit from the knowledge accumulated about antinociceptive effects of BoNTs at the cellular and molecular levels. Considering this, one study looked into the neuron–glia modulation of LPS-induced pain and the effect of BoNT [187]. Results from this study showed that BoNTA reduced the LPS-induced phosphorylation of p38, ERK1/2, and NF-B and blocked the release of pro-inflammatory IL-1, IL-18, and IL-6. Interestingly, it also blocked the release of anti-inflammatory IL-10 in microglia [187]. The authors [187] explained the action of BoNTA in glial cells to be related to the activation of TLR2 and TLR4 receptors. An interesting finding of this study [187] was that the activation of TLR2 in astroglia requires a microglial TLR4 receptor [187]. This can be similarly investigated in itch provocation models to further identify glial roles in itch [188] and the modulatory role of BoNTs on these cells in the peripheral and central nervous system to inhibit itch. For example, we investigated the expression of SNAPs on satellite glial cells (SGCs) in the trigeminal ganglia and found that BoNTA can inhibit the vesicular release of substances from these cells in vitro [113]. Such findings can add value to the identification of peripheral components of itch transmission in sensory ganglia (trigeminal and dorsal root ganglia). Moreover, differences and commonalities can be investigated to determine if the region plays a role in anti-pruritic response within spinal-innervated or trigeminal-innervated areas. This type of investigation can, for example, also address potential mechanisms underlying conditions such as psoriasis, where inverse psoriasis [189] might show a better response to BoNTs compared with psoriasis.

The central effects of BoNTs, following their peripheral administration, have been the subject of debate in the literature, and a consensus has not yet been reached as to their direct or indirect effects [190–192]. A 2018 review by Caleo and Restani [191] summarizes studies that have so far provided evidence of the retrograde transport of BoNTA after its peripheral injection and that this process might contribute to the clinical effects of BoNTA through a direct action on the central circuits [191]. It is yet to be determined if this or other mechanisms can be involved in the effects of centrally mediated itch in clinical conditions such as chronic neuropathic itch [57,193]. Independent of the direct or indirect mechanism of action, it has been proposed that central alterations made by BoNTs can offer therapeutic applications to pathological conditions that are maintained by maladaptive plastic changes, such as neuropathic pain [194,195] or neuropathic itch [57,193], which are both difficult to treat. The accumulation of mechanism-based evidence with high-quality data can potentially lead to the approval of BoNTs for the treatment of itch [43,60,67] most likely in chronic itch states and in troublesome conditions that are resistant to other treatment options. Figure 4 depicts an overview of the potential action points of BoNTs to exert antipruritic effects. Questions marks indicate uncertainty or unknown areas in the available literature, where further investigation is required for clarification.

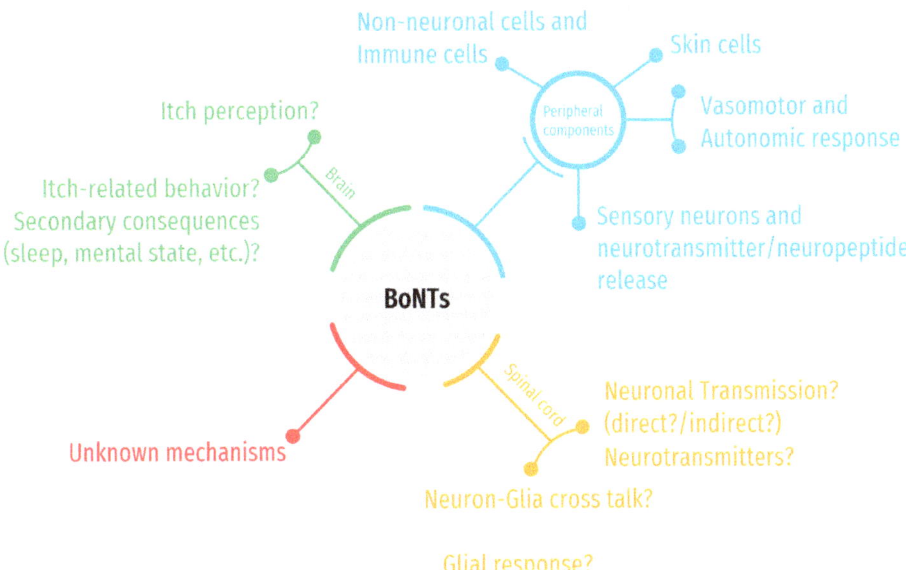

Figure 4. Proposed antipruritic mechanisms of BoNTs. Peripheral components of itch have been studied extensively, and the literature provides evidence of the beneficial effects of BoNTs on these components, such as blocking the release of neurotransmitters and neuropeptides (pruritogen, itch mediator) and blocking the vasomotor and autonomic components of itch. Spinal cord and brain components of itch are being actively investigated, and animal experiments and imaging studies are revealing components of itch mechanisms, transmission, and perception in addition to itch behaviors and emotional response. Limited evidence in the literature points to potential direct/indirect itch inhibitory effects of BoNTs in the central nervous system and these elements (transmission, perception). Some unknown complex mechanisms might also exist that can indirectly lead to antipruritic effects of BoNTs, which are yet to be determined. For details of this simplified schematic, please refer to the text.

5. Conclusions and Future Perspectives

Most of the clinical studies presented in the literature show potential for the beneficial effects of BoNTs against itch and dermatological pathologies with itch as a cardinal symptom. High-level evidence, resulting from well-designed conducted studies, will most likely lead to the enhancement of the grade of recommendation to the point that BoNTs can be approved for their antipruritic efficacy [60,67]. The long-lasting effects of BoNTs after administration appear to be in favor of patient compliance, but the cost might be inhibitory for regular use, in addition to responsiveness to other available compounds with comparative effects to BoNTs. Theoretically, it makes sense to evaluate each pruritus condition prior to the selection of or switching to BoNTs. Safety considerations are also crucial. The overall consensus is that BoNTs are generally safe if used following the recommended dose and interval, but the risks and benefits must be weighted and justified in each case [196]. It is predicted that the present and near-future focus will be directed towards conditions with itch where stronger evidence is available for the choice of BoNTs against itch. This is a positive reason to conduct studies with a proper design. Points to be considered can be moving toward study designs that can provide higher-grade evidence, i.e., randomized controlled trials (RCTs) instead of observational, open-label, or case-based studies. Blinding, and considering a placebo or control arm would minimize bias and enhance the validity of results. This method can also help in determining the existence or proportion of a placebo effect. Power calculation, the effect size for treatment, and sample size determination will result in sufficient study size and limit the potential for false positive or false negative results as a consequence of poor study power or small sample size. A well-defined inclusion and exclusion criteria will also help in recruiting study participants who are the target group for testing the study hypothesis as if BoNTs can exert beneficial antipruritic effects. In this line, multi-center studies that enable the recruitment of a large number of participants can be helpful. Confounding factors, such as gender, age, and ethnic background of participants must be considered and reported to permit the identification of responders versus non-responders. The administration dose and interval of dosing (single or repeated injections), and optimal way of delivery (e.g., intradermal or subcutaneous), must also be justified and included in the study design. Current studies lack a unified description of the location and dosage for the application of BoNTs. In terms of outcome measures, the antipruritic effect seems to be a proper and important primary outcome. However, discrepancies still exist in the application of validated or approved measurement tools. Hence, it is valuable to describe in sufficient detail how the outcomes are measured. Secondary outcomes can also be defined and described. For example, sleep quality can be determined as a secondary outcome for the antipruritic effect as itch disturbs sleep and reduce the quality of sleep. Well-defined study outcomes would allow for comparisons between studies or meta-research (e.g., meta-analysis) in the future. BoNTs injections are considered safe and well-established in the cosmetic field for example for facial wrinkles, however, it is important to consider safety measures in the next trials investigating the antipruritic effects of BoNTs. Collecting information such as patient satisfaction, or the presence of tingling or bleeding during the injections can help in the determination of benefits versus challenges or any potential risks.

The clinical practice reports show the main use of BoNTs for localized chronic pruritus in a diverse range of dermatological conditions such as burns, scars, lichen simplex, and inverse psoriasis. However, the therapeutic response to BoNTs for itch of non-dermatological origin [197] and itch as a systemic disease [198] (e.g., renal, liver, endocrine–metabolic diseases, and hematologic–lymphoproliferative diseases) can also be investigated. BoNTs could be a potential option when multiple symptoms are presented at the same time in patients, for example, in cosmetic and non-cosmetic surgeries, where multiple complex mechanisms including sensory alterations are involved (e.g., wound healing and ulcer treatment [199]. It has even been claimed that *"if there was ever a drug that was likely to affect every cell of the body, this is BoNT"* [200].

Preclinical surrogate models of itch in humans and laboratory animals confirm that BoNTs are capable of preventing itch. Animal studies, in particular, could enrich our understanding of the mechanisms of BoNTs antipruritic effects, and provide evidence on acute and chronic itch prevention using BoNTs. These studies have also provided a platform for the investigation of the peripheral and central effects of BoNTs and their potential antipruritic mechanisms. The current understanding is that BoNTs potentially inhibit itch at several levels, through the blockade of neurotransmitter and neuropeptide release; the blockade of mast cell degranulation; the downregulation of TRPs; and the inhibition of key players in the neuro-immune system, such as cytokines and interleukins involved in itch, as well as the Th_2 cytokine-induced release of itch-promoting substances, such as BNP. Independent of uncertainty around the direct or indirect central role of BoNTs, accumulating evidence highlights the potential role of peripheral and central glia in itch and the potential modulatory role of BoNTs.

Surrogate and preclinical models of itch can help in further studying the roles of neuronal and non-neuronal (e.g., glia) components and their interactions in the development and maintenance of chronic itch, and how BoNTs and other potential targets can prevent or stop it. The field of itch pathogenesis and treatment [130] is active, and new formulations of botulinum toxins with desirable safety profiles and enhanced potency are emerging [201]. The crossover of these two fields offers an exciting horizon for multifold potentials in the future of chronic itch treatment and widening the medical indications of BoNTs.

Funding: No funding was available for this work. No writing assistance was received in the production of this manuscript.

Informed Consent Statement: Not applicable.

Data Availability Statement: A part of this work was presented during the invited plenary lecture, at the 6th International Conference of Basic Science and Clinical aspects of Botulinum and Other Neurotoxins, Toxins 2022, organized by the International Neurotoxin Association in New Orleans, Louisiana, US (27–30 July 2022).

Conflicts of Interest: The author has no relevant affiliations or financial involvement with any organization or entity with a financial interest in, or financial conflict with, the subject matter or materials discussed in the manuscript. This includes employment, consultancies, honoraria, stock ownership or options, expert testimony, grants or patents received or pending, or royalties.

References

1. Erbguth, F.J. Historical notes on botulism, Clostridium botulinum, botulinum toxin, and the idea of the therapeutic use of the toxin. *Mov. Disord.* **2004**, *19* (Suppl. S8), S2–S6. [CrossRef] [PubMed]
2. Tyler, H.R. Botulism. *Arch. Neurol.* **1963**, *9*, 652–660. [CrossRef]
3. Hanchanale, V.S.; Rao, A.R.; Martin, F.L.; Matanhelia, S.S. The Unusual History and the Urological Applications of Botulinum Neurotoxin. *Urol. Int.* **2010**, *85*, 125–130. [CrossRef]
4. Rao, A.K.; Sobel, J.; Chatham-Stephens, K.; Luquez, C. Clinical Guidelines for Diagnosis and Treatment of Botulism, 2021. *MMWR Recomm. Rep.* **2021**, *70*, 1–30. [CrossRef] [PubMed]
5. Erbguth, F.J. From poison to remedy: The chequered history of botulinum toxin. *J. Neural Transm.* **2008**, *115*, 559–565. [CrossRef]
6. Smith, T.; Williamson, C.H.D.; Hill, K.; Sahl, J.; Keim, P. Botulinum Neurotoxin-Producing Bacteria. Isn't It Time that We Called a Species a Species? *mBio* **2018**, *9*, e01469-18. [CrossRef]
7. Jabbari, B. History of Botulinum Toxin Treatment in Movement Disorders. *Tremor Other Hyperkinetic Mov.* **2016**, *6*, 394. [CrossRef]
8. Dhaked, R.K.; Singh, M.K.; Singh, P.; Gupta, P. Botulinum toxin: Bioweapon & magic drug. *Indian J. Med. Res.* **2010**, *132*, 489–503.
9. Lebeda, F.J.; Adler, M.; Dembek, Z.F. Yesterday and Today: The Impact of Research Conducted at Camp Detrick on Botulinum Toxin. *Mil. Med.* **2018**, *183*, 85–95. [CrossRef]
10. Malizio, C.J.; Goodnough, M.C.; Johnson, E.A. Purification of *Clostridium botulinum* Type A Neurotoxin. In *Bacterial Toxins: Methods and Protocols*; Holst, O., Ed.; Humana Press: Totowa, NJ, USA, 2000; pp. 27–39. [CrossRef]
11. Burgen, A.S.; Dickens, F.; Zatman, L.J. The action of botulinum toxin on the neuro-muscular junction. *J. Physiol.* **1949**, *109*, 10–24. [CrossRef]
12. Scott, A.B. Botulinum toxin injection of eye muscles to correct strabismus. *Trans. Am. Ophthalmol. Soc.* **1981**, *79*, 734–770. [PubMed]
13. Chen, S. Clinical uses of botulinum neurotoxins: Current indications, limitations and future developments. *Toxins* **2012**, *4*, 913–939. [CrossRef] [PubMed]

14. Cardoso, F.; Jankovic, J. Clinical use of botulinum neurotoxins. *Curr. Top Microbiol. Immunol.* **1995**, *195*, 123–141. [CrossRef] [PubMed]
15. Cartee, T.V.; Monheit, G.D. An overview of botulinum toxins: Past, present, and future. *Clin. Plast. Surg.* **2011**, *38*, 409–426. [CrossRef] [PubMed]
16. Cheng, C.M.; Chen, J.S.; Patel, R.P. Unlabeled uses of botulinum toxins: A review, part 1. *Am. J. Health Syst. Pharm.* **2006**, *63*, 145–152. [CrossRef]
17. Cheng, C.M.; Chen, J.S.; Patel, R.P. Unlabeled uses of botulinum toxins: A review, part 2. *Am. J. Health Syst. Pharm.* **2006**, *63*, 225–232. [CrossRef]
18. Dressler, D.; Adib Saberi, F. Botulinum Toxin: Mechanisms of Action. *Eur. Neurol.* **2005**, *53*, 3–9. [CrossRef]
19. Pirazzini, M.; Rossetto, O.; Eleopra, R.; Montecucco, C. Botulinum Neurotoxins: Biology, Pharmacology, and Toxicology. *Pharmacol. Rev.* **2017**, *69*, 200–235. [CrossRef]
20. Pirazzini, M.; Montecucco, C.; Rossetto, O. Toxicology and pharmacology of botulinum and tetanus neurotoxins: An update. *Arch. Toxicol.* **2022**, *96*, 1521–1539. [CrossRef]
21. Peck, M.W.; Smith, T.J.; Anniballi, F.; Austin, J.W.; Bano, L.; Bradshaw, M.; Cuervo, P.; Cheng, L.W.; Derman, Y.; Dorner, B.G.; et al. Historical Perspectives and Guidelines for Botulinum Neurotoxin Subtype Nomenclature. *Toxins* **2017**, *9*, 38. [CrossRef]
22. Fung, H.T.; Chan, K.M.; Lam, S.K.T. A review on iatrogenic botulism. *Hong Kong J. Emerg. Med.* **2020**, *27*, 356–367. [CrossRef]
23. Horowitz, B.Z. Botulinum toxin. *Crit. Care Clin.* **2005**, *21*, 825–839. [CrossRef] [PubMed]
24. Jankovic, J. Botulinum toxin in clinical practice. *J. Neurol. Neurosurg. Psychiatry* **2004**, *75*, 951–957. [CrossRef] [PubMed]
25. Bach, K.; Simman, R. The Multispecialty Toxin: A Literature Review of Botulinum Toxin. *Plast. Reconstr. Surg. Glob. Open* **2022**, *10*, e4228. [CrossRef]
26. Heikkilä, H. Botulinum Toxin Treatment in Veterinary Medicine: Clinical Implications. In *Botulinum Toxin Treatment in Surgery, Dentistry, and Veterinary Medicine*; Jabbari, B., Ed.; Springer International Publishing: Cham, Switzerland, 2020; pp. 337–357. [CrossRef]
27. Sławek, J.; Bogucki, A.; Bonikowski, M.; Car, H.; Dec-Ćwiek, M.; Drużdż, A.; Koziorowski, D.; Sarzyńska-Długosz, I.; Rudzińska, M. Botulinum toxin type-A preparations are not the same medications—Clinical studies (Part 2). *Neurol. Neurochir. Pol.* **2021**, *55*, 141–157. [CrossRef]
28. Car, H.; Bogucki, A.; Bonikowski, M.; Dec-Ćwiek, M.; Drużdż, A.; Koziorowski, D.; Rudzińska-Bar, M.; Sarzyńska-Długosz, I.; Sławek, J. Botulinum toxin type-A preparations are not the same medications—Basic science (Part 1). *Neurol. Neurochir. Pol.* **2021**, *55*, 133–140. [CrossRef]
29. Brin, M.F.; James, C.; Maltman, J. Botulinum toxin type A products are not interchangeable: A review of the evidence. *Biologics* **2014**, *8*, 227–241. [CrossRef]
30. Dressler, D.; Benecke, R. Pharmacology of therapeutic botulinum toxin preparations. *Disabil. Rehabil.* **2007**, *29*, 1761–1768. [CrossRef]
31. Choudhury, S.; Baker, M.R.; Chatterjee, S.; Kumar, H. Botulinum Toxin: An Update on Pharmacology and Newer Products in Development. *Toxins* **2021**, *13*, 58. [CrossRef]
32. Spiegel, L.L.; Ostrem, J.L.; Bledsoe, I.O. FDA Approvals and Consensus Guidelines for Botulinum Toxins in the Treatment of Dystonia. *Toxins* **2020**, *12*, 332. [CrossRef]
33. Alster, T.S.; Harrison, I.S. Alternative Clinical Indications of Botulinum Toxin. *Am. J. Clin. Dermatol.* **2020**, *21*, 855–880. [CrossRef] [PubMed]
34. Lewandowski, M.; Świerczewska, Z.; Barańska-Rybak, W. Off-Label Use of Botulinum Toxin in Dermatology—Current State of the Art. *Molecules* **2022**, *27*, 3143. [CrossRef] [PubMed]
35. Phan, K.; Younessi, S.; Dubin, D.; Lin, M.J.; Khorasani, H. Emerging off-label esthetic uses of botulinum toxin in dermatology. *Dermatol. Ther.* **2022**, *35*, e15205. [CrossRef] [PubMed]
36. Argyriou, A.A.; Mitsikostas, D.D.; Mantovani, E.; Vikelis, M.; Tamburin, S. Beyond chronic migraine: A systematic review and expert opinion on the off-label use of botulinum neurotoxin type-A in other primary headache disorders. *Expert Rev. Neurother.* **2021**, *21*, 923–944. [CrossRef] [PubMed]
37. Schulze, J.; Neumann, I.; Magid, M.; Finzi, E.; Sinke, C.; Wollmer, M.A.; Krüger, T.H.C. Botulinum toxin for the management of depression: An updated review of the evidence and meta-analysis. *J. Psychiatr. Res.* **2021**, *135*, 332–340. [CrossRef] [PubMed]
38. Sloan, G.; Alam, U.; Selvarajah, D.; Tesfaye, S. The Treatment of Painful Diabetic Neuropathy. *Curr. Diabetes Rev.* **2022**, *18*, e070721194556. [CrossRef]
39. Kayani, A.M.A.; Silva, M.S.; Jayasinghe, M.; Singhal, M.; Karnakoti, S.; Jain, S.; Jena, R. Therapeutic Efficacy of Botulinum Toxin in Trigeminal Neuralgia. *Cureus* **2022**, *14*, e26856. [CrossRef]
40. Gharib, K.; Mostafa, A.; Elsayed, A. Evaluation of Botulinum Toxin Type A Injection in the Treatment of Localized Chronic Pruritus. *J. Clin. Aesthet Dermatol.* **2020**, *13*, 12–17.
41. Zanchi, M.; Favot, F.; Bizzarini, M.; Piai, M.; Donini, M.; Sedona, P. Botulinum toxin type-A for the treatment of inverse psoriasis. *J. Eur. Acad. Dermatol. Venereol.* **2008**, *22*, 431–436. [CrossRef]
42. Weinfeld, P.K. Successful treatment of notalgia paresthetica with botulinum toxin type A. *Arch. Dermatol.* **2007**, *143*, 980–982. [CrossRef]
43. Gazerani, P. Antipruritic Effects of Botulinum Neurotoxins. *Toxins* **2018**, *10*, 143. [CrossRef] [PubMed]

44. França, K.; Kumar, A.; Fioranelli, M.; Lotti, T.; Tirant, M.; Roccia, M.G. The history of Botulinum toxin: From poison to beauty. *Wien. Med. Wochenschr.* **2017**, *167*, 46–48. [CrossRef] [PubMed]
45. Naik, P.P. Utilities of Botulinum Toxins in Dermatology and Cosmetology. *Clin. Cosmet. Investig. Dermatol.* **2021**, *14*, 1319–1330. [CrossRef] [PubMed]
46. Ikoma, A.; Steinhoff, M.; Ständer, S.; Yosipovitch, G.; Schmelz, M. The neurobiology of itch. *Nat. Rev. Neurosci.* **2006**, *7*, 535–547. [CrossRef]
47. Cevikbas, F.; Wang, X.; Akiyama, T.; Kempkes, C.; Savinko, T.; Antal, A.; Kukova, G.; Buhl, T.; Ikoma, A.; Buddenkotte, J.; et al. A sensory neuron–expressed IL-31 receptor mediates T helper cell–dependent itch: Involvement of TRPV1 and TRPA1. *J. Allergy Clin. Immunol.* **2014**, *133*, 448–460.e7. [CrossRef] [PubMed]
48. Garibyan, L.; Rheingold, C.G.; Lerner, E.A. Understanding the pathophysiology of itch. *Dermatol. Ther.* **2013**, *26*, 84–91. [CrossRef]
49. Jin, S.-Y.; Wang, F. Sensitization Mechanisms of Chronic Itch. *Int. J. Dermatol. Venereol.* **2019**, *2*, 211–215. [CrossRef]
50. Tominaga, M.; Takamori, K. Peripheral itch sensitization in atopic dermatitis. *Allergol. Int.* **2022**, *71*, 265–277. [CrossRef] [PubMed]
51. Legat, F.J. Itch in Atopic Dermatitis—What Is New? *Front. Med.* **2021**, *8*, 644760. [CrossRef]
52. Dong, X.; Dong, X. Peripheral and Central Mechanisms of Itch. *Neuron* **2018**, *98*, 482–494. [CrossRef]
53. Szöllősi, A.G.; Oláh, A.; Lisztes, E.; Griger, Z.; Tóth, B.I. Pruritus: A Sensory Symptom Generated in Cutaneous Immuno-Neuronal Crosstalk. *Front. Pharmacol.* **2022**, *13*, 745658. [CrossRef]
54. Tarikci, N.; Kocatürk, E.; Güngör, Ş.; Topal, I.O.; Can, P.; Singer, R. Pruritus in Systemic Diseases: A Review of Etiological Factors and New Treatment Modalities. *Sci. World J.* **2015**, *2015*, 803752. [CrossRef]
55. Uppal, N.N.; Corona, A.; Fishbane, S. Pruritus in chronic kidney disease. *Curr. Opin. Nephrol. Hypertens.* **2022**, *31*, 435–441. [CrossRef] [PubMed]
56. Bhandari, A.; Mahajan, R. Skin Changes in Cirrhosis. *J. Clin. Exp. Hepatol.* **2022**, *12*, 1215–1224. [CrossRef]
57. Pereira, M.P.; Schmelz, M.; Ständer, S. Mechanisms and therapeutic targets for neuropathic itch. *Curr. Opin. Neurobiol.* **2022**, *75*, 102573. [CrossRef]
58. Heckmann, M.; Heyer, G.; Brunner, B.; Plewig, G. Botulinum toxin type A injection in the treatment of lichen simplex: An open pilot study. *J. Am. Acad. Dermatol.* **2002**, *46*, 617–619. [CrossRef]
59. Swartling, C.; Naver, H.; Lindberg, M.; Anveden, I. Treatment of dyshidrotic hand dermatitis with intradermal botulinum toxin. *J. Am. Acad. Dermatol.* **2002**, *47*, 667–671. [CrossRef]
60. Martina, E.; Diotallevi, F.; Radi, G.; Campanati, A.; Offidani, A. Therapeutic Use of Botulinum Neurotoxins in Dermatology: Systematic Review. *Toxins* **2021**, *13*, 120. [CrossRef]
61. Dreyfus, I.; Maza, A.; Rodriguez, L.; Merlos, M.; Texier, H.; Rousseau, V.; Sommet, A.; Mazereeuw-Hautier, J. Botulinum toxin injections as an effective treatment for patients with intertriginous Hailey-Hailey or Darier disease: An open-label 6-month pilot interventional study. *Orphanet J. Rare Dis.* **2021**, *16*, 93. [CrossRef] [PubMed]
62. González, C.; Franco, M.; Londoño, A.; Valenzuela, F. Breaking paradigms in the treatment of psoriasis: Use of botulinum toxin for the treatment of plaque psoriasis. *Dermatol. Ther.* **2020**, *33*, e14319. [CrossRef] [PubMed]
63. Campanati, A.; Martina, E.; Giuliodori, K.; Consales, V.; Bobyr, I.; Offidani, A. Botulinum Toxin Off-Label Use in Dermatology: A Review. *Ski. Appendage Disord.* **2017**, *3*, 39–56. [CrossRef]
64. Kim, Y.S.; Hong, E.S.; Kim, H.S. Botulinum Toxin in the Field of Dermatology: Novel Indications. *Toxins* **2017**, *9*, 403. [CrossRef] [PubMed]
65. Burns, P.B.; Rohrich, R.J.; Chung, K.C. The levels of evidence and their role in evidence-based medicine. *Plast. Reconstr. Surg.* **2011**, *128*, 305–310. [CrossRef]
66. Maari, C.; Marchessault, P.; Bissonnette, R. Treatment of notalgia paresthetica with botulinum toxin A: A double-blind randomized controlled trial. *J. Am. Acad. Dermatol.* **2014**, *70*, 1139–1141. [CrossRef]
67. Boozalis, E.; Sheu, M.; Selph, J.; Kwatra, S.G. Botulinum toxin type A for the treatment of localized recalcitrant chronic pruritus. *J. Am. Acad. Dermatol.* **2018**, *78*, 192–194. [CrossRef] [PubMed]
68. Akhtar, N.; Brooks, P. The use of botulinum toxin in the management of burns itching: Preliminary results. *Burns* **2012**, *38*, 1119–1123. [CrossRef] [PubMed]
69. Pérez-Pérez, L.; García-Gavín, J.; Allegue, F.; Caeiro, J.L.; Fabeiro, J.M.; Zulaica, A. Notalgia paresthetica: Treatment using intradermal botulinum toxin A. *Actas Dermosifiliogr.* **2014**, *105*, 74–77. [CrossRef] [PubMed]
70. Wallengren, J.; Bartosik, J. Botulinum toxin type A for neuropathic itch. *Br. J. Dermatol.* **2010**, *163*, 424–426. [CrossRef]
71. Kavanagh, G.M.; Tidman, M.J. Botulinum A toxin and brachioradial pruritus. *Br. J. Dermatol.* **2012**, *166*, 1147. [CrossRef]
72. Salardini, A.; Richardson, D.; Jabbari, B. Relief of intractable pruritus after administration of botulinum toxin A (botox): A case report. *Clin. Neuropharmacol.* **2008**, *31*, 303–306. [CrossRef]
73. Rai, S.; Stephanides, S.; Jamieson, L.A.; Smyth, J.V.; Madan, V. Pruritic axillary papules. *Clin. Exp. Dermatol.* **2012**, *37*, 928–930. [CrossRef] [PubMed]
74. Saber, M.; Brassard, D.; Benohanian, A. Inverse psoriasis and hyperhidrosis of the axillae responding to botulinum toxin type A. *Arch. Dermatol.* **2011**, *147*, 629–630. [CrossRef] [PubMed]
75. Ho, D.; Jagdeo, J. Successful botulinum toxin (onabotulinumtoxinA) treatment of Hailey-Hailey disease. *J. Drugs Dermatol.* **2015**, *14*, 68–70. [PubMed]
76. Toyama, S.; Tominaga, M.; Takamori, K. Treatment Options for Troublesome Itch. *Pharmaceuticals* **2022**, *15*, 1022. [CrossRef]

77. Klager, S.; Kumar, M.G. Treatment of pruritus with botulinum toxin in a pediatric patient with Fox-Fordyce disease. *Pediatr. Dermatol.* **2021**, *38*, 950–951. [CrossRef]
78. Sohrabi, C.; Goutos, I. The use of botulinum toxin in keloid scar management: A literature review. *Scars Burn Heal.* **2020**, *6*, 2059513120926628. [CrossRef]
79. Huang, S.H.; Wu, K.W.; Lo, J.J.; Wu, S.H. Synergic Effect of Botulinum Toxin Type A and Triamcinolone Alleviates Scar Pruritus by Modulating Epidermal Hyperinnervation: A Preliminary Report. *Aesthet. Surg. J.* **2021**, *41*, NP1721–NP1731. [CrossRef]
80. Khattab, F.M. Evaluation of Botulinum Toxin A as an Optional Treatment for Atopic Dermatitis. *J. Clin. Aesthet. Dermatol.* **2020**, *13*, 32–35.
81. Jin, J.; Sklar, G.E.; Min Sen Oh, V.; Chuen Li, S. Factors affecting therapeutic compliance: A review from the patient's perspective. *Ther. Clin. Risk Manag.* **2008**, *4*, 269–286. [CrossRef]
82. Fabbri, M.; Leodori, G.; Fernandes, R.M.; Bhidayasiri, R.; Marti, M.J.; Colosimo, C.; Ferreira, J.J. Neutralizing Antibody and Botulinum Toxin Therapy: A Systematic Review and Meta-analysis. *Neurotox. Res.* **2016**, *29*, 105–117. [CrossRef]
83. Gazerani, P.; Pedersen, N.S.; Drewes, A.M.; Arendt-Nielsen, L. Botulinum toxin type A reduces histamine-induced itch and vasomotor responses in human skin. *Br. J. Dermatol.* **2009**, *161*, 737–745. [CrossRef] [PubMed]
84. Meunier, F.A.; Schiavo, G.; Molgó, J. Botulinum neurotoxins: From paralysis to recovery of functional neuromuscular transmission. *J. Physiol. Paris* **2002**, *96*, 105–113. [CrossRef]
85. Johnson, E.A.; Montecucco, C. Chapter 11 Botulism. In *Handbook of Clinical Neurology*; Elsevier: Amsterdam, The Netherlands, 2008; Volume 91, pp. 333–368.
86. Scott, A.B. Development of botulinum toxin therapy. *Dermatol. Clin.* **2004**, *22*, 131–133. [CrossRef]
87. Multani, I.; Manji, J.; Hastings-Ison, T.; Khot, A.; Graham, K. Botulinum Toxin in the Management of Children with Cerebral Palsy. *Pediatric Drugs* **2019**, *21*, 261–281. [CrossRef]
88. Rossetto, O.; Pirazzini, M.; Montecucco, C. Botulinum neurotoxins: Genetic, structural and mechanistic insights. *Nat. Rev. Microbiol.* **2014**, *12*, 535–549. [CrossRef]
89. Gardner, A.P.; Barbieri, J.T. Light Chain Diversity among the Botulinum Neurotoxins. *Toxins* **2018**, *10*, 268. [CrossRef]
90. Dressler, D. Chapter 17 Botulinum toxin mechanisms of action. In *Supplements to Clinical Neurophysiology*; Hallett, M., Phillips, L.H., Schomer, D.L., Massey, J.M., Eds.; Elsevier: Amsterdam, The Netherlands, 2004; Volume 57, pp. 159–166.
91. Kumar, R.; Dhaliwal, H.P.; Kukreja, R.V.; Singh, B.R. The Botulinum Toxin as a Therapeutic Agent: Molecular Structure and Mechanism of Action in Motor and Sensory Systems. *Semin. Neurol.* **2016**, *36*, 10–19. [CrossRef]
92. Matak, I.; Bölcskei, K.; Bach-Rojecky, L.; Helyes, Z. Mechanisms of Botulinum Toxin Type A Action on Pain. *Toxins* **2019**, *11*, 459. [CrossRef]
93. Zhang, T.; Adatia, A.; Zarin, W.; Moitri, M.; Vijenthira, A.; Chu, R.; Thabane, L.; Kean, W. The efficacy of botulinum toxin type A in managing chronic musculoskeletal pain: A systematic review and meta analysis. *Inflammopharmacology* **2011**, *19*, 21–34. [CrossRef]
94. Muñoz Lora, V.R.M.; Del Bel Cury, A.A.; Jabbari, B.; Lacković, Z. Botulinum Toxin Type A in Dental Medicine. *J. Dent. Res.* **2019**, *98*, 1450–1457. [CrossRef]
95. Lippi, L.; de Sire, A.; Folli, A.; D'Abrosca, F.; Grana, E.; Baricich, A.; Carda, S.; Invernizzi, M. Multidimensional Effectiveness of Botulinum Toxin in Neuropathic Pain: A Systematic Review of Randomized Clinical Trials. *Toxins* **2022**, *14*, 308. [CrossRef] [PubMed]
96. Shaterian, N.; Shaterian, N.; Ghanaatpisheh, A.; Abbasi, F.; Daniali, S.; Jahromi, M.J.; Sanie, M.S.; Abdoli, A. Botox (OnabotulinumtoxinA) for Treatment of Migraine Symptoms: A Systematic Review. *Pain Res. Manag.* **2022**, *2022*, 3284446. [CrossRef] [PubMed]
97. Frampton, J.E.; Silberstein, S. OnabotulinumtoxinA: A Review in the Prevention of Chronic Migraine. *Drugs* **2018**, *78*, 589–600. [CrossRef] [PubMed]
98. Luvisetto, S.; Gazerani, P.; Cianchetti, C.; Pavone, F. Botulinum Toxin Type a as a Therapeutic Agent against Headache and Related Disorders. *Toxins* **2015**, *7*, 3818–3844. [CrossRef] [PubMed]
99. Ramachandran, R.; Yaksh, T.L. Therapeutic use of botulinum toxin in migraine: Mechanisms of action. *Br. J. Pharmacol.* **2014**, *171*, 4177–4192. [CrossRef] [PubMed]
100. Do, T.P.; Hvedstrup, J.; Schytz, H.W. Botulinum toxin: A review of the mode of action in migraine. *Acta Neurol. Scand.* **2018**, *137*, 442–451. [CrossRef]
101. Martinelli, D.; Arceri, S.; Tronconi, L.; Tassorelli, C. Chronic migraine and Botulinum Toxin Type A: Where do paths cross? *Toxicon* **2020**, *178*, 69–76. [CrossRef]
102. Cairns, B.E.; Gazerani, P. Botulinum neurotoxin A for chronic migraine headaches: Does it work and how? *Pain Manag.* **2014**, *4*, 377–380. [CrossRef]
103. Kumar, R. Therapeutic use of botulinum toxin in pain treatment. *Neuronal Signal.* **2018**, *2*, Ns20180058. [CrossRef]
104. Burstein, R.; Blumenfeld, A.M.; Silberstein, S.D.; Manack Adams, A.; Brin, M.F. Mechanism of Action of OnabotulinumtoxinA in Chronic Migraine: A Narrative Review. *Headache J. Head Face Pain* **2020**, *60*, 1259–1272. [CrossRef]
105. Liu, T.; Ji, R.R. New insights into the mechanisms of itch: Are pain and itch controlled by distinct mechanisms? *Pflügers Arch. Eur. J. Physiol.* **2013**, *465*, 1671–1685. [CrossRef] [PubMed]

106. Davidson, S.; Giesler, G.J. The multiple pathways for itch and their interactions with pain. *Trends Neurosci.* **2010**, *33*, 550–558. [CrossRef]
107. Sharif, B.; Ase, A.R.; Ribeiro-da-Silva, A.; Séguéla, P. Differential Coding of Itch and Pain by a Subpopulation of Primary Afferent Neurons. *Neuron* **2020**, *106*, 940–951.e4. [CrossRef]
108. Anzelc, M.; Burkhart, C.G. Pain and Pruritus: A study of their similarities and differences. *Int. J. Dermatol.* **2020**, *59*, 159–164. [CrossRef]
109. Gazerani, P.; Au, S.; Dong, X.; Kumar, U.; Arendt-Nielsen, L.; Cairns, B.E. Botulinum neurotoxin type A (BoNTA) decreases the mechanical sensitivity of nociceptors and inhibits neurogenic vasodilation in a craniofacial muscle targeted for migraine prophylaxis. *Pain* **2010**, *151*, 606–616. [CrossRef]
110. Gazerani, P.; Pedersen, N.S.; Staahl, C.; Drewes, A.M.; Arendt-Nielsen, L. Subcutaneous Botulinum toxin type A reduces capsaicin-induced trigeminal pain and vasomotor reactions in human skin. *Pain* **2009**, *141*, 60–69. [CrossRef]
111. Gazerani, P.; Staahl, C.; Drewes, A.M.; Arendt-Nielsen, L. The effects of Botulinum Toxin type A on capsaicin-evoked pain, flare, and secondary hyperalgesia in an experimental human model of trigeminal sensitization. *Pain* **2006**, *122*, 315–325. [CrossRef]
112. da Silva, L.B.; Kulas, D.; Karshenas, A.; Cairns, B.E.; Bach, F.W.; Arendt-Nielsen, L.; Gazerani, P. Time course analysis of the effects of botulinum neurotoxin type A on pain and vasomotor responses evoked by glutamate injection into human temporalis muscles. *Toxins* **2014**, *6*, 592–607. [CrossRef] [PubMed]
113. da Silva, L.B.; Poulsen, J.N.; Arendt-Nielsen, L.; Gazerani, P. Botulinum neurotoxin type A modulates vesicular release of glutamate from satellite glial cells. *J. Cell. Mol. Med.* **2015**, *19*, 1900–1909. [CrossRef] [PubMed]
114. Bittencourt da Silva, L.; Karshenas, A.; Bach, F.W.; Rasmussen, S.; Arendt-Nielsen, L.; Gazerani, P. Blockade of glutamate release by botulinum neurotoxin type A in humans: A dermal microdialysis study. *Pain Res. Manag.* **2014**, *19*, 126–132. [CrossRef] [PubMed]
115. Durham, P.L.; Cady, R.; Cady, R. Regulation of calcitonin gene-related peptide secretion from trigeminal nerve cells by botulinum toxin type A: Implications for migraine therapy. *Headache J. Head Face Pain* **2004**, *44*, 35–42. [CrossRef] [PubMed]
116. Purkiss, J.; Welch, M.; Doward, S.; Foster, K. Capsaicin-stimulated release of substance P from cultured dorsal root ganglion neurons: Involvement of two distinct mechanisms. *Biochem. Pharmacol.* **2000**, *59*, 1403–1406. [CrossRef]
117. Lora, V.R.; Clemente-Napimoga, J.T.; Abdalla, H.B.; Macedo, C.G.; Canales, G.T.; Barbosa, C.M. Botulinum toxin type A reduces inflammatory hypernociception induced by arthritis in the temporomadibular joint of rats. *Toxicon* **2017**, *129*, 52–57. [CrossRef]
118. Cui, M.; Khanijou, S.; Rubino, J.; Aoki, K.R. Subcutaneous administration of botulinum toxin A reduces formalin-induced pain. *Pain* **2004**, *107*, 125–133. [CrossRef] [PubMed]
119. Zhang, X.; Strassman, A.M.; Novack, V.; Brin, M.F.; Burstein, R. Extracranial injections of botulinum neurotoxin type A inhibit intracranial meningeal nociceptors' responses to stimulation of TRPV1 and TRPA1 channels: Are we getting closer to solving this puzzle? *Cephalalgia* **2016**, *36*, 875–886. [CrossRef] [PubMed]
120. Gfrerer, L.; Xu, W.; Austen, W.; Ashina, S.; Melo-Carrillo, A.; Longhi, M.S.; Adams, A.M.; Houle, T.; Brin, M.F.; Burstein, R. OnabotulinumtoxinA alters inflammatory gene expression and immune cells in chronic headache patients. *Brain* **2022**, *145*, 2436–2449. [CrossRef]
121. Ständer, S.; Steinhoff, M.; Schmelz, M.; Weisshaar, E.; Metze, D.; Luger, T. Neurophysiology of pruritus: Cutaneous elicitation of itch. *Arch. Dermatol.* **2003**, *139*, 1463–1470. [CrossRef]
122. Wimalasena, N.K.; Milner, G.; Silva, R.; Vuong, C.; Zhang, Z.; Bautista, D.M.; Woolf, C.J. Dissecting the precise nature of itch-evoked scratching. *Neuron* **2021**, *109*, 3075–3087.e2. [CrossRef]
123. Song, J.; Xian, D.; Yang, L.; Xiong, X.; Lai, R.; Zhong, J. Pruritus: Progress toward Pathogenesis and Treatment. *Biomed Res. Int.* **2018**, *2018*, 9625936. [CrossRef]
124. Sanders, K.M.; Akiyama, T. The vicious cycle of itch and anxiety. *Neurosci. Biobehav. Rev.* **2018**, *87*, 17–26. [CrossRef] [PubMed]
125. Lee, J.S.; Han, J.S.; Lee, K.; Bang, J.; Lee, H. The peripheral and central mechanisms underlying itch. *BMB Rep.* **2016**, *49*, 474–487. [CrossRef] [PubMed]
126. Papoiu, A.D.; Coghill, R.C.; Kraft, R.A.; Wang, H.; Yosipovitch, G. A tale of two itches. Common features and notable differences in brain activation evoked by cowhage and histamine induced itch. *Neuroimage* **2012**, *59*, 3611–3623. [CrossRef]
127. Han, L.; Dong, X. Itch mechanisms and circuits. *Annu. Rev. Biophys.* **2014**, *43*, 331–355. [CrossRef] [PubMed]
128. Sun, S.; Dong, X. Trp channels and itch. *Semin. Immunopathol.* **2016**, *38*, 293–307. [CrossRef]
129. Guo, C.J.; Grabinski, N.S.; Liu, Q. Peripheral Mechanisms of Itch. *J. Investig. Dermatol.* **2022**, *142*, 31–41. [CrossRef]
130. Sutaria, N.; Adawi, W.; Goldberg, R.; Roh, Y.S.; Choi, J.; Kwatra, S.G. Itch: Pathogenesis and treatment. *J. Am. Acad. Dermatol.* **2022**, *86*, 17–34. [CrossRef]
131. Yosipovitch, G.; Rosen, J.D.; Hashimoto, T. Itch: From mechanism to (novel) therapeutic approaches. *J. Allergy Clin. Immunol.* **2018**, *142*, 1375–1390. [CrossRef]
132. Hoeck, E.A.; Marker, J.B.; Gazerani, P.; Andersen, H.H.; Arendt-Nielsen, L. Preclinical and human surrogate models of itch. *Exp. Dermatol.* **2016**, *25*, 750–757. [CrossRef]
133. Akdogan, N.; Uysal, P.I.; Oktem, A.; Karabulut, E.; Hayran, Y.; Yalcin, B. Factors Affecting the Etiology of Intractable Pruritus in Hospitalized Patients without Primary Skin Lesions. *İstanbul Med. J* **2019**, *20*, 224–230. [CrossRef]
134. Reddy, K.S.; Naidu, M.U.; Rani, P.U.; Rao, T.R. Human experimental pain models: A review of standardized methods in drug development. *J. Res. Med. Sci.* **2012**, *17*, 587–595.

135. Quesada, C.; Kostenko, A.; Ho, I.; Leone, C.; Nochi, Z.; Stouffs, A.; Wittayer, M.; Caspani, O.; Brix Finnerup, N.; Mouraux, A.; et al. Human surrogate models of central sensitization: A critical review and practical guide. *Eur. J. Pain* **2021**, *25*, 1389–1428. [CrossRef]
136. Chizh, B.A.; Priestley, T.; Rowbotham, M.; Schaffler, K. Predicting therapeutic efficacy—Experimental pain in human subjects. *Brain Res. Rev.* **2009**, *60*, 243–254. [CrossRef] [PubMed]
137. Sycha, T.; Samal, D.; Chizh, B.; Lehr, S.; Gustorff, B.; Schnider, P.; Auff, E. A lack of antinociceptive or antiinflammatory effect of botulinum toxin A in an inflammatory human pain model. *Anesth. Analg.* **2006**, *102*, 509–516. [CrossRef]
138. Yuan, T.; Li, J.; Shen, L.; Zhang, W.; Wang, T.; Xu, Y.; Zhu, J.; Huang, Y.; Ma, C. Assessment of Itch and Pain in Animal Models and Human Subjects. *Adv. Exp. Med. Biol.* **2016**, *904*, 1–22. [CrossRef]
139. Fostini, A.C.; Girolomoni, G. Experimental elicitation of itch: Evoking and evaluation techniques. *J. Dermatol. Sci.* **2015**, *80*, 13–17. [CrossRef]
140. Ikoma, A.; Handwerker, H.; Miyachi, Y.; Schmelz, M. Electrically evoked itch in humans. *Pain* **2005**, *113*, 148–154. [CrossRef] [PubMed]
141. Fukuoka, M.; Miyachi, Y.; Ikoma, A. Mechanically evoked itch in humans. *Pain* **2013**, *154*, 897–904. [CrossRef] [PubMed]
142. Simone, D.A.; Alreja, M.; LaMotte, R.H. Psychophysical studies of the itch sensation and itchy skin ("alloknesis") produced by intracutaneous injection of histamine. *Somatosens. Mot. Res.* **1991**, *8*, 271–279. [CrossRef]
143. Simone, D.A.; Ngeow, J.Y.; Whitehouse, J.; Becerra-Cabal, L.; Putterman, G.J.; LaMotte, R.H. The magnitude and duration of itch produced by intracutaneous injections of histamine. *Somatosens. Res.* **1987**, *5*, 81–92. [CrossRef] [PubMed]
144. LaMotte, R.H.; Shimada, S.G.; Green, B.G.; Zelterman, D. Pruritic and nociceptive sensations and dysesthesias from a spicule of cowhage. *J. Neurophysiol.* **2009**, *101*, 1430–1443. [CrossRef] [PubMed]
145. Papoiu, A.D.; Tey, H.L.; Coghill, R.C.; Wang, H.; Yosipovitch, G. Cowhage-induced itch as an experimental model for pruritus. A comparative study with histamine-induced itch. *PLoS ONE* **2011**, *6*, e17786. [CrossRef] [PubMed]
146. Sikand, P.; Shimada, S.G.; Green, B.G.; LaMotte, R.H. Similar itch and nociceptive sensations evoked by punctate cutaneous application of capsaicin, histamine and cowhage. *Pain* **2009**, *144*, 66–75. [CrossRef] [PubMed]
147. Sikand, P.; Dong, X.; LaMotte, R.H. BAM8-22 peptide produces itch and nociceptive sensations in humans independent of histamine release. *J. Neurosci.* **2011**, *31*, 7563–7567. [CrossRef]
148. Liu, Q.; Sikand, P.; Ma, C.; Tang, Z.; Han, L.; Li, Z.; Sun, S.; LaMotte, R.H.; Dong, X. Mechanisms of itch evoked by β-alanine. *J. Neurosci.* **2012**, *32*, 14532–14537. [CrossRef]
149. Weisshaar, E.; Ziethen, B.; Gollnick, H. Can a serotonin type 3 (5-HT3) receptor antagonist reduce experimentally-induced itch? *Inflamm. Res.* **1997**, *46*, 412–416. [CrossRef]
150. Nattkemper, L.A.; Stull, C.; Lavery, M.; Valdes-Rodriguez, R.; McGregory, M.; Ramsey, R.; Chen, Y.; Mochizuki, H.; Yosipovitch, G. Prolonged antipruritic effect of botulinum toxin type A on cowhage-induced itch. In *ACTA DERMATO-VENEREOLOGICA*; 2017; Volume 97, p. 1021. Available online: https://www.medicaljournals.se/acta/content_files/files/pdf/97/8/IFSIAbstracts2017.pdf (accessed on 9 October 2022).
151. Arthur, R.P.; Shelley, W.B. The role of proteolytic enzymes in the production of pruritus in man. *J. Investig. Dermatol.* **1955**, *25*, 341–346. [CrossRef]
152. Cevikbas, F.; Lerner, E.A. Physiology and Pathophysiology of Itch. *Physiol. Rev.* **2020**, *100*, 945–982. [CrossRef]
153. Hersini, K.J.; Melgaard, L.; Gazerani, P.; Petersen, L.J. Microdialysis of inflammatory mediators in the skin: A review. *Acta Derm. Venereol.* **2014**, *94*, 501–511. [CrossRef]
154. Barré-Sinoussi, F.; Montagutelli, X. Animal models are essential to biological research: Issues and perspectives. *Future Sci. OA* **2015**, *1*, FSO63. [CrossRef]
155. Singh, V.P.; Pratap, K.; Sinha, J.; Desiraju, K.; Bahal, D.; Kukreti, R. Critical evaluation of challenges and future use of animals in experimentation for biomedical research. *Int. J. Immunopathol. Pharmacol.* **2016**, *29*, 551–561. [CrossRef]
156. Kahremany, S.; Hofmann, L.; Gruzman, A.; Cohen, G. Advances in Understanding the Initial Steps of Pruritoceptive Itch: How the Itch Hits the Switch. *Int. J. Mol. Sci.* **2020**, *21*, 4883. [CrossRef] [PubMed]
157. Joglekar, G.V.; Bhide, M.B.; Balwani, J.H. An experimental method for screening antipruritic agents. *Br. J. Dermatol.* **1963**, *75*, 117–119. [CrossRef] [PubMed]
158. Gmerek, D.E.; Cowan, A. An animal model for preclinical screening of systemic antipruritic agents. *J. Pharmacol. Methods* **1983**, *10*, 107–112. [CrossRef] [PubMed]
159. Ko, M.C.; Naughton, N.N. An experimental itch model in monkeys: Characterization of intrathecal morphine-induced scratching and antinociception. *Anesthesiology* **2000**, *92*, 795–805. [CrossRef] [PubMed]
160. Kuraishi, Y.; Nagasawa, T.; Hayashi, K.; Satoh, M. Scratching behavior induced by pruritogenic but not algesiogenic agents in mice. *Eur. J. Pharmacol.* **1995**, *275*, 229–233. [CrossRef]
161. Yamaguchi, T.; Nagasawa, T.; Satoh, M.; Kuraishi, Y. Itch-associated response induced by intradermal serotonin through 5-HT2 receptors in mice. *Neurosci. Res.* **1999**, *35*, 77–83. [CrossRef]
162. Inan, S.; Cowan, A. Kappa opioid agonists suppress chloroquine-induced scratching in mice. *Eur. J. Pharmacol.* **2004**, *502*, 233–237. [CrossRef] [PubMed]
163. Shimada, S.G.; Shimada, K.A.; Collins, J.G. Scratching behavior in mice induced by the proteinase-activated receptor-2 agonist, SLIGRL-NH2. *Eur. J. Pharmacol.* **2006**, *530*, 281–283. [CrossRef] [PubMed]

164. Cowan, A.; Lyu, R.M.; Chen, Y.H.; Dun, S.L.; Chang, J.K.; Dun, N.J. Phoenixin: A candidate pruritogen in the mouse. *Neuroscience* **2015**, *310*, 541–548. [CrossRef]
165. Suto, H.; Matsuda, H.; Mitsuishi, K.; Hira, K.; Uchida, T.; Unno, T.; Ogawa, H.; Ra, C. NC/Nga mice: A mouse model for atopic dermatitis. *Int. Arch. Allergy Immunol.* **1999**, *120*, 70–75. [CrossRef]
166. Sakai, K.; Sanders, K.M.; Youssef, M.R.; Yanushefski, K.M.; Jensen, L.; Yosipovitch, G.; Akiyama, T. Mouse model of imiquimod-induced psoriatic itch. *Pain* **2016**, *157*, 2536–2543. [CrossRef]
167. Miyamoto, T.; Nojima, H.; Shinkado, T.; Nakahashi, T.; Kuraishi, Y. Itch-associated response induced by experimental dry skin in mice. *Jpn. J. Pharmacol.* **2002**, *88*, 285–292. [CrossRef] [PubMed]
168. Aoki, K.R. Review of a proposed mechanism for the antinociceptive action of botulinum toxin type A. *Neurotoxicology* **2005**, *26*, 785–793. [CrossRef] [PubMed]
169. Bach-Rojecky, L.; Lacković, Z. Antinociceptive effect of botulinum toxin type a in rat model of carrageenan and capsaicin induced pain. *Croat. Med. J.* **2005**, *46*, 201–208. [PubMed]
170. Reyes-Long, S.; Alfaro-Rodríguez, A.; Cortes-Altamirano, J.L.; Lara-Padilla, E.; Herrera-Maria, E.; Romero-Morelos, P.; Salcedo, M.; Bandala, C. The Mechanisms of Action of Botulinum Toxin Type A in Nociceptive and Neuropathic Pathways in Cancer Pain. *Curr. Med. Chem.* **2021**, *28*, 2996–3009. [CrossRef] [PubMed]
171. Ramachandran, R.; Marino, M.J.; Paul, S.; Wang, Z.; Mascarenhas, N.L.; Pellett, S.; Johnson, E.A.; DiNardo, A.; Yaksh, T.L. A Study and Review of Effects of Botulinum Toxins on Mast Cell Dependent and Independent Pruritus. *Toxins* **2018**, *10*, 134. [CrossRef]
172. Shimizu, T.; Shibata, M.; Toriumi, H.; Iwashita, T.; Funakubo, M.; Sato, H.; Kuroi, T.; Ebine, T.; Koizumi, K.; Suzuki, N. Reduction of TRPV1 expression in the trigeminal system by botulinum neurotoxin type-A. *Neurobiol. Dis.* **2012**, *48*, 367–378. [CrossRef]
173. Apostolidis, A.; Popat, R.; Yiangou, Y.; Cockayne, D.; Ford, A.P.; Davis, J.B.; Dasgupta, P.; Fowler, C.J.; Anand, P. Decreased sensory receptors P2X3 and TRPV1 in suburothelial nerve fibers following intradetrusor injections of botulinum toxin for human detrusor overactivity. *J. Urol.* **2005**, *174*, 977–982. [CrossRef]
174. Cochrane, D.E.; Douglas, W.W. Calcium-induced extrusion of secretory granules (exocytosis) in mast cells exposed to 48/80 or the ionophores A-23187 and X-537A. *Proc. Natl. Acad. Sci. USA* **1974**, *71*, 408–412. [CrossRef]
175. Hirokawa, N.; Heuser, J.E. Structural evidence that botulinum toxin blocks neuromuscular transmission by impairing the calcium influx that normally accompanies nerve depolarization. *J. Cell Biol.* **1981**, *88*, 160–171. [CrossRef]
176. Cao, L.F.; Si, M.; Huang, Y.; Chen, L.H.; Peng, X.Y.; Qin, Y.Q.; Liu, T.T.; Zhou, Y.; Liu, T.; Luo, W.F. Long-term anti-itch effect of botulinum neurotoxin A is associated with downregulation of TRPV1 and TRPA1 in the dorsal root ganglia in mice. *Neuroreport* **2017**, *28*, 518–526. [CrossRef] [PubMed]
177. Amalia, S.N.; Uchiyama, A.; Baral, H.; Inoue, Y.; Yamazaki, S.; Fujiwara, C.; Sekiguchi, A.; Yokoyama, Y.; Ogino, S.; Torii, R.; et al. Suppression of neuropeptide by botulinum toxin improves imiquimod-induced psoriasis-like dermatitis via the regulation of neuroimmune system. *J. Dermatol. Sci.* **2021**, *101*, 58–68. [CrossRef]
178. Girolomoni, G.; Strohal, R.; Puig, L.; Bachelez, H.; Barker, J.; Boehncke, W.H.; Prinz, J.C. The role of IL-23 and the IL-23/T(H) 17 immune axis in the pathogenesis and treatment of psoriasis. *J. Eur. Acad. Dermatol. Venereol.* **2017**, *31*, 1616–1626. [CrossRef]
179. Jabeen, M.; Boisgard, A.S.; Danoy, A.; El Kholti, N.; Salvi, J.P.; Boulieu, R.; Fromy, B.; Verrier, B.; Lamrayah, M. Advanced Characterization of Imiquimod-Induced Psoriasis-Like Mouse Model. *Pharmaceutics* **2020**, *12*, 789. [CrossRef]
180. Ostrowski, S.M.; Belkadi, A.; Loyd, C.M.; Diaconu, D.; Ward, N.L. Cutaneous Denervation of Psoriasiform Mouse Skin Improves Acanthosis and Inflammation in a Sensory Neuropeptide-Dependent Manner. *J. Investig. Dermatol.* **2011**, *131*, 1530–1538. [CrossRef]
181. Saraceno, R.; Kleyn, C.E.; Terenghi, G.; Griffiths, C.E. The role of neuropeptides in psoriasis. *Br. J. Dermatol.* **2006**, *155*, 876–882. [CrossRef]
182. Meng, J.; Li, Y.; Fischer, M.J.M.; Steinhoff, M.; Chen, W.; Wang, J. Th2 Modulation of Transient Receptor Potential Channels: An Unmet Therapeutic Intervention for Atopic Dermatitis. *Front. Immunol.* **2021**, *12*, 696784. [CrossRef]
183. Meng, J.; Ovsepian, S.V.; Wang, J.; Pickering, M.; Sasse, A.; Aoki, K.R.; Lawrence, G.W.; Dolly, J.O. Activation of TRPV1 mediates calcitonin gene-related peptide release, which excites trigeminal sensory neurons and is attenuated by a retargeted botulinum toxin with anti-nociceptive potential. *J. Neurosci.* **2009**, *29*, 4981–4992. [CrossRef]
184. Meng, J.; Wang, J.; Steinhoff, M.; Dolly, J.O. TNFα induces co-trafficking of TRPV1/TRPA1 in VAMP1-containing vesicles to the plasmalemma via Munc18–1/syntaxin1/SNAP-25 mediated fusion. *Sci. Rep.* **2016**, *6*, 21226. [CrossRef]
185. Meng, Q.T.; Liu, X.Y.; Liu, X.T.; Liu, J.; Munanairi, A.; Barry, D.M.; Liu, B.; Jin, H.; Sun, Y.; Yang, Q.; et al. BNP facilitates NMB-encoded histaminergic itch via NPRC-NMBR crosstalk. *Elife* **2021**, *10*, e71689. [CrossRef]
186. Agarwal, R.; Schmidt, J.J.; Stafford, R.G.; Swaminathan, S. Mode of VAMP substrate recognition and inhibition of Clostridium botulinum neurotoxin F. *Nat. Struct. Mol. Biol.* **2009**, *16*, 789–794. [CrossRef]
187. Rojewska, E.; Piotrowska, A.; Popiolek-Barczyk, K.; Mika, J. Botulinum Toxin Type A—A Modulator of Spinal Neuron-Glia Interactions under Neuropathic Pain Conditions. *Toxins* **2018**, *10*, 145. [CrossRef]
188. Andersen, H.H.; Arendt-Nielsen, L.; Gazerani, P. Glial Cells are Involved in Itch Processing. *Acta Derm. Venereol.* **2016**, *96*, 723–727. [CrossRef]
189. Micali, G.; Verzì, A.E.; Giuffrida, G.; Panebianco, E.; Musumeci, M.L.; Lacarrubba, F. Inverse Psoriasis: From Diagnosis to Current Treatment Options. *Clin. Cosmet. Investig. Dermatol.* **2019**, *12*, 953–959. [CrossRef]

190. Weise, D.; Weise, C.M.; Naumann, M. Central Effects of Botulinum Neurotoxin—Evidence from Human Studies. *Toxins* **2019**, *11*, 21. [CrossRef]
191. Caleo, M.; Restani, L. Direct central nervous system effects of botulinum neurotoxin. *Toxicon* **2018**, *147*, 68–72. [CrossRef]
192. Luvisetto, S. Botulinum Neurotoxins in Central Nervous System: An Overview from Animal Models to Human Therapy. *Toxins* **2021**, *13*, 751. [CrossRef]
193. Pereira, M.P.; Wiegmann, H.; Agelopoulos, K.; Ständer, S. Neuropathic Itch: Routes to Clinical Diagnosis. *Front. Med.* **2021**, *8*, 641746. [CrossRef]
194. Colloca, L.; Ludman, T.; Bouhassira, D.; Baron, R.; Dickenson, A.H.; Yarnitsky, D.; Freeman, R.; Truini, A.; Attal, N.; Finnerup, N.B.; et al. Neuropathic pain. *Nat. Rev. Dis. Primers* **2017**, *3*, 17002. [CrossRef]
195. Finnerup, N.B.; Kuner, R.; Jensen, T.S. Neuropathic Pain: From Mechanisms to Treatment. *Physiol. Rev.* **2021**, *101*, 259–301. [CrossRef]
196. Witmanowski, H.; Błochowiak, K. The whole truth about botulinum toxin—A review. *Allergol. Postępy Dermatol. Alergol.* **2020**, *37*, 853–861. [CrossRef]
197. Kremer, A.E.; Mettang, T.; Weisshaar, E. Non-dermatological Challenges of Chronic Itch. *Acta Derm. Venereol.* **2020**, *100*, adv00025. [CrossRef]
198. Hashimoto, T.; Yosipovitch, G. Itching as a systemic disease. *J. Allergy Clin. Immunol.* **2019**, *144*, 375–380. [CrossRef]
199. Winayanuwattikun, W.; Vachiramon, V. Botulinum Toxin Type A for the Treatment of Skin Ulcers: A Review Article. *Toxins* **2022**, *14*, 406. [CrossRef]
200. Grando, S.A.; Zachary, C.B. The non-neuronal and nonmuscular effects of botulinum toxin: An opportunity for a deadly molecule to treat disease in the skin and beyond. *Br. J. Dermatol.* **2018**, *178*, 1011–1019. [CrossRef]
201. Dressler, D.; Johnson, E.A. Botulinum toxin therapy: Past, present and future developments. *J. Neural Transm.* **2022**, *129*, 829–833. [CrossRef]

toxins MDPI

Review

Botulinum Neurotoxins beyond Neurons: Interplay with Glial Cells

Siro Luvisetto

Institute of Biochemistry and Cell Biology, National Research Council of Italy, Via Ercole Ramarini 32, Monterotondo Scalo, 00015 Rome, Italy; siro.luvisetto@cnr.it

Abstract: In recent years, numerous studies have highlighted the significant use of botulinum neurotoxins (BoNTs) in the human therapy of various motor and autonomic disorders. The therapeutic action is exerted with the selective cleavage of specific sites of the SNARE's protein complex, which plays a key role in the vesicular neuroexocytosis which is responsible for neural transmission. The primary target of the BoNTs' action is the peripheral neuromuscular junction (NMJ), where, by blocking cholinergic neurons releasing acetylcholine (ACh), they interfere with neural transmission. A great deal of experimental evidence has demonstrated that BoNTs are also effective in blocking the release of other neurotransmitters or neuromodulators, such as glutamate, substance-P, and CGRP, and they can interfere with the function of glial cells, both at the peripheral and central level. The purpose of this review is to provide an update on the available experimental data from animal models that suggest or confirm the direct interactions between BoNTs and glial cells. From the data collected, it appears evident that, through mechanisms that are not yet fully understood, BoNTs can block the activation of spinal glial cells and their subsequent release of pro-inflammatory factors. BoNTs are also able to promote peripheral regeneration processes after nerve injury by stimulating the proliferation of Schwann cells. The data will be discussed in consideration of the possible therapeutic implications of the use of BoNTs on those pathological conditions where the contribution of glial cell activation is fundamental, such as in peripheral and central neuropathies.

Keywords: botulinum; glia; peripheral nervous system; central nervous system; animal models

Key Contribution: This review constitutes an update on research concerning the possible interactions between botulinum neurotoxins and both peripheral and spinal glial cells, mainly Schwann cells, microglia, and astrocytes. The data covered in this review can help us to understand the mechanisms underlying the interaction between BoNTs and glial cells. A better understanding of these mechanisms may form the basis for the further use of BoNTs for therapeutic purposes.

Citation: Luvisetto, S. Botulinum Neurotoxins beyond Neurons: Interplay with Glial Cells. *Toxins* **2022**, *14*, 704. https://doi.org/10.3390/toxins14100704

Received: 7 September 2022
Accepted: 12 October 2022
Published: 13 October 2022

1. Introduction

Botulinum neurotoxins (BoNTs) are produced by clostridium botulinum in the subtypes of seven serotypes, named from A to G [1,2], in addition to the recently characterized serotype X, and a certain number of chimeric neurotoxins [3,4]. In this paper, the different types of BoNTs will be reported using the acronym BoNT/Y, where Y stands for serotypes A to X. If the subtype is not specified, then the acronym BoNTs will be used in a generic manner; where necessary, the commercial name will be used.

All BoNTs have similar structure, consisting of two chains (L-chain, 50 kDa, and H-chain, 100 kDa) that are linked by a single disulfide bridge, with the L-chain being the catalytic domain and the H-chain being the receptor-binding and translocation domain [5]. BoNTs act as powerful blockers of synaptic vesicle fusion at the peripheral neuromuscular junction (NMJ), where they block the release of acetylcholine (ACh). Inhibition of vesicular ACh release is achieved via the cleavage of SNARE proteins, which constitute the protein complex essential for the vesicular release evoked by the action potential at the NMJ.

Different BoNT serotypes cleave different SNARE proteins: BoNT/B, D, F, G, and X cleave VAMP at single and different sites; BoNT/C cleaves both syntaxin and SNAP-25, whereas BoNT/A and E cleave SNAP-25 at different sites [6].

In recent years, many basic scientific studies, and a great deal of clinical evidence, demonstrated that, apart from the canonical anticholinergic effects at the NMJ, BoNTs are effective in inhibiting the ACh release at sites other than the NMJ, in addition to inhibiting the release of neurotransmitters other than ACh [7]. Currently, there is no doubt that BoNTs can block the release of excitatory neurotransmitters and neuropeptides, such as glutamate, substance-P, and CGRP. These substances, along with certain actions, are strongly involved in pain modulation [8]. Moreover, the finding that BoNTs may block their release provided studies, that were conducted on both animal models and in clinics [9,10], with the impetus to suggest the use of BoNTs as an analgesic for chronic pain conditions that did not respond to other analgesic drugs; therefore, BoNTs have been suggested as a third line analgesic treatment [11].

Many studies have provided evidence for the fact that, apart from the canonical action on neurons, BoNTs may interact also with glial cells, both in the central (CNS) and peripheral (PNS) nervous system. Historically, the first evidence for an interaction between BoNTs and glial cells, came from a series of in vitro studies in cultured astrocytes. Jeftinija et al. [12] demonstrated that the pre-treatment of cultured astrocytes with BoNT/A or BoNT/C decreased both the baseline and the bradykinin-evoked release of glutamate. Verderio et al. [13] demonstrated that BoNT/B and BoNT/F are internalized to culture astrocytes, whereas Araque et al. [14] demonstrated that a microinjection of cultured astrocytes with light chain BoNT/B strongly reduced SNARE protein-dependent glutamate releases. Moreover, it was demonstrated that BoNT/A also reduced the extracellular high K^+-induced increase in glutamate that was released from the astrocytes [15]. Other in vitro studies showed that BoNT/A blocked the uridine triphosphate-stimulated ATP release from both cultured astrocytes that were isolated from rat cortexes [16] and Schwann cells (SCs) that were cultured from the sciatic nerve [17]. An additional effect of BoNT/A on cultured SCs was found by Marinelli et al. [18]; in ex vivo experiments, Marinelli et al. [18] found that, after a peripheral injection of BoNT/A in mice that were subjected to a ligature of the sciatic nerve, cleaved SNAP-25 co-localized with astrocytes. This finding has been considered as strong evidence for the possible transcytosis of BoNT/A from neuronal cells to astrocytes. Blocking the release of astrocytic glutamate from BoNT/A may contribute to the reduction of pain.

As the aim of this review is to give a description of the current findings on the interaction between BoNTs and glial cells, further evidence will be discussed below. Given that there may be a possible therapeutic application of BoNTs, regarding pathologies in which glial cells are deeply involved in the onset and maintenance of the disease, particular emphasis will be given to the evidence concerning the direct effects of BoNTs on glial activation for the treatment of neuropathic conditions and/or neuronal lesions. Finally, given the fact that the last review on this subject dates back to 2018 [19], this descriptive review aims to update the current literature, starting from 2018, with the aim to stimulate future studies on this important topic.

2. Search Strategy and Inclusion Criteria

The online search was conducted on the databases PubMed/MEDLINE, ScienceDirect, Scopus, and Web of Science, and it included articles from January 2018 up to September 2022. The search on the databases were performed using all paired combinations of the term 'botulinum' with (AND) one of the following terms: 'glia', 'glial', 'microglia', 'macroglia', 'astroglia', 'astrocytes', 'oligodendrocyte', 'Schwann', 'satellite glia', 'ependymal', 'radial glia', and 'enteric glia'. The experimental articles included both animal-based and clinical studies; the only studies which were excluded were those that were conducted with botulinum C3, a toxin which secretes the enzyme C3, that selectively causes the mono-ADP-ribosylation of the small GTPases Rho-A, -B, and -C, thus inhibiting cell–Rho signaling.

Such exclusions are justified given the fact that the mechanism of action for the botulinum toxin C3 is completely different from that of classical BoNTs, which act on SNARE proteins; this is the focus of the present review.

3. Glia

Glia are non-neuronal cells existing both in CNS and PNS (Figure 1). The main difference between glia and neurons, is that glial cells do not produce action potentials, and thus, for many years, their function was considered to be exclusively structural, providing only functional and protective support for neurons. For this reason, they were named "glia", which is a Greek translation for "glue". The "gluing" function of glia is now considered to be minor, and it has been widely recognized that glial cells are actively involved in the modulation of the neuronal environment, and the regulation of many neuronal functions, such as the nerve firing discharge, brain plasticity, the immune and inflammatory response, the formation of the myelin sheath of axons, and recovery from nerve injuries [20]. Different types of glial cells play different specific roles.

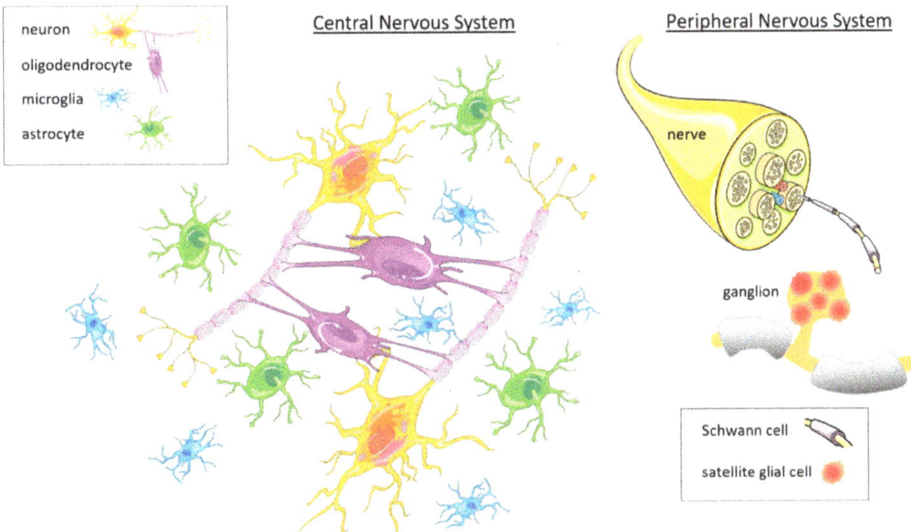

Figure 1. Diagram of the main types of glial cells in the central and peripheral nervous system. In the CNS, glial cells include astrocytes, oligodendrocytes, and microglia, whereas in the PNS, the glial cells include SCs and SGCs. Other glial cells such as ependymal, radial, and enteric glial cells are not depicted in this diagram. Parts of this figure were composed using pictures from Servier Medical Art (accessed on 18 July 2022 from http://smart.servier.com), a free service provided by Les Laboratoires Servier (accessed on 18 July 2022 from http://www.servier.com). Servier Medical Art is licensed under a Creative Commons Attribution 3.0 Unported License (accessed on 18 July 2022 from https://creativecommons.org/licenses/by/3.0/).

Glial cell types belong to two main categories: macroglia and microglia. Macroglia are involved in the regulation and optimization of neuronal function, whereas microglia have phagocytic properties and help make the neuronal environment safer. Macroglia exist in four main forms in the CNS, namely, the oligodendrocytes, astrocytes, ependymal cells, and radial glia. In the PNS, they exist in three main forms, namely, the Schwann cells, satellite glial cells, and enteric glia.

Oligodendrocytes are cells that line axons in the CNS with their cell membrane, which forms the myelin sheath, thus allowing electrical signals to propagate along axons more efficiently [21,22]. A single oligodendrocyte provides insulation for multiple neurons. Oligodendrocytes also support the metabolic needs of the axons in nerve cells [23].

Ependymal cells form layers that delimit the space in brain ventricles and the spinal cord with a continuous sheet of epithelium [24,25]. The main function of ependymal cells is the production of cerebrospinal fluid [26]. Since they are ciliated cells, they help to distribute neurotransmitters and hormones, and they also contribute to osmotic control within the brain via the regulation of glucose and the uptake of ions [24].

Radial glia comprise a subgroup of glial cells that include Bergmann and Muller cells. Radial glia are only found in specific areas of the CNS [27], namely, the cerebellum (Bergmann cells) and retina (Muller cells), where they modulate neurotransmission and optimize how information is processed. During brain development, radial glia also function as neuronal stem cells [28].

Astrocytes, also called astroglia, are the most abundant type of macroglia in the CNS [29]. They regulate the external chemical environment of neurons by recycling the neurotransmitters that are released during synaptic transmission [30]. Astrocytes are also involved in vasoconstriction and vasodilation, which are actions that occur after substances such as arachidonic acid have been produced, as their metabolites are vasoactive. They also connect neurons to blood vessels and help to maintain the permeability of the blood–brain barrier, where they sense the levels of glucose and ions and regulate their flow into, or out of, the brain [31].

Schwann cells (SCs) in the PNS operate similarly to the oligodendrocytes in the CNS. They provide myelination to nerve axons and they modulate the extracellular fluid [32]. Unlike oligodendrocytes in the CNS, where a single oligodendrocyte myelinates multiple axons, a single axon in the PNS is myelinated by multiple SCs. SCs not only have a myelinating function, but they also exhibit phagocytic activity. In fact, together with the infiltrating macrophages, they help to clear cellular debris after neuronal lesions, thus they favor the regrowth of PNS neurons [33].

Satellite glial cells (SGCs) are small, flattened cells found in the PNS, where they surround sensory and autonomic ganglia. These cells, which help to regulate the external chemical environment, are highly sensitive to injury and inflammation, and they appear to contribute to pathological states, such as acute and chronic pain [34]. In some respects, their functions are similar to the astrocytes in the CNS [35,36]. In sensory neurons, SGCs regulate K^+ levels, and the neurons respond to the evoked potentials [37].

Enteric glial cells are found in the lining of the intestines, among the enteric ganglia in the digestive system, in smooth muscle layers, and in gut mucosa. They have many roles that are related to homeostasis and the muscular digestive processes in the enteric nervous system [38]. They are implicated in peristalsis, and they encourage contact between the different cells of the intestinal wall [39]. Enteric glia are also involved in the pathogenesis of inflammatory diseases concerning the enteric nervous system, such as inflammatory bowel diseases, and of functional gastrointestinal diseases, such as irritable bowel syndrome [40].

Microglia are found in all regions of the brain and spinal cord. Microglia are specialized macrophages that are capable of phagocytosis, which eliminates cellular debris and toxins, thus protecting CNS neurons [41]. Microglia are small, relative to macroglia, with changing shapes and oblong nuclei [42,43]. As macrophages, microglia hold fundamental roles in terms of nervous tissue immunity and inflammatory responses [44]. They are also implicated in many other processes that are involved in the optimization of different brain circuits which enable cognitive development. In fact, microglia perform their "cleaning action" by phagocyting previously formed synapses that are no longer useful [45].

4. Interactions between BoNTs, Microglia, and Astrocytes

The main evidence for the interactions between BoNTs, astrocytes, and microglia comes from studies on pathological pain in animal models. Pathological pain is characterized by an amplified response to normally harmless stimuli and an amplified response to acute pain. With conditions that cause pathological pain, which results from a dysfunction in the neuronal activity of sensory neurons, the activation of spinal glial cells, microglia, and astrocytes, contributes to the development and maintenance of chronic pain [46–49]. Glial cells are activated by the neuronal release of neuromediators, including substance P, glutamate, and fractalkine. The activated glia may release other mediators that, via a feedback action on glia and neurons, produce an amplification of the pain signals. Critical mediators that sustain the amplification of pain have been demonstrated to be pro-inflammatory cytokines [50,51]. In a previous review, Rojewska et al. [19] provided a detailed analysis of the evidence that shows a modulatory interaction between BoNT/A and microglia, astrocytes, and neurons under neuropathic pain conditions; they had a particular interest in clarifying how BoNT/A may affect spinal neuron–glial interactions. Starting with the review of Rojewska et al. [19], which gives a comprehensive review on these topics before 2018, the current chapter of this review aims to provide an update on the effects of BoNTs on microglia and/or astrocytes by using the findings of newer studies.

Before analyzing the interaction between BoNTs and microglia in detail, some preliminary fixed points must be addressed. It is well recognized that, under pathophysiological conditions, microglia can act in two different forms: a classic pro-inflammatory phenotype and an alternative anti-inflammatory phenotype [52,53]. As a pro-inflammatory phenotype, microglia release TNF-α, IL-1β, IL-12, IL-23, and pro-inflammatory cytokines, which exacerbate inflammation and tissue injury. In contrast, as an anti-inflammatory phenotype, they release TGF-β, IL-4, IL-10, Il-13, VEGF, BDNF, PDGF, anti-inflammatory cytokines, and growth factors, which suppress inflammation and promote tissue recovery, respectively [47]. Differentiation of microglia towards the pro-inflammatory phenotype often occurs during neuropathic pain, and a transition from a pro-inflammatory to an anti-inflammatory phenotype may represent an innovative therapeutic strategy for relieving neuropathic pain [19,47].

In a neuropathic pain model that involved a chronic constrictive injury (CCI) to the sciatic nerve of a rat, Gui et al. [54] found that the subcutaneous injection of BoNT/A (10–20 U/kg Botox® into the metatarsal surface; three days after CCI) promoted the polarization of a microglial to become an anti-inflammatory phenotype. This finding correlated with the decreased expression of the microglial purinergic P2X7 receptor, along with the increased mechanical withdrawal threshold and thermal withdrawal latency. The reduced expression of P2X7 receptors was also confirmed by in vitro assays performed in a microglial cell line stimulated by lipopolysaccharide (LPS). The exact mechanism by which BoNT/A reduced the expression of P2X7 receptors, and thus, the mechanism that caused the microglia to become an anti-inflammatory phenotype, remains unknown. In another report from the same laboratory [55], the authors confirmed that BoNT/A attenuated CCI-induced neuropathic pain in rats by slowing the release of pro-inflammatory factors from activated microglia, as well as mitigating the expression of SNAP-23. Reducing the SNAP-23 expression in microglia occurs by targeting toll-like receptor 2 (TLR2), and its adaptor protein, MyD88. Toll-like receptors are normally expressed in immune and glial cells, where they regulate innate and adaptive immunity.

It is worth considering that the Botox dose used in references [54,55] appear particularly high. In fact, the dose of 20 U/Kg would represent a dose of 1500 units for a 75 Kg adult, which is exceptionally high (15 vials of 100 U per vial). In light of this, translating the results in references [54,55] from rats to humans seems questionable; however, it should be also noted that doses of BoNT/A in animal models cannot be simply converted into therapeutic doses for humans on the basis of weight ratios, rather, they must be appropriately chosen on the basis of toxicity criteria.

The role of TLR2-mediated neuroinflammation was also evidenced by Chen et al. [56], who demonstrated that the unilateral subcutaneous facial injection of BoNT/A (0.18 U of Lanzhou manufactured BoNT/A into the whisker pad), in a trigeminal neuralgia model induced by CCI of the distal infraorbital nerve in mice, attenuated bilateral trigeminal neuropathic pain behaviors and inhibited the upregulation of microglia in TLR2 expression.

Altogether, these results are confirmed by the findings of Piotroska et al. [57], who revealed that BoNT/A inhibits the expression of pro-inflammatory factors through the modulation of NF-kB, p38, and ERK1/2. Moreover, it interacts with the TLR2/MyD88 signaling pathway, thus resulting in the decreased expression of SNAP-23 in LPS stimulated microglial cells. The results from Piotroska et al. [57] are in line with the results of Hepp et al. [58], who reported that SNAP-23 replaces SNAP-25 in microglia and oligodendrocytes. The effects of BoNT/A on SNAP-23 seem to contrast with the natural molecular targets of BoNT/A in neuronal cells (i.e., SNAP-25); however, when analyzing the structural features of SNAP-25, and its non-neuronal SNAP-23 isoforms, which include the murine mSNAP-23 and human hSNAP-23, Vaydianathan et al. [59] found that BoNT/A was effectively able to cleave the non-neuronal mSNAP-23, but not hSNAP-23. Additionally, BoNT/E was more efficient than BoNT/A in cleaving mSNAP-23. Notably, if BoNT/A was only able to block microglial mSNAP-23, this finding would pose an important limitation, in that it would highlight the difficulty in translating these results to a therapeutic setting for humans, from a pharmacological perspective (i.e., regarding the possibility of interacting BoNT/A with microglia as an alternative method to treat chronic pain). Further research is needed to clarify these points.

In a rat model involving a spinal cord injury (SCI), Yu et al. [60] observed that the combined application of BoNT/A (injection of two doses of 1.25 U of Botox® around the SCI site and forelimb muscle) and minocycline, an antibiotic agent that also has anti-inflammatory properties, synergistically reduced neuropathic pain and apoptosis by inactivating the glial cells. In further detail, the authors found that the combination of BoNT/A and minocycline promotes the expression of the SIRT1 cell signaling pathway, inactivates the NF-κB, P53, and PI3K/AKT signaling pathways, and attenuates an inflammatory response and oxidative stress. These combined effects greatly enhance the therapeutic effect of the two drugs.

Feng et al. [61] showed that a single intraplantar, or the intrathecal, pre-administration of BoNT/A (0.5–1 U/kg of Botox®), in a rat that was subjected to a partial sciatic nerve ligation (PSNL) pain model, significantly prevented PSNL-induced allodynia and thermal hyperalgesia, together with a reduced upregulation of pro-inflammatory cytokines in the spinal cord, dorsal horn, and dorsal root ganglions (DRGs). In order to determine the direct effect of BoNT/A on microglia and/or astrocytes, the authors also performed an in vitro experiment on the LPS-activated glial cells that were treated with BoNT/A (1–2 U/mL of Botox®). They found that BoNT/A significantly inhibited the activation of LPS-activated microglia and reduced the release of TNF-α, IL-6, Il-1β, iNOS, and MIP-1α, without any effect on the astrocytes' activation. This latter result appears to be in contrast with findings that detect the cl-SNAP25 immunoreaction after treatment with BoNT/A, both in LPS-activated astrocytes [56] and in spinal astrocytes, either in CCI pain models [18,62] or in spinal cord injury models [63]. Interestingly, Finocchiaro et al. [64] reported a strong reduction in the activation of spinal astrocytes, a study which also involved CCI mice that were treated with BoNT/B (intraplantar injection of 7.5 pg/mouse of 150 KDa of purified BoNT/B). Conversely, no difference in the expression of resting and activated microglia were observed in CCI mice treated with BoNT/B. The discrepancies observed in the different BoNTs serotypes may depend on the different targets of the toxins, namely, SNAP-25 for BoNT/A and VAMP-2 for BoNT/B, and the different expressions of these targets in neuronal and non-neuronal cells.

The possible interaction between BoNTs and glial cells was not only analyzed in the context of neuropathic pain, but also in the context of inflammatory pain. In a chronic inflammatory pain model, which involved an intraarticular injection of a solution of complete Freund's adjuvant (CFA) into the ankle joint cavity of the left leg of a rat, Shi et al. [65] observed that an intraarticular injection of BoNT/A (5–10 U/Kg of Botox® into ankle articular cavity) reduced CFA-induced pain-related behaviors in a dose dependent manner. Similar behavioral effects were achieved by blocking the activation of spinal microglia and reducing TNF-α. Furthermore, the authors found that the effect of BoNT/A on spinal microglial activation was associated with the inhibition of spinal microglial P2X4R–P38MAPK intracellular signaling pathways. Similarly, in a model that involved antigen-induced arthritis of the temporomandibular joint (TMJ) in rats, which was induced by injecting and emulsifying CFA and methylated serum albumin, Munoz-Lora et al. [66,67] found that the intra-TMJ injection of BoNT/A (7 U/Kg of Botox®, or 14 U/Kg of Dysport®) was able to reduce the P2X7/Cathepsin-S/Fractalkine microglia-activated pathway in the trigeminal subnucleus caudalis. Moreover, BoNT/A also reduced the protein level of IL-1β and TNF-α.

In all the studies presented thus far, the exact mechanism by which the peripherally injected BoNT/A reaches the spinal cord, where it may block both neuronal synaptic release and spinal glial activation, is not yet completely understood. As has been suggested in many studies that have used animal models [68], a direct central effect of the peripheral administration of BoNT/A is conceivable as a consequence of its retrograde transport along the axons of sensory neurons and its subsequent transcytosis to neuronal and non-neuronal spinal cells, where it can block both the release of neurotransmitters and the activation of spinal glia cells. It should be noted that, although the retrograde transport of the toxin can be evoked as a mechanism by which the peripheral toxin can reach the spinal cord in animal models, for obvious reasons, it is desirable that this does not happen in humans. The retrograde transport of the toxin from the peripheral injection site, which is an uncontrollable event, is an undesirable adverse effect, and in practical medicine, every effort is aimed at ensuring that this event does not occur. In light of this, it is unthinkable to consider the possibility of translating this type of mechanism, which concerns the toxin–glial cell interaction, for use in a clinical setting. Nonetheless, this problem could be circumvented by synthesizing new chimera toxins, which, if they are successfully designed to recognize specific receptors, may selectively target glial cells. In recent years, the development of engineered toxins has become the subject of intense research in the field of botulinum toxins, and it is desirable that this continues further [69–71].

5. Interaction between BoNTs and Myelin Forming Cells

Oligodendrocytes and SCs are the myelin-forming cells within CNS and PNS, respectively. A recent paper on the therapeutic potential of BoNT/A in counteracting paralysis and neuropathic pain, which used a model involving spinal cord injuries (SCI) in mice, analyzed the possible interaction between BoNT/A and oligodendrocytes [63]. Oligodendrocytes are highly susceptible to spinal cord damage that is induced by traumatic injury, and they easily undergo apoptosis as consequence of SCI [72]. Accordingly, Vacca et al. [63] observed a massive expression of the apoptotic marker Caspase 3, which is partially co-localized with oligodendrocytes, at the epicenter of the spinal impact. An intrathecal injection of BoNT/A (15 pg/mouse of 150 KDa purified BoNT/A in spinal cord) significantly reduced Caspase 3 expression, thus indicating protection against apoptotic processes. In parallel to this, the spared tissue was also abnormally myelinated, with a consequent reduction in axonal conduction. As a reactive response, to induce the healing of the injured spinal cord, the surviving oligodendrocytes produce a myelin basic protein (MBP) [73]. Vacca et al. [63] found an increased expression of MBP in saline-treated SCI mice, whereas in BoNT/A-treated SCI mice, the MBP was approximatively at the same level as the naïve mice, thus indicating that there was support for the reduced reactivity of oligodendrocytes due to a minor degree of degeneration.

In the last decade, convincing evidence that supports the interactive effect of BoNT/A on SCs proliferation has also been obtained, mainly in CCI neuropathic mice. Marinelli et al. [74] showed that the intraplantar injection of BoNT/A (15 pg of 150 KDa of purified BoNT/A/paw) in CCI mice prompts the functional recovery of injured hindlimbs, accelerates regenerative processes, and enhances the expression of proliferative cells in sciatic nerve tissue. In a subsequent study that was performed in naïve mice, the mice underwent an intraplantar injection of BoNT/A (15 pg of 150 KDa of purified BoNT/A/paw). Marinelli et al. [18] detected cl-SNAP-25 along the nociceptive pathway, which started at the hind paw skin and ended at the spinal cord. In particular, in the sciatic nerve, cl-SNAP-25 co-localized with a glial fibrillar acidic protein, which is a protein expressed in unmyelinated fibers and in dedifferentiated SCs after injury [75]. To determine if BoNT/A may directly interact with the proliferative state of SCs, Marinelli et al. [18] performed an in vitro experiment and found that the ACh released from SCs was reduced using BoNT/A, thus suggesting that BoNT/A may not only cleave SNAP-25 in neuronal cells, but also in SCs. In accordance with this finding, the presence of SNAP-25 in SCs has been demonstrated by Barden et al. [76] who found a receptor cluster located in SCs, which was composed by SNAP-25, together with other SNARE proteins, N-type calcium channels, and P2X receptors. The ACh release from SCs, which was reduced by BoNT/A, may affect the proliferation of SCs. With this in mind, it should be noted that low levels of ACh in the environment that surrounds the SCs, stimulates SC proliferation, whereas high levels of ACh arrests SC proliferation [77,78]. After CCI in vivo, the ACh that is released from the SCs, and from axons undergoing myelination, may be blocked by BoNT/A, thus causing a consequent reduction of ACh levels in the microenvironment that surrounds the regenerating nerve. This finding was confirmed by Cobianchi et al. [79], who observed that a single intranerve injection of BoNT/A (15 pg of 150 KDa of purified BoNT/A/paw) in mice stimulates the regeneration of myelin fibers and the speed of axonal elongation after an injury to the peripheral nerve; this occurs due to the activation and proliferation of SCs. Similar results were obtained by Seo et al. [80], who observed an increased proliferation of SCs after an intranerve injection of BoNT/A (7 U/Kg of Botox®) in the crushed sciatic nerve of rats. It is notable that, in contrast to what was observed with BoNT/A, BoNT/B reduces the proliferation of SCs; this could be explained by considering the different targets of the two BoNTs [64].

6. Interactions between BoNTs and Other Glial Cells

Regarding the SGCs, Da Silva et al. [81] demonstrated that trigeminal SGCs express the docking proteins, SNAP-25 and SNAP-23. Moreover, authors observed that a pre-treatment with different concentrations of BoNT/A (medium containing 0.1, 1, 10, and 100 pM of Botox®) inhibited, in a concentration-dependent manner, the ionomycin-evoked release of glutamate from cultured SGCs, thus indicating a possible direct interaction between BoNT/A and SCGs through the cleavage of SNAP-25 and/or SNAP-23, which would inhibit the vesicular release of glutamate. With this in mind, SGCs appear as potential targets for future therapeutic options in the treatment of chronic pain [82]. Another in vitro study on cultured trigeminal ganglions (TG) demonstrated a direct interaction between BoNT/A and sensory mechanisms [83]. Accordingly, the authors observed the expression of both SNAP-25 and BoNTs receptor SV2-A in TG. The expression of SNAP-25 ed SV2-A was also observed in sphenopalatine ganglions that were isolated from rats and human autopsies [84]. Interestingly, a mixed expression was observed between rats and humans. In rats, SV2-A was expressed in SGCs and SNAP-25 in neurons, whereas the opposite was observed in humans. The significance of the differential expressions between SNAP25 and SV2-A remain unclear. Among the other glial cells, namely, ependymal, radial, and enteric glial cells, no evidence of a possible interaction between BoNTs and these cells has been presented until now.

7. Summary

In this review, the most recent evidence that highlight a possible interaction between BoNTs, mainly BoNT/A, and glial cells was presented and discussed. Two main conclusions can be drawn from the cited articles, which might be considered as the strengths of this review. First, it is confirmed that BoNT/A is able to block the release of neuroactive substances, not only from neurons, but also from glial cells. This action can contribute, for example, to the analgesic activity of BoNT/A for chronic pain, during which the activation of microglia and astrocytes, secondary to the onset of painful phenomenology, contributes to the amplification and maintenance of chronic pain itself. Second, the interaction between BoNT/A and glial cells is not only confined to microglia and astrocytes. In fact, although there are still only a few studies on this matter, BoNT/A can interact with the glial cells that are responsible for the reconstruction of the myelin sheath of neural axons, such as SCs and oligodendrocytes, and thus the gradual restoration of the myelin function may occur even after it has been subjected to traumatic injury.

Although the strengths of this review are undeniable, at the same time, it has some weaknesses. First of all, unfortunately, there are still few publications on this topic. This is why the explanation of the results presented in this paper can sometimes appear incomplete and controversial. Secondly, since the effects of BoNTs on glial cells are very complex, and due to the diversity of the models used and the diversity of the glial cells themselves, it is impossible, at this stage, to formulate a unifying mechanism of action. Despite these weaknesses, this review still aims to establish a starting point for stimulating further research on the interaction between BoNTs and glial cells in order to elucidate the mechanism of action, which is still far from being defined.

8. Concluding Remarks and Future Perspectives

Although it still seems utopian to think of using BoNTs as agents that are capable of blocking the activation and release of pro-inflammatory substances from glial cells in humans, the door is open in terms of conducting more research on this topic. Certainly, this constitutes a field of investigation that is worth being continuously explored in order to clarify questions, mainly regarding the mechanism of action of BoNTs, which are still unsolved; for example, a fundamental contribution should come from a definitive clarification of the hypothesized retrograde transport of BoNTs. In particular, it would be useful to precisely clarify the following questions:

1. Are all the effects of peripheral BoNTs on central spinal glial cells solely due to direct interactions that occur after the retrograde transport of BoNTs?
2. How should this retrograde transport be controlled so that it does not cause adverse effects after the BoNTs are injected?
3. Which receptors on the membrane of the glial cell allow BoNTs to enter the glial cell and exert their proteolytic effect?
4. Is the mechanism of action of BoNTs inside the glial cells, as has been widely demonstrated at the neuronal level? Are there specific internal glial cell targets involved?

These are just some of the questions that future studies on this topic should answer. On a final note, it is worth considering that, once again, this fascinating molecule, the botulinum toxin, never ceases to amaze with its multiple applications. Given that the therapeutic use of this molecule began in the 70s, with the pioneering work of Dr. Alan B. Scott and colleagues, who applied BoNT/A in the ophthalmological field, there are no other molecules in nature that have shown such a variety of effects and therapeutic uses, both registered and non-registered. Surely, in the future, BoNTs will continue to surprise.

Funding: This research received no external funding.

Institutional Review Board Statement: Not applicable.

Informed Consent Statement: Not applicable.

Data Availability Statement: Not applicable.

Conflicts of Interest: The author declares no conflict of interest.

References

1. Montecucco, C.; Rasotto, M.B. On botulinum neurotoxin variability. *mBio* **2015**, *6*, e02131. [CrossRef]
2. Rossetto, O.; Pirazzini, M.; Montecucco, C. Botulinum neurotoxins: Genetic, structural and mechanistic insights. *Nat. Rev. Microbiol.* **2014**, *12*, 535–549. [CrossRef]
3. Peck, M.W.; Smith, T.J.; Anniballi, F.; Austin, J.W.; Bano, L.; Bradshaw, M.; Cuervo, P.; Cheng, L.W.; Derman, Y.; Dorner, B.G.; et al. Historical perspectives and guidelines for botulinum neurotoxin subtype nomenclature. *Toxins* **2017**, *9*, 38. [CrossRef]
4. Steward, L.; Brin, M.F.; Brideau-Andersen, A. Novel Native and Engineered Botulinum Neurotoxins. In *Botulinum Toxin Therapy*; Handbook of Experimental Pharmacology; Springer: Cham, Switzerland, 2021; Volume 263, pp. 63–89.

5. Rossetto, O.; Pirazzini, M.; Fabris, F.; Montecucco, C. Botulinum Neurotoxins: Mechanism of Action. *Handb. Exp. Pharm.* **2021**, *263*, 35–47.
6. Pirazzini, M.; Rossetto, O.; Eleopra, R.; Montecucco, C. Botulinum Neurotoxins: Biology, Pharmacology and Toxicology. *Pharm. Rev.* **2017**, *69*, 200–235. [CrossRef]
7. Pavone, F.; Luvisetto, S. Botulinum Neurotoxin for Pain Management: Insights from Animal Models. *Toxins* **2010**, *2*, 2890–2913. [CrossRef]
8. Yan, M.F.; Loh, Y.C.; Tan, C.S.; Khadijah Adam, S.; Abdul Manan, N.; Basir, R. General Pathways and the Major Nuerotransmitters Involved in Pain Regulation. *Int. J. Mol. Sci.* **2018**, *19*, 2164.
9. Park, J.; Park, H.J. Botulinum Toxin for the Treatement of Neuropathic Pain. *Toxins* **2017**, *9*, 260. [CrossRef]
10. Lackovic, Z. Botulinum Toxin and Pain. *Handb. Exp. Pharm.* **2021**, *263*, 251–264.
11. Attal, N. Pharmacological treatments of neuropathic pain: The latest recommendations. *Rev. Neurol.* **2019**, *175*, 46–50. [CrossRef]
12. Jeftinija, S.D.; Jeftinija, K.V.; Stefanovic, G. Cultured astrocytes express proteins involved in vesicular glutamate release. *Brain Res.* **1997**, *750*, 41–47. [CrossRef]
13. Verderio, C.; Coco, S.; Rossetto, O.; Montecucco, C.; Matteoli, M. Internalization and proteolytic action of botulinum toxins in CNS neurons and astrocytes. *J. Neurochem.* **1999**, *73*, 372–379. [CrossRef]
14. Araque, A.; Li, N.; Doyle, R.T.; Haydon, P.G. SNARE protein-dependent glutamate release from astrocytes. *J. Neurosci.* **2000**, *20*, 666–673. [CrossRef]
15. Yaguchi, T.; Nishizaki, T. Extracellular high K+ stimulates glutamate release from astrocytes by activating voltage-dependent calcium channels. *J. Cell Physiol.* **2010**, *225*, 512–518. [CrossRef]
16. Abdipranoto, A.; Liu, G.J.; Werry, E.L.; Bennett, M.R. Mechanisms of secretion of ATP from cortical astrocytes triggered by uridine triphosphate. *Neuroreport* **2003**, *14*, 2177–2178. [CrossRef]
17. Liu, G.J.; Werry, E.L.; Bennett, M.R. Secretion of ATP from Schwann cells in response to uridine triphosphate. *Eur. J. Neurosci.* **2005**, *21*, 151–160. [CrossRef]
18. Marinelli, S.; Vacca, V.; Ricordy, R.; Uggenti, C.; Tata, A.M.; Luvisetto, S.; Pavone, F. The analgesic effect on neuropathic pain of retrogradely transported botulinum neurotoxin A involves Schwann cells and astrocytes. *PLoS ONE* **2012**, *7*, e47977. [CrossRef]
19. Allen, N.J.; Lyons, D.A. Glia as architects of central nervous system formation and function. *Science* **2018**, *362*, 181–185. [CrossRef]
20. Barateiro, A.; Brites, D.; Fernandes, A. Oligodendrocyte Development and Myelination in Neurodevelopment: Molecular Mechanisms in Health and Disease. *Curr. Pharm. Des.* **2016**, *22*, 656–679. [CrossRef]
21. Zhou, B.; Zhu, Z.; Ransom, B.R.; Tong, X. Oligodendrocyte lineage cells and depression. *Mol. Psychiatry* **2021**, *26*, 103–117. [CrossRef]
22. Yamazaki, Y. Oligodendrocyte Physiology Modulating Axonal Excitability and Nerve Conduction. *Adv. Exp. Med. Biol.* **2019**, *1190*, 123–144.
23. Del Bigio, M.R. Ependymal cells: Biology and pathology. *Acta Neuropathol.* **2010**, *119*, 55–73. [CrossRef]
24. Moreno-Manzano, V. Ependymal cells in the spinal cord as neuronal progenitors. *Curr. Opin. Pharm.* **2020**, *50*, 82–87. [CrossRef]
25. Seo, J.S.; Mantas, J.; Svenningsson, P.; Greengard, P. Ependymal cells-CSF flow regulates stress-induced depression. *Mol. Psychiatry* **2021**, *26*, 7308–7315. [CrossRef]
26. Barry, D.S.; Pakan, J.M.; McDermott, K.W. Radial glial cells: Key organisers in CNS development. *Int. J. Biochem. Cell Biol.* **2014**, *46*, 76–79. [CrossRef]
27. Kriegstein, A.; Alvarez-Buylla, A. The glial nature of embryonic and adult neural stem cells. *Annu. Rev. Neurosci.* **2009**, *32*, 149–184. [CrossRef]
28. Sofronien, M.V.; Vinters, H.V. Astrocytes: Biology and pathology. *Acta Neuropathol.* **2010**, *119*, 7–35. [CrossRef]
29. Araque, A. Astrocytes process synaptic information. *Neuron Glia Biol.* **2008**, *4*, 3–10. [CrossRef]
30. Abbott, N.J.; Rönnbäck, L.; Hansson, E. Astrocyte-endothelial interactions at the blood-brain barrier. *Nat. Rev. Neurosci.* **2006**, *7*, 41–53. [CrossRef]
31. Salzer, J.L. Schwann cell meylination. *Cold Spring Harb. Perspect. Biol.* **2015**, *7*, a020529. [CrossRef]
32. Jessen, K.R.; Mirsky, R.; Lloyd, A.C. Schwann Cells: Development and Role in Nerve Repair. *Cold Spring Harb. Perspect. Biol.* **2015**, *7*, a020487. [CrossRef] [PubMed]
33. Ohara, P.T.; Vit, J.P.; Bhargava, A.; Romero, M.; Sundberg, C.; Charles, A.C.; Jasmin, L. Gliopathic pain: When satellite glial cells go bad. *Neuroscientist* **2009**, *15*, 450–463. [CrossRef]
34. Hanani, M.; Verkhrastky, A. Satellite Glial Cells and Astrocytes, a Comparative Review. *Neurochem. Res.* **2021**, *46*, 2525–2537. [CrossRef] [PubMed]
35. Hanani, M. How Is Peripheral Injury Signaled to Satellite Glial Cells in Sensory Ganglia? *Cells* **2022**, *11*, 512. [CrossRef]
36. Huang, L.Y.; Gu, Y.; Chen, Y. Communication between neuronal somata and satellite glial cells in sensory ganglia. *Glia* **2013**, *61*, 1571–1581. [CrossRef] [PubMed]
37. Ochoa-Cortes, F.; Turco, F.; Linan-Rico, A.; Soghomonyan, S.; Whitaker, E.; Wehner, S.; Cuomo, R.; Christofi, F.L. Enteric Glial Cells: A New Frontier in Neurogastroenterology and Clinical Target for Inflammatory Bowel Diseases. *Inflamm. Bowel Dis.* **2016**, *22*, 433–449. [CrossRef]
38. Seguella, L.; Gulbransen, B.D. Enteric glial biology, intercellular signalling and roles in gastrointestinal disease. *Nat. Rev. Gastroenterol. Hepatol.* **2021**, *18*, 571–587. [CrossRef]

39. Sharkey, K.A. Emerging roles for enteric glia in gastrointestinal disorders. *J. Clin. Investig.* **2015**, *125*, 918–925. [CrossRef]
40. Prinz, M.; Jung, S.; Priller, J. Microglia Biology: One Century of Evolving Concepts. *Cell* **2019**, *179*, 292–311. [CrossRef]
41. Masuda, T.; Sankowski, R.; Staszewski, O.; Prinz, M. Microglia Heterogeneity in the Single-Cell Era. *Cell Rep.* **2020**, *30*, 1271–1281. [CrossRef]
42. Cserép, C.; Pósfai, B.; Dénes, Á. Shaping Neuronal Fate: Functional Heterogeneity of Direct Microglia-Neuron Interactions. *Neuron* **2021**, *109*, 222–240. [CrossRef] [PubMed]
43. Borst, K.; Dumas, A.A.; Prinz, M. Microglia: Immune and non-immune functions. *Immunity* **2021**, *54*, 2194–2208. [CrossRef] [PubMed]
44. Paolicelli, R.C.; Ferretti, M.T. Function and Dysfunction of Microglia during Brain Development: Consequences for Synapses and Neural Circuits. *Front. Synaptic Neurosci.* **2017**, *9*, 9. [CrossRef] [PubMed]
45. Watkins, L.R.; Milligan, E.D.; Maier, S.F. Glial activation: A driving force for pathological pain. *Trends Neurosci.* **2001**, *24*, 450–455. [CrossRef]
46. Austin, P.J.; Moalem-Taylor, G. The neuro-immune balance in neuropathic pain: Involvement of inflammatory immune cells, immune-like glial cells and cytokines. *J. Neuroimmunol.* **2010**, *229*, 26–50. [CrossRef]
47. Mika, J.; Zychowska, M.; Popiolek-Barczyk, K.; Rojewska, E.; Przewlocka, B. Importance of glial activation in neuropathic pain. *Eur. J. Pharmacol.* **2013**, *716*, 106–119. [CrossRef]
48. Tsuda, M. Microglia in the spinal cord and neuropathic pain. *J. Diabetes Investig.* **2016**, *7*, 17–26. [CrossRef]
49. Sommer, C.; Schäfers, M. Mechanisms of neuropathic pain: The role of cytokines. *Drug Discov. Today Dis. Mech.* **2004**, *1*, 441–448. [CrossRef]
50. Kiguchi, N.; Kobayashi, Y.; Kishioka, S. Chemokines and cytokines in neuroinflammation leading to neuropathic pain. *Curr. Opin. Pharmacol.* **2012**, *12*, 55–61. [CrossRef] [PubMed]
51. Rojewska, E.; Piotrowska, A.; Popiolek-Barczyk, K.; Mika, J. Botulinum Toxin Type A-A Modulator of Spinal Neuron-Glia Interactions under Neuropathic Pain Conditions. *Toxins* **2018**, *10*, 145. [CrossRef]
52. Martinez, F.O.; Gordon, S. The M1 and M2 paradigm of macrophage activation: Time for reassessment. *F1000Prime Rep.* **2014**, *6*, 13. [CrossRef] [PubMed]
53. Kanazawa, M.; Ninomiya, I.; Hatakeyama, M.; Takahashi, T.; Shimohata, T. Microglia and Monocytes/Macrophages Polarization Reveal Novel Therapeutic Mechanism against Stroke. *Int. J. Mol. Sci.* **2017**, *18*, 2135. [CrossRef] [PubMed]
54. Gui, X.; Wang, H.; Wu, L.; Tian, S.; Wang, X.; Zheng, H.; Wu, W. Botulinum toxin type A promotes M2 polarization and suppresses chronic constriction injury-induced neuropathic pain through the P2X7 receptor. *Cell Biosci.* **2020**, *10*, 45. [CrossRef] [PubMed]
55. Wang, X.; Tian, S.; Wang, H.; Liu, P.; Zheng, H.; Wu, L.; Liu, Q.; Wu, W. Botulinum toxin type A alleviates neuropathic pain and suppresses inflammatory cytokines release from microglia by targeting TLR2/MyD88 and SNAP23. *Cell Biosci.* **2020**, *10*, 141. [CrossRef]
56. Chen, W.J.; Niu, J.Q.; Chen, Y.T.; Deng, W.J.; Xu, Y.Y.; Liu, J.; Luo, W.F.; Liu, T. Unilateral facial injectionn of Botulinum neurotoxin A attenuates bilateral trigeminal neuropathic pain and anxiety-like behaviors through inhibition of TLR2-mediated neuroinflammation in mice. *J. Headache Pain* **2021**, *22*, 38. [CrossRef]
57. Piotrowska, A.; Popiolek-Barczyk, K.; Pavone, F.; Mika, J. Comparison of the Expression Changes after Botulinum Toxin Type A and Minocycline Administration in Lipopolysaccharide-Stimulated Rat Microglial and Astroglial Cultures. *Front. Cell. Infect. Microbiol.* **2017**, *7*, 141. [CrossRef]
58. Hepp, R.; Perraut, M.; Chasserot-Golaz, S.; Galli, T.; Aunis, D.; Langley, K.; Grant, N.J. Cultured glial cells express the SNAP-25 analogue SNAP-23. *Glia* **1999**, *27*, 181–187. [CrossRef]
59. Vaidyanathan, V.V.; Yoshino, K.; Jahnz, M.; Dörries, C.; Bade, S.; Nauenburg, S.; Niemann, H.; Binz, T. Proteolysis of SNAP-25 isoforms by botulinum neurotoxin types A, C, and E: Domains and amino acid residues controlling the formation of enzyme-substrate complexes and cleavage. *J. Neurochem.* **1999**, *72*, 327–337. [CrossRef]
60. Yu, Z.; Liu, J.; Sun, L.; Wang, Y.; Meng, H. Combination of Botulinum Toxin and Minocycline Ameliorates Neuropathic Pain through Antioxidant Stress and Anti-inflammation via Promoting SIRT1 Pathway. *Front. Pharm.* **2021**, *11*, 602417. [CrossRef]
61. Feng, X.; Xiong, D.; Li, J.; Xiao, L.; Xie, W.; Qiu, Y. Direct Inhibition of Microglia Activation by Pretreatment with Botulinum Neurotoxin A for the Prevention of Neuropathic Pain. *Front. Neurosci.* **2021**, *15*, 760403. [CrossRef]
62. Vacca, V.; Marinelli, S.; Luvisetto, S.; Pavone, F. Botulinum toxin A increases analgesic effects of morphine, counters development of morphine tolerance and modulates glia activation and μ opioid receptor expression in neuropathic mice. *Brain Behav. Immun.* **2013**, *32*, 40–50. [CrossRef] [PubMed]
63. Vacca, V.; Madaro, L.; De Angelis, F.; Proietti, D.; Cobianchi, S.; Orsini, T.; Puri, P.L.; Luvisetto, S.; Pavone, F.; Marinelli, S. Revealing the therapeutic potential of Botulinum neurotoxin type A in counteracting paralysis and neuropathic pain in spinally injured mice. *Toxins* **2020**, *12*, 491. [CrossRef] [PubMed]
64. Finocchiaro, A.; Marinelli, S.; De Angelis, F.; Vacca, V.; Luvisetto, S.; Pavone, F. Botulinum Toxin B Affects Neuropathic Pain but Not Functional Recovery after Peripheral Nerve Injury in a Mouse Model. *Toxins* **2018**, *10*, 128. [CrossRef] [PubMed]
65. Shi, X.; Gao, C.; Wang, L.; Chu, X.; Shi, Q.; Yang, H.; Li, T. Botulinum toxin type A ameliorates adjuvant-arthritis pain by inhibiting microglial activation-mediated neuroinflammation and intracellular molecular signaling. *Toxicon* **2020**, *178*, 33–40. [CrossRef]

66. Munoz-Lora, V.R.M.; Abdalla, H.B.; Del Bel Cury, A.A.; Clemente-Napimoga, J.T. Modulatory effect of botulinum toxin type A on the microglia P2X7/CatS/FKN activated-pathway in antigen-induced arthritis of the temporomandibular joint of rats. *Toxicon* **2020**, *187*, 116–121. [CrossRef]
67. Munoz-Lora, V.R.M.; Dugonjic Okrosa, A.; Matak, I.; Del Bel Cury, A.A.; Kalinichev, M.; Lackovic, Z. Antinociceptive Actions of Botulinum Toxin A1 on Immunogenic Hypersensitivity in Temporomandibular Joint of Rats. *Toxins* **2022**, *14*, 161. [CrossRef]
68. Luvisetto, S. Botulinum Neurotoxins in Central Nervous System: An Overview from Animal Models to Human Therapy. *Toxins* **2021**, *13*, 751. [CrossRef]
69. Ferrari, E.; Gu, C.; Niranjan, D.; Restani, L.; Rasetti-Escargueil, C.; Obara, I.; Geranton, S.M.; Arsenault, J.; Goetze, T.A.; Harper, C.B.; et al. Synthetic Self-Assembling Clostridial Chimera for Modulation of Sensory Functions. *Bioconjug. Chem.* **2013**, *24*, 1750–1759. [CrossRef]
70. Webb, R.P. Engineering of Botulinum Neurotoxins for Biomedical Applications. *Toxins* **2018**, *10*, 231. [CrossRef]
71. Fonfria, E.; Elliott, M.; Beard, M.; Chaddock, J.A.; Krupp, J. Engineering Botulinum Toxins to Improve and Expand Targeting and SNARE Cleavage Activity. *Toxins* **2018**, *10*, 278. [CrossRef]
72. Li, N.; Leung, G.K.K. Oligodendrocyte precursor cells in spinal cord injury: A review and update. *BioMed Res. Int.* **2015**, *2015*, 235195. [CrossRef] [PubMed]
73. Bartholdi, D.; Schwab, M.E. Oligodendroglial reaction following spinal cord injury in rat: Transient upregulation of MBP mRNA. *Glia* **1998**, *23*, 278–284. [CrossRef]
74. Marinelli, S.; Luvisetto, S.; Cobianchi, S.; Makuch, W.; Obara, I.; Mezzaroma, E.; Caruso, M.; Straface, E.; Przewlocka, B.; Pavone, F. Botulinum neurotoxin type A counteracts neuropathic pain and facilitates fucntional recovery after peripheral nerve injury in animal models. *Neuroscience* **2010**, *171*, 316–328. [CrossRef] [PubMed]
75. Jessen, K.R.; Mirsky, R. The origin and development of glial cells in peripheral nerves. *Nat. Rev. Neurosci.* **2005**, *6*, 671–682. [CrossRef]
76. Barden, J.A.; Cottee, L.J.; Bennett, M.R. Vesicle-associated proteins and P2X receptor clusters at single sympathetic varicosities in mouse vas deferens. *J. Neurocytol.* **1999**, *28*, 469–480. [CrossRef]
77. Loreti, S.; Ricordy, R.; De Stefano, M.E.; Augusto-Tocco, G.; Tata, M.A. Acetylcholine inhibits cell cycle progression in rat Schwann cells by activation of the M2 receptor subtype. *Neuron Glia Biol.* **2007**, *3*, 269–279. [CrossRef]
78. Uggenti, C.; De Stefano, M.E.; Costantino, M.; Loreti, S.; Pisano, A.; Avallone, B.; Talora, C.; Magnaghi, V.; Tata, A.M. M2 muscarinic receptor activation regulates Schwann cell differentiation and myelin organization. *Dev. Neurobiol.* **2014**, *74*, 676–691. [CrossRef]
79. Cobianchi, S.; Jaramillo, J.; Luvisetto, S.; Pavone, F.; Navarro, X. Botulinum neurotoxin A promotes functional recovery after peripheral nerve injury by increasing regeneration of myelinated fibers. *Neuroscience* **2017**, *359*, 82–91. [CrossRef]
80. Seo, M.; Lim, D.; Kim, S.; Kim, T.; Kwon, B.S.; Nam, K. Effect of Botulinum Injection and Extracorporeal Shock Wave Therapy on Nerve Regeneration in Rats with Experimentally Induced Sciatic Nerve Injury. *Toxins* **2021**, *13*, 879. [CrossRef]
81. da Silva, L.B.; Poulsen, J.N.; Arendt-Nielsen, L.; Gazerani, P. Botulinum neurotoxin type A modulates vesicular release of glutamate from satellite glial cells. *J. Cell Mol. Med.* **2015**, *19*, 1900–1909. [CrossRef]
82. Gazerani, P. Satellite Glial Cells in Pain Research: A Targeted Viewpoint of Potential and Future Directions. *Front. Pain Res.* **2021**, *2*, 646068. [CrossRef] [PubMed]
83. Edvinson, J.; Warfvinge, K.; Edvinsson, L. Modulation of inflammatory mediators in the trigeminal ganglion by botulinum neurotoxin type A: An organ culture study. *J. Headache Pain* **2015**, *16*, 73. [CrossRef] [PubMed]
84. Steinberg, A.; Frederiksen, S.D.; Blixt, F.W.; Warfvinge, K.; Edvinsson, L. Expression of messenger molecules and receptors in rat and human sphenopalatine ganglion indicating therapeutic targets. *J. Headache Pain* **2016**, *17*, 78. [CrossRef] [PubMed]

toxins

Article

Frequency of Hemorrhagic Side Effects of Botulinum Neurotoxin Treatment in Patients with Blepharospasm and Hemifacial Spasm on Antithrombotic Medication

Fiona Carolin Wenninger [1,2] and Bettina Wabbels [1,*]

[1] Department of Ophthalmology, University Hospital of Bonn, Ernst-Abbe-Str. 2, D-53127 Bonn, Germany
[2] Department of Neurology, University Hospital of Münster, Albert-Schweitzer-Campus 1, D-48149 Münster, Germany
* Correspondence: bettina.wabbels@ukbonn.de; Tel.: +49-(0)228-287-15612; Fax: +49-(0)228-287-14692

Abstract: The aim of this study was to investigate the frequency of hemorrhagic side effects of botulinum neurotoxin A injections (BoNT/A) for the treatment of benign essential blepharospasm (BEB) and hemifacial spasm (HFS) in patients taking antithrombotic drugs (ATD). A total of 140 patients were included (female: 65%; BEB: 75%; mean age: 70 ± 12 years). According to their current antithrombotic medication, participants were either assigned to the ATD group (41%), or to the control group (59%). The ATD group was further divided into subgroups depending on the medication administered: acetylsalicylic acid, ADP receptor antagonists, direct oral anticoagulants, vitamin-K antagonists, or dual antiplatelet therapy. The frequency of hemorrhagic side effects was recorded by retrospective analysis of past treatments as documented in the patient's file set in relation to the number of past treatments (hematoma frequency of past treatments, HF_{retro}) as well as by a prospective survey capturing the side effects of one single treatment (hematoma frequency of actual treatment, HF_{actual}). There was no significant difference in hematoma frequency between the ATD group and the control group, neither for past (HF_{retro}: ATD: 2%; 45/2554; control: 4%; 109/2744) nor for the current BoNT/A treatments (HF_{actual}: ATD: 30%; 16/53; control: 31%; 22/72). Even between ATD subgroups, hematoma frequency did not differ significantly. Overall, hemorrhagic side effects of the BoNT/A treatment for BEB and HFS were mild and non-disabling.

Keywords: hematoma; botulinum neurotoxin; blepharospasm; hemifacial spasm; antithrombotics; anticoagulants

Key Contribution: Botulinum neurotoxin A injections for the treatment of benign essential blepharospasm and hemifacial spasm are not associated with more frequent hemorrhagic side effects in patients on antithrombotic drugs; regardless of the agent; than in patients without these drugs.

Citation: Wenninger, F.C.; Wabbels, B. Frequency of Hemorrhagic Side Effects of Botulinum Neurotoxin Treatment in Patients with Blepharospasm and Hemifacial Spasm on Antithrombotic Medication. *Toxins* **2022**, *14*, 769. https://doi.org/10.3390/toxins14110769

Received: 27 September 2022
Accepted: 2 November 2022
Published: 7 November 2022

1. Introduction

Benign essential blepharospasm (BEB), a focal dystonia, is characterized by increased blinking due to overactivity of the orbicularis oculi muscle. In the course of time the involuntary muscle contractions may be prolonged, causing functional blindness [1–4], and may spread to neighboring muscles (Meiges syndrome) [5,6]. Up until now the underlying cause of the disease remains unknown. Depending on the geographical region, the prevalence of BEB varies between 1.7 and 13.3 per 100,000 [7,8]. Women are predominantly found among those affected. The first symptoms usually appear between the fifth and seventh decade of life [9,10].

Hemifacial spasm (HFS) is characterized by mostly unilateral, involuntary contractions of the facial muscles innervated by the facial nerve. These can appear as slight muscle tremors, significant twitching or long-lasting muscle spasms of individual facial muscle groups or half of the face [11–14]. HFS usually occurs due to compression of the facial nerve [15] caused by neighboring arterial vessels of the brainstem and cerebellum as well as masses, cysts, or bony abnormalities [16–18]. According to Rosenstengel et al. (2012), approximately 8000–9000 people in Germany suffer from HFS [19], corresponding to a prevalence of 9.9–11.8 per 100,000 inhabitants. Women are particularly affected [19,20], as are people over the age of 40 years [20,21].

The unsatisfactory effects of numerous oral drugs [13,22] and a significantly higher complication rate of surgical interventions [1,13] make regular injections of BoNT/A the therapy of first choice for BEB and HSF [22,23]. For this purpose, three different preparations of BoNT/A are currently available in Germany: onabotulinum toxin (Botox®, Allergan Pharmaceuticals, Dublin, Ireland), incobotulinum toxin (Xeomin®, Merz Pharmaceuticals GmbH, Frankfurt/Main, Germany) and abobotulinum toxin (Dysport®, Ipsen Pharma, Paris, France). To treat BEB and HFS, a combined subcutaneous-intramuscular injection of the toxin in affected muscles is carried out. Since the paralytic effect is only temporary, the treatment must be repeated at regular intervals of usually eight to twelve weeks [24].

Compared to the high therapeutic effect, adverse effects of BoNT/A treatment occur rarely, are mostly mild, and are only of short duration [13,25]. This also applies to subcutaneous bleeding and hematomas around the injection sites. However, there is concern that patients on antithrombotic medication are at higher risk of subcutaneous bleeding and hematoma. Antithrombotic therapy is used to prevent or treat thrombosis in patients in a variety of clinical situations of different cardiovascular conditions [26], and includes two major classes of drugs: anticoagulants and platelet aggregation inhibitors. Antiplatelet agents prevent or delay the activation and aggregation of platelets and include acetylsalicylic acid (ASA) and ADP receptor antagonists (ADP-RA) [27,28], while a combination of platelet aggregation inhibitor drug groups is referred to as dual antiplatelet therapy (DAPT). Anticoagulants prevent or delay the formation of a red thrombus through activated coagulation factors and include vitamin K antagonists (VKA) and direct oral anticoagulants (DOAC).

According to the manufacturers of vitamin K antagonists (VKA), the subcutaneous-intramuscular injection of BoNT/A in patients undergoing therapy with these preparations is contraindicated "due to the risk of massive bleeding" (direction for use Marcumar® 3 mg tablets, MEDA Pharma GmbH & Co. KG, Bad Homburg, Germany, 2018; direction for use Phenpro.-ratiopharm® 3 mg tablets, ratiopharm GmbH, Ulm, Germany 2018). The respective directions for use of other antithrombotics do not include this contraindication explicitly but include a warning about hematoma. However, due to lifestyle and demographic changes with an aging population, the number of patients with antithrombotic medication is expected to increase. At the same time, the number of studies investigating the risk of bleeding after BoNT/A injection in patients on antithrombotic medication is limited to date. The aim of this study was to assess the frequency and severity of hemorrhagic side effects of BoNT/A treatment for BEB and HFS in patients on antithrombotic medication. Special attention was paid to the medical relevance of these adverse effects and their influence on the physical and psychological well-being of the affected patients.

2. Results

2.1. Demographic Data

A total of 140 patients were included in the study between May and November 2019. Overall, the mean (±SD) age of the participants was 70 ± 12 [37;91] years, with 65% (n = 91) being women. Most participants underwent BoNT/A treatment for BEB (75%, 105/140). In median, participants had already been treated at the University Eye Clinic Bonn for 9 [0;33] years. The number of past BoNT/A treatments, including the current treatment, varied between two and 155 treatments, with a median of 35 treatments per study participant. For the current treatment, 70% (73/105) of the participants suffering from BEB were treated with Xeomin®, whereas Botox® was mostly used to treat HFS (69%, 24/35). The median total dose of the BoNT/A preparation administered for the treatment of BEB was 35.0 [7.5;75.0] units of Botox® or 30.0 [5.0;85.0] units of Xeomin®.

A median of 12.5 [7.5;30.0] units of Botox® and 20.0 [7.5;65.0] units of Xeomin® were administered for the treatment of HFS. Antithrombotic medication was taken by 41% (58/140) of the participants.

Table 1 shows demographic and treatment characteristics of patients in the control group and the ATD group. Apart from age ($p < 0.01$) and the applied BoNT/A preparation ($p = 0.04$), there were no significant differences between groups. The most frequently used antithrombotic drug was ASA (52%, 30/58). Almost every fourth participant on antithrombotic medication claimed to be on DOAC (24%, 14/58). The regular intake of VKA, ADP-RA or DAPT was much less common. Table 2 shows demographic and treatment characteristics of patients in the ATD group subgroups. Except for gender distribution ($p = 0.01$), there was no significant difference between the subgroups of the ATD group.

Table 1. Comparison of demographic and therapy-related characteristics of the participants in the control and ATD group.

Characteristics		Control Group (n = 82)	ATD Group (n = 58)
Number of participants	n (%)	82 (59)	58 (41)
Age (years)	mean ± SD	66 ± 11	74 ± 10
Female	n (%)	58 (71)	33 (57)
BEB	n (%)	60 (73)	45 (78)
Xeomin®	n (%)	55 (67)	29 (50)
Number of treatments	median [R]	29 [2;155]	41 [2;137]

ATD, antithrombotic drug; BEB, benign essential blepharospasm.

Table 2. Comparison of demographic and therapy-related characteristics of participants taking antithrombotic medication according to subgroup.

Characteristics		ASA	DOAC	VKA	ADP-RA	DAPT
Number of participants	n (%)	30 (52)	14 (24)	8 (14)	4 (7)	2 (4)
Age (years)	mean ± SD	73 ± 10	77 ± 9	73 ± 16	82 ± 5	73 ± 5
Female	n(%)	14 (47)	12 (59)	3 (38)	4 (100)	0 (0)
BEB	n(%)	22 (73)	11 (79)	7 (88)	3 (75)	2 (100)
Xeomin®	n(%)	16 (67)	6 (43)	4 (40)	2 (50)	1 (50)
Number of treatments	median [R]	41 [2;107]	46 [2;120]	29 [2;107]	25 [17;77]	6 [4;8]
Treatment duration (years)	median [R]	12 [0;33]	14 [0;24]	13 [2;32]	5 [4;26]	2 [2;2]

ASA, acetylsalicylic acid; ADP-RA, adenosine diphosphate receptor antagonists; DAPT, dual antiplatelet therapy; DOAC, direct oral anticoagulants; VKA, vitamin K antagonists.

2.2. Hematoma Frequency

More than one-third of patients had at least once experienced a hematoma after treatment with BoNT/A in the past (HF_{retro}) in both the control and the ATD group (control: 39%, 31/80; ATD: 38%, 22/58). Adjusting for the number of past BoNT/A treatments, HF_{retro} averaged 3% (154/5298, [0.0;100]) of treatments, with a mean of 4% (109/2744), [0.0;100] within the control group and 2% (45/2554, [0.0:33.3] in the ATD group (Table 3). The files of two participants in the control group were incomplete regarding the history of side effects and were not included in the analysis.

Table 3. HF_{retro} and HF_{actual} based on the number BoNT/A treatments (T) and the number of following hematoma events (H) in control and ATD group as well as ADT subgroups.

Group		Past Treatments				Current Treatment		
		T	H	HF_{retro}		T	H	HF_{actual}
	n (%)	n	n	Mean %	[R %]	n (%)	n	(%)
Control	80 (58) [a]	2744	109	4	[0.0;100]	72 (58) [b]	22	31
ATD	58 (42)	2554	45	2	[0.0;33.3]	53 (42)	16	30
ASA	30 (52)	1231	24	2	[0.0;6.0]	27 (51) [b]	11	41
DOAC	14 (24)	730	7	1	[0.0;6.0]	13 (25) [b]	1	8
VKA	8 (14)	443	13	0	[0.0;14.5]	7 (13) [b]	1	14
ADP-RA	4 (7)	140	0	0	[0.0;0.0]	4 (8)	2	50
DAPT	2 (7)	10	1	0	[0.0;33.3]	2 (4)	1	50
total	138 (100) [a]	5298	154	3	[0.0;100]	125 (100)	38	30

[a] Due to missing information, two participants of the control group were not included in the determination of the absolute number of past hematoma events and HF_{retro}. [b] 10 participants of the control group, three participants on ASA, and one each on DOAC or VKA did not return the questionnaire and were therefore not included in the determination HF_{actual}. ASA, acetylsalicylic acid; ADP-RA, adenosine diphosphate receptor antagonists; DAPT, dual antiplatelet therapy; DOAC, direct oral anticoagulants; VKA, vitamin K antagonists.

Overall, the proportion of patients with hematoma after current BoNT/A treatment (HF_{actual}) was 30.4% (38/125). Participants in both the control and ATD group were equally affected by hematoma after current BoNT/A treatment (HF_{actual}: control: 31%, 22/72; ATD: 30%, 16/53). Within the subgroups ADP-RA and DAPT, half of all participants suffered a hematoma after current BoNT/A injection (ADP-RA: 50%, 2/4; DAPT: 50%, 1/2), whereas for the participants who regularly took a DOAC the HF_{actual} was 8% (1/13). Despite the evidence of these tendencies, a significant connection between the intake of certain antithrombotic drugs and the occurrence of a hematoma after current treatment could not be demonstrated (Table 3). A total of 15 participants (control: 10; ATD: 5) did not complete the questionnaire on current BoNT/A treatment and were not included in the analysis.

2.3. *Hematoma Intensity*

The majority of participants (42%, 15/36) reported that the hematoma occurred immediately after treatment or on the same day (47%, 17/36). Only 11% (4/36) of the patients with hematoma, all of whom belonged to the ATD group, stated that the hematoma appeared one or more days after the BoNT/A injection. The hematomas were visible for a mean period (±SD) of 9.9 ± 7.4 [3.0;33.0] days. There was no significant difference between the control and ATD group or between the ATD subgroups regarding the duration of the hematoma. Two participants in the control group did not provide information on the duration of the hematoma and were therefore not included in this analysis.

More than half of the participants stated that the hematoma was punctiform and ≤1.5 cm in diameter (57%, 21/37). The other participants reported more pronounced hematomas, however mostly ≤2.5 cm in diameter (24%, 9/37). The size of the hematoma drawn in a facial image varied between two and 800 mm^2 per participant, with a mean (±SD) of 118 ± 203 mm^2. There was no significant difference between the control and ATD group or between the ATD subgroups regarding the size of the hematoma based on drawings (Figure 1). One participant in the control group did not provide information on the size of the hematoma and was therefore not included in this analysis.

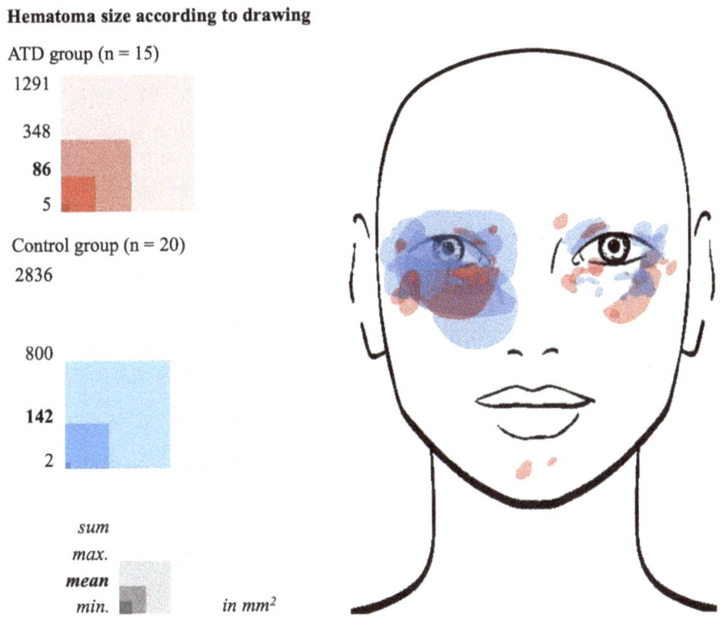

Figure 1. Depiction of hematomas drawn in a facial image after current BoNT/A injection according to their size and localization. Overview of mean, maximum and minimum of total hematoma area per participant and sum of total hematoma area per group in comparison of the control and ATD groups.

Overall, 58 individual hematomas were drawn by 35 participants. Accordingly, a mean of 1.6 ± 0.8 [1;4] hematomas occurred per affected participant. The number of hematomas following current BoNT/A injection differed significantly (p = 0.02) between control and ATD groups. Participants in the control group reported a mean of 1.4 ± 0.7 [1;3] individual hematomas, while participants in the ATD group suffered from a mean of 2.0 ±0.9 [1;4] hematomas. However, no significant difference in the number of hematomas after current BoNT/A treatment was observed between ATD subgroups.

2.4. Hematoma Consequence

On a visual analogue scale (VAS) the mean (±SD) impairment caused by the hematoma was reported as 1.4 ± 2.2 [0.0;7.6], with most of the participants (49%, 16/33) feeling completely unaffected (VAS: 0). There was no significant difference between the ATD group (VAS: 1.4 ± 2.3 [0.0;7.0]) and the control group (VAS: 1.4 ± 2.2 [0;7.6.0]) or between ATD subgroups. Cosmetic reasons (21%, 7/33), pain, or a feeling of pressure (9%, 3/33) were given as reasons for the impairment caused by the hematoma. To deal with the disturbing side effects, 27% (10/37) of those affected took further measures such as cooling affected areas (22%, 8/37). Five participants (control: 4; ATD: 1) did not provide information about impairment and were therefore not included in this analysis.

2.5. Non-Hemorrhaghic Side Effects

Regarding past BoNT/A treatments in 84% (118/138) of patients records other side effects than hematoma were noted. The most common were ptosis (34%, 47/139), lacrimation (33%, 46/139), and unspecified visual disturbances (27%, 38/139). For two patients, this information was not available.

Regarding the current BoNT/A treatment, most participants (60%, 74/124) reported no side effects apart from hematoma. The most common included tearing (6%, 7/124) and burning, pain, and dry eye sensation (4% each, 5/124).

3. Discussion

Our study shows that BoNT/A injections for the treatment of BEB and HFS are not associated with more frequent hemorrhagic side effects in patients on ATD, regardless of the agent, than in patients without—both in the retrospective analysis of past BoNT/A treatments and in the prospective survey on current BoNT/A injections. Except the number of single hematomas per patient, there is also no significant difference in the severity of hematomas occurring after BoNT/A treatment between patients on ATD, regardless of the agent, and patients without.

However, the retrospectively determined hematoma frequency of past BoNT/A treatments (HF_{retro}) was significantly lower than of the current BoNT/A injection (HF_{actual})—both in the overall cohort of participants ($p < 0.01$) and within the control ($p < 0.01$) and ATD group ($p < 0.01$).

Regarding past BoNT/A treatments (HF_{retro}), overall, a hematoma frequency of 3% was determined. Various studies on the effectiveness and safety of different BoNT/A preparations for the treatment of BEB and HFS revealed comparable results (Table 4). Bentivoglio et al. (2009) retrospectively determined a hematoma frequency of 3.2% (43/1341) after treatment of BEB [29]. Jankovic and colleagues (2011) also observed a comparable hematoma frequency in patients with HFS [30]. In contrast, the hematoma frequency of 25.0% (16/64), determined by Wabbels et al. (2010) was considerably higher and corresponds to HFS_{actual} (30%) in our study [31].

Table 4. Overview of the hematoma frequency (HF) determined in past studies from the number of observed treatments (T) and subsequently among the patients (P) In descending order.

Subject of Study	P [n]	T [n]	H [n]	HF [%]	Methods [a]	Reference
Hematoma frequency after current treatment of BEB and HFS (our study)	125	125	38	30	questionnaire ≤14 days explicit	(HF_{actual})
Hematoma frequency after past treatments of BEB and HFS (our study)	138	5298	154	3	patient's record following session open	(HF_{retro})
Efficacy and safety of BOTOX® versus Xeomin® for the treatment of BEB	64	64	16	25,0	Interview ≥4 weeks open	Wabbels et al., 2010 [31]
Comparison of the effectiveness of preseptal and pretarsal injections of BOTOX® for the treatment of BEB and HFS	40	80	4	5,0	Report ≥1 month open	Lolekha et al., 2017 [32]
Hematoma frequency of BoNT/A treatment of BEB or HFS taking phenprocoumon versus control group	28	437	19	4,4	patient's record open	Schrader et al., 2018 [33]
Clinic features and treatment options of BEB including treatment with BoNT/A	151	-	≥ 5 [b]	≥3,3 [c]	patient's record open	Grandas et al., 1988 [1]
Efficacy and safety of BOTOX® versus Dysport® in the treatment of BEB	-	1341	43	3,2	patient's record open	Bentivoglio et al., 2009 [29]
Efficacy and safety of long-term BoNT/A treatment of BEB	234	10,632	340	3,1	patient's record following session open	Wabbels et al., 2022 [34]
Efficacy and safety of treating HFS with Xeomin® versus placebo	108	108	3	2,8	-	Jankovic et al., 2011 [30]
Efficacy and safety of BoNT/A treatment of BEB and HFS	83	241	6	2,5	report open	Jankovic et al., 1990 [35]
Efficacy and safety of BoNT/A treatment of BEB and HFS	112	212	≥2 [b]	≥1,8 [c]	report ≥3–6 days open	Park et al., 1993 [36]

Table 4. *Cont.*

Subject of Study	P [n]	T [n]	H [n]	HF [%]	Methods [a]	Reference
Efficacy and safety of BoNT/A treatment of BEB and HFS	106	1028	10	1,0	questionnaire following session open	Hsiung et al., 2002 [37]
Efficacy and safety of BoNT/A treatment of BEB and HFS	131	920	≥7 [b]	≥0,8 [c]	patient's record open	Cillino et al., 2010 [38]
Efficacy and safety of BoNT/A treatment of BEB and HFS	32	1421	≥2 [b]	≥0,1 [c]	patient's record following session open	Ababneh et al., 2014 [39]
Efficacy and safety of treating HFS with BoNT/A versus placebo	288	10,701	0	0,0	patient's record open	Kollewe et al., 2015 [40]

[a] Data source, time of data collection after BoNT/A treatment and questioning technique used (open/explicit). [b] Number of participants with at least one hematoma after BoNT/A treatment. [c] Missing information on the number of hematoma events, therefore it is not possible to determine the exact hematoma frequency. BEB, benign essential blepharospasm; BoNT/A, botulinumtoxin A; HFS, hemifacial spasm.

The heterogeneous data on frequency of hematoma after BoNT/A treatment within our study and within the literature (Table 4) could be explained by different study designs. Presumably, small, only briefly visible and non-burdening hematomas are often forgotten and therefore only remembered when asked explicitly and in time, as in the case of determination of HF_{actual}. In contrast to this, determination of the HF_{retro} as well as the hematoma frequency in numerous other studies was carried out by an open-ended questioning about side effects of past BoNT/A treatments without particular attention to hemorrhagic complications. In addition, the data collection usually took place several weeks after the treatment in question. It can therefore be assumed that, both in determination of the HF_{retro} in our study and the hematoma frequency in comparable studies that followed a retrospective study design, fewer hematomas were recorded than actually occurred.

Given the large difference between HF_{actual} and Hf_{retro}, we consider it unlikely that the higher hematoma frequency of current BoNT/A treatments was randomly generated by case number differences (current treatment: n = 125; past treatments: n = 5298). Since participants received the current BoNT/A injection by different physicians, a physician-related increased hematoma frequency can also be ruled out.

To date, very few studies investigated the hematoma frequency after BoNT/A treatment in patients on antithrombiotic medication. Schrader and colleagues (2018) retrospectively determined the hematoma frequency of BoNT/A treatment for BEB, HFS, cervical dystonia and stroke-related spasticity in patients on phenprocoumon. After a total of 231 and 206 BoNT/A treatments for BEB and HFS, respectively, there was no significant difference in hematoma frequency between patients on phenprocoumon and their matched controls (BEB: 5.2% vs. 2.6%; HFS: 3.9% vs. 2.9%). These results largely correspond to the HF_{retro} of the control and VKA group in our study. In contrast, the HF_{actual} of both groups in our study is considerably higher. The comparison of these results confirms the already discussed assumption that an open-ended and delayed questioning of patients, as performed when determining the hematoma frequency in the study of Schrader et al. and the HF_{retro} in our study, underestimates the real hematoma frequency.

Furthermore, Jagatsinh and George (2012) investigated the safety of different BoNT/A preparations for the treatment of spastic disorders on warfarin [41]. After a total of 103 intramuscular injections, none of the 14 participants registered hemorrhagic complications. The hematoma frequency to be calculated would therefore be 0%, and corresponds to the HF_{retro} of the VKA group in our study. Again, the HF_{actual} of the VKA group determined in our study is considerably higher than the hematoma frequency determined by Jagatsinh and George. However, in view of the small number of participants on phenprocoumon in our study, the comparison of the results is only possible to a very limited extent. In addition, the comparability of the study results might be limited by the different pharmacological properties of the VKA warfarin and phenprocoumon [24,42–44].

There are no studies on the frequency of hematomas after BoNT/A treatment on other antithrombotics such as DOAC, ADP-RA or ASA. However, according to the direction for use for the preparations in question, hematomas or bleeding generally after medical interventions and injections or punctures occur "rarely" to "frequently" ($\geq$1/10,000 to <1/10) depending on the respective preparation. This corresponds to a hematoma frequency between $\geq$0.01% and <10.0%, and is therefore comparable to the HF_{retro} determined in our study to be between 0% (ADP-RA) and 2% (ASA).

4. Conclusions

In view of the results of our study, pausing antithrombotic medication with ASA, VKA or DOAC in the context of BoNT/A treatment for BEB and HFS does not seem justified. However, since only a small number of patients on ADP-RA and DAPT were included in our study, subsequent studies are necessary to be able to make a recommendation for these agents. For future studies, it should be considered that timing and questioning technique may lead to significant differences in the reported frequency and description of hemorrhagic side effects.

Overall, our study showed that hemorrhagic side effects of the BoNT/A treatment for BEB and HFS are mild and non-disabling.

5. Materials and Methods

The study was approved by the local ethics committee of the "Rheinische Friedrichs-Wilhelms-Universität Bonn" and has been performed in accordance with the ethical standards of the Declaration of Helsinki and its later amendments. All participants provided a signed informed consent form for participation in this study. Participation did not affect the patient's medical care. Antithrombotic treatment was continued as prescribed and therefore was not influenced by the study.

Patients with BEB or HFS undergoing regular treatment with BoNT/A at the University Eye Clinic in Bonn were consecutively included in the study. According to their medication, the participants were assigned to the following study groups: the control group (no intake of antithrombotic drugs), the ATD group (regular intake of antithrombotic drugs), and their subgroups regarding the antithrombotic agent (ASA, ADP-RA, DAPT, DOAC or VKA).

The frequency of hematomas caused by BoNT/A treatment for BEB and HFS was recorded for two separate periods: in a retrospective analysis of past BoNT/A treatments (HF_{retro}), and in a prospective survey on one single BoNT/A treatment (HF_{actual}).

The HF_{retro} was calculated for each participant and each study group from the absolute number of past BoNT/A treatments that caused a hematoma in relation to the total number of BoNT/A treatments in our clinic. Information on adverse events including hematoma following past BoNT/A treatments was collected from patient records. Adverse events are queried in a standardized manner at every follow-up visit in our clinic (about 8–12 weeks after the respective treatment) and documented in the patient file. The patient records were also used to collect data on demographics, diagnosis (BEB or HFS), specifics of the BoNT/A therapy performed in our eye clinic (duration, number of injections administered, type of BoNT/A), comorbidities, and medication.

The HF_{actual} was calculated for each study group from the number of participants who suffered a hematoma after a single, currently performed BoNT/A treatment in relation to the total number of participants in the group. The mentioned BoNT/A treatment was performed by different physicians according to the patient's current treatment regimen. In order to ensure that the injection was carried out as usual and without special care, the attending physician was not informed about the patient's participation in the study. Post-treatment, participants were given a questionnaire to assess the occurrence, size, location, onset, and duration of hematoma, as well as the occurrence of non-hemorrhagic adverse events after this single BoNT/A injection. To assess the hematoma size in an objective way, participants were asked to compare the hematoma to a 1 cent coin (about

1.5 cm in diameter, 207 mm^2), a 2 euro coin (about 2.5 cm in diameter, 520 mm^2) or a quarter of the face. In addition, the area of hematomas drawn in the facial image was calculated by creating representative polygons using the SketchAndClac software (Dobbs, Elliott M. "www.SketchAndCalc.com" (accessed on 12th November 2021). SketchAndCalc. Elliott M Dobbs, 20 February 2011. Web, version 4.1.8.11.). A visual analogue scale (VAS) ranging from 0 (no impairment) to 10 (worst impairment) was used to assess how much the hematoma bothered the patients. Patients were asked to complete the questionnaire within two weeks of the treatment in question and to return it to our clinic.

Statistical analysis of the pseudonymized data was performed using the statistical program IBM SPSS Statistics, version 26 (IBM Corp., Armonk, NY, USA). Results were expressed as mean, standard deviation, minimum, median, quartiles, and maximum. For determining statistical significance between groups, the Mann-Whitney-Test (control vs. ATD group) and the Kruskall-Wallis-Test (between ATD subgroups) were performed ($p < 0.05$ considered statistically significant).

Author Contributions: Conception and design of the study, B.W.; data acquisition, F.C.W.; statistical analysis and interpretation of the data, F.C.W. and B.W.; draft of the paper and critical revision, F.C.W. and B.W. All authors have read and agreed to the published version of the manuscript.

Funding: This research received no external funding.

Institutional Review Board Statement: The study has been performed in accordance with the ethical standards of the Declaration of Helsinki and its later amendments, was approved by the local ethics committee of the "Rheinische Friedrichs-Wilhelms-Universität Bonn"(Approved code: 127/19; approved date: 11 June 2019), and all participants provided a signed informed consent form for participation in this study.

Informed Consent Statement: Informed consent was obtained from all subjects involved in the study.

Data Availability Statement: Data and materials can be provided by the authors upon reasonable request. The original data is included in paper files.

Acknowledgments: The authors would like to thank Eva Stetzer for assistance with the preparation of the manuscript.

Conflicts of Interest: F.C.W. has no relevant financial or non-financial interests to disclose. B.W. received travel grants and speaker honoraria from Desitin and Merz and research funding from Allergan and Merz.

References

1. Grandas, F.; Elston, J.; Quinn, N.; Marsden, C.D. Blepharospasm: A review of 264 patients. *J. Neurol. Neurosurg. Psychiatry* **1988**, *51*, 767–772. [CrossRef] [PubMed]
2. Jankovic, J.; Havins, W.E.; Wilkins, R.B. Blinking and Blepharospasm. *JAMA* **1982**, *248*, 3160–3164. [CrossRef] [PubMed]
3. Jankovic, J.; Orman, J. Blepharospasm: Demographic and clinical survey of 250 patients. *Ann. Ophthalmol.* **1984**, *16*, 371–376.
4. Evalls-Sole, J.; Edefazio, G. Blepharospasm: Update on Epidemiology, Clinical Aspects, and Pathophysiology. *Front. Neurol.* **2016**, *7*, 45. [CrossRef]
5. Abbruzzese, G.; Berardelli, A.; Girlanda, P.; Marchese, R.; Martino, D.; Morgante, F.; Avanzino, L.; Colosimo, C.; Defazio, G. Long-term assessment of the risk of spread in primary late-onset focal dystonia. *J. Neurol. Neurosurg. Psychiatry* **2008**, *79*, 392–396. [CrossRef] [PubMed]
6. Ba, E.M.W.; Hershey, T.; Karimi, M.; Racette, B.; Tabbal, S.D.; Mink, J.W.; Paniello, R.C.; Perlmutter, J.S. Relative risk of spread of symptoms among the focal onset primary dystonias. *Mov. Disord.* **2006**, *21*, 1175–1181. [CrossRef]
7. Defazio, G.; Abbruzzese, G.; Livrea, P.; Berardelli, A. Epidemiology of primary dystonia. *Lancet Neurol.* **2004**, *3*, 673–678. [CrossRef]
8. DeFazio, G.; Berardelli, A.; Hallett, M. Do primary adult-onset focal dystonias share aetiological factors? *Brain* **2007**, *130*, 1183–1193. [CrossRef]
9. Defazio, G.; Hallett, M.; Jinnah, H.A.; Conte, A.; Berardelli, A. Blepharospasm 40 Years Later. *Mov. Disord.* **2017**, *32*, 498–509. [CrossRef]
10. The Epidemiological Study of Dystonia in Europe (ESDE) Collaborative Group. A prevalence study of primary dystonia in eight European countries. *J. Neurol.* **2000**, *247*, 787–792. [CrossRef]

11. Digre, K.; Corbett, J.J. Hemifacial spasm: Differential diagnosis, mechanism, and treatment. *Adv. Neurol.* **1988**, *49*, 151–176. [PubMed]
12. Tan, E.K.; Jankovic, J. Bilateral hemifacial spasm: A report of five cases and a literature review. *Mov. Disord.* **1999**, *14*, 345–349. [CrossRef]
13. Wang, A.; Jankovic, J. Hemifacial spasm: Clinical findings and treatment. *Muscle Nerve* **1998**, *21*, 1740–1747. [CrossRef]
14. Wartenberg, R. Hemifacial spasm. *Trans. Am. Neurol. Assoc.* **1950**, *51*, 276–277. [PubMed]
15. Nielsen, V.K. Electrophysiology of the facial nerve in hemifacial spasm: Ectopic/ephaptic excitation. *Muscle Nerve* **1985**, *8*, 545–555. [CrossRef] [PubMed]
16. Adler, C.H.; Zimmerman, R.A.; Savino, P.J.; Bernardi, B.; Bosley, T.M.; Sergott, R.C. Hemifacial spasm: Evaluation by magnetic resonance imaging and magnetic resonance tomographic angiography. *Ann. Neurol.* **1992**, *32*, 502–506. [CrossRef]
17. Gálvez-Jiménez, N.; Hanson, M.R.; Desai, M. Unusual causes of hemifacial spasm. *Semin. Neurol.* **2001**, *21*, 75–83. [CrossRef]
18. Jannetta, P.J.; Abbasy, M.; Maroon, J.C.; Ramos, F.M.; Albin, M.S. Etiology and definitive microsurgical treatment of hemifacial spasm. Operative techniques and results in 47 patients. *J. Neurosurg.* **1977**, *47*, 321–328. [CrossRef]
19. Rosenstengel, C.; Matthes, M.; Baldauf, J.; Fleck, S.; Schroeder, H. Hemifacial spasm: Conservative and surgical treatment options. *Dtsch. Arztebl. Int.* **2012**, *109*, 667–673. [CrossRef]
20. Auger, R.G.; Whisnant, J.P. Hemifacial Spasm in Rochester and Olmsted County, Minnesota, 1960 to 1984. *Arch. Neurol.* **1990**, *47*, 1233–1234. [CrossRef]
21. Nilsen, B.; Le, K.-D.; Dietrichs, E. Prevalence of hemifacial spasm in Oslo, Norway. *Neurology* **2004**, *63*, 1532–1533. [CrossRef] [PubMed]
22. Albanese, A.; Barnes, M.P.; Bhatia, K.P.; Fernandez-Alvarez, E.; Filippini, G.; Gasser, T.; Krauss, J.K.; Newton, A.; Rektor, I.; Savoiardo, M.; et al. A systematic review on the diagnosis and treatment of primary (idiopathic) dystonia and dystonia plus syndromes: Report of an EFNS/MDS-ES Task Force. *Eur. J. Neurol.* **2006**, *13*, 433–444. [CrossRef] [PubMed]
23. Kenney, C.; Jankovic, J. Botulinum toxin in the treatment of blepharospasm and hemifacial spasm. *J. Neural Transm.* **2007**, *115*, 585–591. [CrossRef] [PubMed]
24. Kaltenmaier, M.; Vanselow, K.; Rollnik, J.; Maschke, M. Therapie des essentiellen Blepharospasmus mit Botulinumtoxin. *Fortschritte Neurol. Psychiatr.* **2019**, *87*, 355–360. [CrossRef]
25. Naumann, M.; Jankovic, J. Safety of botulinum toxin type A: A systematic review and meta-analysis. *Curr. Med. Res. Opin.* **2004**, *20*, 981–990. [CrossRef]
26. Favaloro, E.J. Anticoagulant and antithrombotic therapy: Globally applied according to local geographical selection criiteria. *Ann. Blood.* **2020**, *5*, 22. [CrossRef]
27. Antithrombotic Trialists' (ATT) Collaboration; Baigent, C.; Blackwell, L.; Collins, R.; Emberson, J.; Godwin, J.; Peto, R.; Buring, J.; Hennekens, C.; Kearney, P.; et al. Aspirin in the primary and secondary prevention of vascular disease: Collaborative meta-analysis of individual participant data from randomised trials. *Lancet* **2009**, *373*, 1849–1860. [CrossRef]
28. Paikin, J.S.; Eikelboom, J.W. Aspirin. *Circulation* **2012**, *125*, e439-42. [CrossRef]
29. Bentivoglio, A.R.; Fasano, A.; Ialongo, T.; Soleti, F.; Fermo, S.L.; Albanese, A. Fifteen-Year Experience in Treating Blepharospasm with Botox or Dysport: Same Toxin, Two Drugs. *Neurotox. Res.* **2009**, *15*, 224–231. [CrossRef]
30. Jankovic, J.; Comella, C.; Hanschmann, A.; Grafe, S. Efficacy and safety of incobotulinumtoxinA (NT 201, Xeomin) in the treatment of blepharospasm-A randomized trial. *Mov. Disord.* **2011**, *26*, 1521–1528. [CrossRef]
31. Wabbels, B.; Reichel, G.; Fulford-Smith, A.; Wright, N.; Roggenkämper, P. Double-blind, randomised, parallel group pilot study comparing two botulinum toxin type A products for the treatment of blepharospasm. *J. Neural Transm.* **2011**, *118*, 233–239. [CrossRef] [PubMed]
32. Schrader, C.; Ebke, M.; Saberi, F.A.; Dressler, D. Botulinum toxin therapy in patients with oral anticoagulation: Is it safe? *J. Neural Transm.* **2017**, *125*, 173–176. [CrossRef] [PubMed]
33. Jagatsinh, Y.; George, J. Audit of Safety of Intramuscular Botulinum Toxin Injections among Patients Receiving Warfarin Anticoagulation Therapy. *Indian J. Phys. Med. Rehabil.* **2012**, *23*, 101–104. [CrossRef]
34. Wabbels, B.; Fimmers, R.; Roggenkämper, P. Ultra-Long-Term Therapy of Benign Essential Blepharospasm with Botulinumtoxin A—30 Years of Experience in a Tertiary Care Center. *Toxins* **2022**, *14*, 120. [CrossRef] [PubMed]
35. Jankovic, J.; Schwartz, K.; Donovan, D.T. Botulinum toxin treatment of cranial-cervical dystonia, spasmodic dysphonia, other focal dystonias and hemifacial spasm. *J. Neurol. Neurosurg. Psychiatry* **1990**, *53*, 633–639. [CrossRef]
36. Park, Y.C.; Lim, J.K.; Lee, D.K.; Yi, S.D. Botulinum a toxin treatment of hemifacial spasm and blepharospasm. *J. Korean Med. Sci.* **1993**, *8*, 334–340. [CrossRef]
37. Hsiung, G.-Y.; Das Mbbs, S.; Ranawaya, R.; Lafontaine, A.-L.; Suchowersky, O. Long-term efficacy of botulinum toxin A in treatment of various movement disorders over a 10-year period. *Mov. Disord.* **2002**, *17*, 1288–1293. [CrossRef]
38. Cillino, S.; Raimondi, G.; Guépratte, N.; Damiani, S.; Cillino, M.; Di Pace, F.; Casuccio, A. Long-term efficacy of botulinum toxin A for treatment of blepharospasm, hemifacial spasm, and spastic entropion: A multicentre study using two drug-dose escalation indexes. *Eye* **2009**, *24*, 600–607. [CrossRef]
39. Ababneh, O.H.; Cetinkaya, A.; Kulwin, D.R. Long-term efficacy and safety of botulinum toxin A injections to treat blepharospasm and hemifacial spasm. *Clin. Exp. Ophthalmol.* **2013**, *42*, 254–261. [CrossRef]

40. Kollewe, K.; Mohammadi, B.; Köhler, S.; Pickenbrock, H.; Dengler, R.; Dressler, D. Blepharospasm: Long-term treatment with either Botox®, Xeomin® or Dysport®. *J. Neural Transm.* **2015**, *122*, 427–431. [CrossRef]
41. Kelly, J.G.; O'Malley, K. Clinical Pharmacokinetics of Oral Anticoagulants. *Clin. Pharmacokinet.* **1979**, *4*, 1–15. Available online: http://link.springer.com/10.2165/00003088-197904010-00001 (accessed on 3 June 2019). [CrossRef] [PubMed]
42. Palareti, G.; Legnani, C. Warfarin Withdrawal. *Clin. Pharmacokinet.* **1996**, *30*, 300–313. [CrossRef] [PubMed]
43. Ufer, M. Comparative Pharmacokinetics of Vitamin K Antagonists. *Clin. Pharmacokinet.* **2005**, *44*, 1227–1246. [CrossRef] [PubMed]
44. Lolekha, P.; Choolam, A.; Kulkantrakorn, K. A comparative crossover study on the treatment of hemifacial spasm and blepharospasm: Preseptal and pretarsal botulinum toxin injection techniques. *Neurol. Sci.* **2017**, *38*, 2031–2036. [CrossRef]

toxins

MDPI

Article

Prediction Model for Identifying Computational Phenotypes of Children with Cerebral Palsy Needing Neurotoxin Treatments

Carlo M. Bertoncelli [1,2,3,*], Michal Latalski [4], Domenico Bertoncelli [1,3], Sikha Bagui [1], Subhash C. Bagui [1], Dechelle Gautier [2] and Federico Solla [2]

1. Department of Computer Science, Hal Marcus College of Science & Engineering, University of West Florida, Pensacola, FL 32514, USA
2. EEAP H Germain and Department of Pediatric Orthopaedic Surgery, Lenval Foundation, University Pediatric Hospital of Nice, 06000 Nice, France
3. Department of Information Engineering Computer Science and Mathematics, Computer Science and Mathematics, University of L'Aquila, 67100 L'Aquila, Italy
4. Children Orthopaedic Department, Medical University, 20-059 Lublin, Poland

* Correspondence: carlo.bertoncelli@univ-cotedazur.fr

Abstract: Factors associated with neurotoxin treatments in children with cerebral palsy (CP) are poorly studied. We developed and externally validated a prediction model to identify the prognostic phenotype of children with CP who require neurotoxin injections. We conducted a longitudinal, international, multicenter, double-blind descriptive study of 165 children with CP (mean age 16.5 $\pm$ 1.2 years, range 12–18 years) with and without neurotoxin treatments. We collected functional and clinical data from 2005 to 2020, entered them into the BTX-PredictMed machine-learning model, and followed the guidelines, "Transparent Reporting of a Multivariable Prediction Model for Individual Prognosis or Diagnosis". In the univariate analysis, neuromuscular scoliosis ($p = 0.0014$), equines foot ($p < 0.001$) and type of etiology (prenatal > peri/postnatal causes, $p = 0.05$) were linked with neurotoxin treatments. In the multivariate analysis, upper limbs ($p < 0.001$) and trunk muscle tone disorders ($p = 0.02$), the presence of spasticity ($p = 0.01$), dystonia ($p = 0.004$), and hip dysplasia ($p = 0.005$) were strongly associated with neurotoxin injections; and the average accuracy, sensitivity, and specificity was 75%. These results have helped us identify, with good accuracy, the clinical features of prognostic phenotypes of subjects likely to require neurotoxin injections.

Keywords: prediction model; neurotoxin treatment; cerebral palsy

Key Contribution: BTX-PredictMed identified clinical features of prognostic phenotypes of subjects likely to require neurotoxin injections; the clinical features being neuromuscular scoliosis, equines foot, type of etiology, upper limbs, trunk muscle tone disorders, the presence of spasticity, dystonia, hip dysplasia.

Citation: Bertoncelli, C.M.; Latalski, M.; Bertoncelli, D.; Bagui, S.; Bagui, S.C.; Gautier, D.; Solla, F. Prediction Model for Identifying Computational Phenotypes of Children with Cerebral Palsy Needing Neurotoxin Treatments. *Toxins* **2023**, *15*, 20. https://doi.org/10.3390/toxins15010020

Received: 3 November 2022
Revised: 19 December 2022
Accepted: 20 December 2022
Published: 28 December 2022

1. Introduction

Cerebral palsy (CP) comprises a group of non-progressive motor control and posture disorders due to brain damage during the early stages of development. Clinical manifestations include involuntary movements or gait abnormalities, movement alterations (loss of tone or spasticity of the trunk and limbs with exaggerated reflexes), and abnormal posture [1]. The progression of dynamic contracture into fixed contracture is an issue of paramount importance for the effective use of botulinum toxins.

Muscle hyperactivity can be effectively reduced by injecting botulinum toxins [2]. Over the past 25 years, botulinum toxins have emerged as the most widely used medical intervention in children with CP. Botulinum toxins reduce muscle strength and tone, with a small, short-term improvement in walking and function. The lack of knowledge on pathophysiology and mechanisms leading from hypertonia to contractures explains the complexity of CP. In addition, little is known about the most commonly used treatment, botulinum toxin (BTX) [2].

Electronic Medical Record (EMR) data are very useful in identifying health outcomes. They contain rich clinical information, including laboratory test results, vital signs, discharge summaries, progress notes, and radiologic and pathologic images and reports, among other information.

Computational phenotyping is the creation of computer-processable algorithms to identify individuals with specific health conditions, diseases, or clinical events from EMR data [3,4].

Large-scale EMRs are an obvious data source for clinical phenotype discovery research. However, EMRs are designed primarily for clinical care, and some effort is required to adapt them as a data source for research. EMRs have been employed in clinical [5,6] and genomic [7,8] research using expert domain knowledge to devise phenotype specifications that identify clinical cohorts of interest manually.

One of the key changes needed to achieve precision and personalized medicine is to let the data speak for themselves, to tell us what the phenotypes are, abandoning the use of historical clinical descriptions of each disease. This view is supported in recent studies, indicating that long-recognized diseases such as spasticity in CP are not single entities but collections of many different phenotypes that may or may not coincide with historical disease boundaries [2].

The analysis of EMR with machine-learning methods could help to clarify these issues.

Machine learning (ML) is a contemporary artificial intelligence discipline for analyzing complex data. ML employs algorithms to find patterns in data that are not obvious to humans. Regression and logistic regression (LR) are among the first supervised ML algorithms for creating predictive health models [9–11]. ML algorithms are considered supervised if the output classes are labeled (e.g., BTX treatment, yes/no). Thus, supervised ML prediction models can help identify patients who will undergo BTX treatment.

Prediction models can predict the probability (Prob) of a condition (e.g., BTX treatment) being present [12–14]. In supervised ML algorithms, the output is obtained from labeled training samples. Through the training examples, the program learns a function (e.g., logistic regression) that will predict new incoming patients with unknown conditions.

In previous studies, we developed [1] and validated [15] "PredictMed," a supervised ML model to predict neuromuscular scoliosis and hip dysplasia [10,16], gastrostomy placement [17], and identify factors associated with intellectual disability [18] and autism spectrum disorder [19] in individuals with CP. PredictMed has also proven effective in predicting osteoarthritis in young adults using statistical data mining and machine learning [9].

In the present study, we implemented and externally validated the BTX-PredictMed ML model to predict prognostic phenotypes of children with CP needing BTX treatment.

Following the development of a prediction model, external validation is strongly recommended, that is, to evaluate the model's performance on other participant data not used for model development [13,20–22]. External validation requires that predictions about outcomes be made using the original model for each subject in the new dataset, and be compared with observed outcomes [23]. In this study, we applied the same predictive model to patients from different centers and countries to evaluate the model's performance through external validation.

From a clinical perspective, a reliable predictive model to identify the phenotype of children with CP who need BTX treatment would allow healthcare providers to recruit and schedule patients more efficiently, thereby reducing medical costs.

From a research perspective, this validated predictive model is easily adaptable and could be used in different fields of medicine.

2. Results

We developed and externally validated BTX-PredictMed, a statistical machine-learning model to identify factors associated with neurotoxin treatments for children with CP from two European centers.

This is the first study highlighting the influence of neuromuscular scoliosis, truncal tone disorders, and type of etiology as features constituting the prognostic phenotypes of CP children needing BTX treatment. This study also confirmed previous results [2,24–29] about spasticity, equines foot, hip dysplasia, and manual ability as clinical features predicting the need for BTX treatment for this population.

In univariate analysis, the factors linked with neurotoxins treatments were:

- Neuromuscular scoliosis: $p = 0.0013$, Odds ratio (OR) = 2.7;
- Equines foot: $p < 0.001$, OR = 4.1;
- Type of etiology: prenatal > peri/postnatal causes, $p = 0.05$, OR = 0.53.

Factors linked with neurotoxins treatments in multivariate analysis were:

- Upper limbs ability, $p < 0.001$, OR = 3;
- Trunk muscle tone disorders, $p = 0.02$, OR = 1.9;
- The presence of spasticity, $p = 0.01$, OR = 2;
- Dystonia, $p = 0.004$, OR = 5.3;
- Hip dysplasia, $p = 0.005$, OR = 4.

The multivariate analysis had an accuracy of 76%, sensitivity of 67%, specificity of 81%, and an average of 75%.

3. Discussion

The present study has shown that children with CP with equines foot, hip dysplasia, and dystonia were four to five times more likely to undergo neurotoxin treatments compared to a similar group that lacked these clinical features. Low levels of MACS (OR = 3), the presence of spasticity (OR = 2), neuromuscular scoliosis (OR = 2.9), and truncal tone disorders (OR = 1.9), are also strong predictors of phenotypes of CP children needing BTX treatment.

On the opposite side, prenatal etiology appears to be less related to BTX injections than perinatal or postnatal (OR < 1).

As expected, spastic foot is the prognostic phenotype's main feature.

Since the nineties, botulinum treatment of spastic equines foot has been recommended [30–32]. In young children with CP and high GMFCS, multilevel injection of BTX can be used for focal treatment of spasticity, particularly for the lower extremity [30].

Injection and distal injection were significantly related to a more significant gain in gross motor function in the younger age group [30]. The functional hindrance of the spastic equines foot is often the initial obstacle noted by parents and specialists.

BTX injections into the gastrocnemius and soleus muscles have been effective regardless of the duration of treatment and the number of sessions. However, cerebral palsy, the patient's age, and the impairment level [33] influence this efficacy. In CP children with upper and lower limb spasticity, there is increasing evidence of the time-limited beneficial effect of BTX in decreasing muscle tone. Decreased muscle tone in the lower limbs may translate into improved ambulation in children with CP with spastic equino-varus [30].

We have noticed that children with CP and lower MACS scores (better manual skills) will be more likely to undergo neurotoxin treatment. We, therefore, assume that casts, orthoses and/or orthopedic surgery, should be preferred in case of severe joint deformity related to strong, long-lasting spasticity [33].

Despite the association of BTX and hip dysplasia being classic and widely recommended in the literature [28,29,34], high-quality evidence of the prevention of hip displacement is lacking. From a physiopathologic mechanism perspective, BTX treatment is recommended [27] in patients with initial subluxation or with strong spasticity and concomitant risk factors of dislocation before radiologic evidence of subluxation [16,17]; the high odds ratio confirms this trend.

We noted that dystonia was strongly associated with BTX treatment, even if this practice is sparsely recommended in the literature [27]. Unfortunately, our data do not allow increasing evidence pro or contra to this indication. We plan further research on this topic in the future.

There is a pronounced trend toward BTX treatment in CP children with spasticity [2,25–29]. The present study specifies the phenotype subtypes: about half of the patients with diplegia received BTX, and roughly one-third of patients with hemiplegia, triplegia, or quadriplegia.

Despite this, we have highlighted the link between BTX treatments, neuromuscular scoliosis, and truncal tone disorder. BTX is poorly described for neuromuscular scoliosis treatment [25], and data support its inefficiency [26]. With regard to truncal tone disorder, this subject is nearly absent from the literature.

According to the present research, prenatal etiology appears to be less associated with BTX than perinatal or postnatal. Since the prediction model scored a low odds ratio and there are no studies on this, future research is needed to confirm this finding.

Patients with focal/segmental disabling spasticity are ideal candidates for BTX treatment, and there is a growing need for the early selection of suitable candidates [35]. An algorithm has been implemented to support managing adult patients with disabling spasticity by aiding patient selection for BTX treatments [35]. Identifying the prognostic phenotype through BTX-PredictMed could be a reliable support for identifying and selecting younger patients.

Similar results in two separate groups and the totality of patients confirm the validity of the prediction model, which can also be applied to other fields of medical research.

The availability of a robust predictive algorithm would allow healthcare providers to better understand, prevent and manage orthopedic deformations during growth by delaying or avoiding fixed contractures. It would also facilitate early consultation with a qualified neurotoxin specialist, improving the overall quality of care for patients and families. The personalization of therapies would also lead to a reduction in costs.

3.1. Tolerance and Precautions

Side effects were rare and mostly transient: cramps, pain, or hematoma at the injection site, rashes, and pseudo-influenza syndromes. We have not noted any contraindications, such as amyotrophic lateral sclerosis or major respiratory occlusion. Other problems, such as myasthenia gravis, Lambert-Eaton syndrome, or amyotrophic lateral sclerosis, have not been found either. Adherence to good practice recommendations, such as adherence to low doses in the first injection and delays between injections, remains the best prevention [24].

3.2. Limitations

Limitations of this study include the limited number of patients and the retrospective analysis. This resulted in high specificity and accuracy of diagnosis, while sensitivity was moderate.

The study of the predictive performance of the algorithm by increasing the number of patients and independent variables (>15) will be our next issue.

For the current study, we had a limited cohort of patients to study (hundreds). We plan to study and fine-tune the model on a much larger number (thousands) to confirm and possibly improve the model's predictive performance. In this regard, PredictMed has already achieved excellent results in predicting osteoarthritis in adults using statistical data mining and machine learning [9]. At this stage, we plan to calibrate this model on a much larger database to test its potential overfitting (e.g., by studying a receiver operating characteristic curve) due to the limited number of patients and with respect to a large number of features and independent variables [10]. We also plan to use Lasso (L1) and

Ridge (L2) regularization techniques to improve PredictMed model and implement clinical decision support systems, a tool supporting professionals in making medical decisions.

4. Materials and Methods

4.1. Study Design

This longitudinal, multicenter, multinational study was conducted between June 2005 and June 2021. For model implementation, data collection and assessments were conducted in the last six months of 2017, while data analysis began in June 2018 and lasted 24 months.

External validation followed the guidelines of the "Transparent Reporting of a multivariable prediction model for Individual Prognosis or Diagnosis" (TRIPOD) Statement [13–15].

We compared two groups of CP children with and without BTX treatment in a double-blind study. These CP children, treated in specialized units, had severe motor disorders and cognitive impairment. The development data showed no differences in setting, eligibility criteria, outcome, and predictors.

The flow diagram of study participants for analysis is shown in Figure 1. All consented to be enrolled in the study (Figure 1).

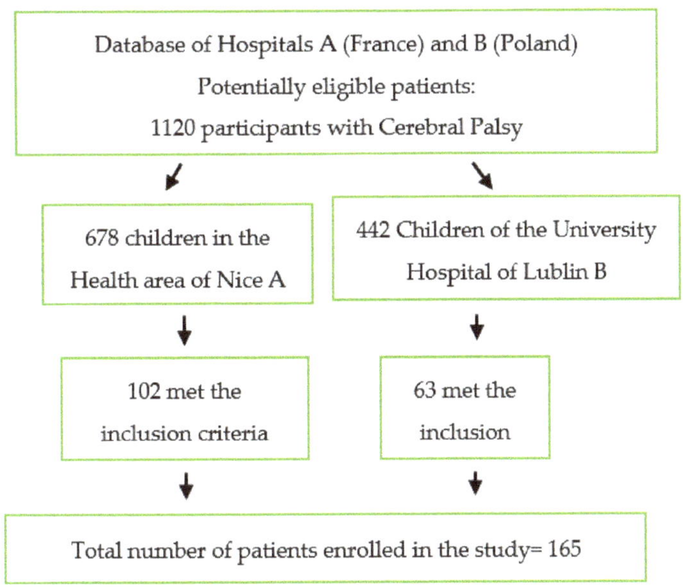

Figure 1. Flow diagram of study participants for analysis.

The mean age was 15.7 years (range: 12–18 years; standard deviation [SD] 1.8). The mean follow-up was 5.1 years (range: 3–12 years). 165 patients (91 male, 74 female) assessed between June 2005 and June 2020 were included (Table 1). There were no dropouts during the trial period.

Table 1. Clinical presentation according to the presence or absence of Neurotoxin Treatments.

Patients Profile	Pediatric Hospital A			Children Hospital B			Multicenter A + B
	Neurotoxin Treatments			Neurotoxin Treatments			
	Yes (%)	No (%)	Total (%)	Yes (%)	No (%)	Total (%)	Total (%)
Patients n. (%)	17 (17)	85 (83)	102 (100)	49 (77)	14 (23)	63 (100)	165 (100)
Male	35 (58)	25 (42)	60 (100)	22 (70)	9 (30)	31 (100)	91 (55)
Female	23 (55)	19 (45)	42 (100)	18 (58)	14 (42)	32 (100)	74 (45)
Average age (mean, SD)	16.4 (1.8)	16.8 (1.8)	16.6 (1.8)	15.8 (1.8)	16.0 (1.8)	15.9 (1.8)	16.2 (1.8)
Spasticity, n. (%)	16 (21)	59 (79)	75 (100)	34 (48)	20 (52)	54 (100)	129 (78)
Hemiplegia	2 (22)	7 (78)	9 (100)	3 (75)	1 (25)	4 (100)	13 (8)
Diplegia	1 (6)	15 (94)	16 (100)	20 (86)	3 (14)	23 (100)	39 (24)
Tri/quadriplegia	13 (26)	37 (74)	50 (100)	11 (41)	16 (59)	27 (100)	77 (68)
Dystonia n. (%)	10 (71)	4 (29)	14 (100)	8 (66)	4 (36)	12 (100)	26 (16)
Well-controlled Epilepsy, n. (%)	10 (20)	40 (80)	50 (100)	23 (79)	6 (21)	29 (100)	79 (48)
Intractable Epilepsy	4 (18)	18 (82)	22 (100)	10 (77)	3 (23)	13 (100)	35 (20)
No Epilepsy	3 (10)	27 (90)	30 (100)	16 (76)	5 (24)	21 (100)	51 (31)
Severe Scoliosis (%)	23 (59)	16 (41)	39 (100)	16 (53)	14 (47)	30 (100)	69 (41)
Equines Foot (%)	31 (75)	10 (25)	41 (100)	21 (75)	7 (25)	28 (100)	69 (41)
Hip Dysplasia (%)	18 (56)	14 (44)	32 (100)	13 (59)	9 (41)	22 (100)	54 (38)
Truncal tone disorder (%)	11 (21)	42 (79)	53 (100)	29 (74)	10 (26)	39 (100)	92 (56)
Ante-natal Causes	10 (16)	54 (84)	64 (100)	21 (84)	4 (16)	25 (100)	89 (54)
Perinatal Causes	4 (14)	25 (86)	29 (100)	24 (72)	9 (28)	33 (100)	62 (37)
Postnatal Causes	3 (34)	6 (66)	9 (100)	4 (80)	1 (20)	5 (100)	14 (9)

4.2. Botulin Toxin Clinical Use

BTX treatment specialists selected children requiring BTX injections based on their clinical and functional characteristics. Once information on possible side effects was provided, the patient's and/or parents' approval was explicitly expressed. The date of injection, doses and muscles treated, and pain assessment through a visual analog scale, were recorded.

The specialties Dysport, Botox, Neurobloc, and Xeomin, were available and used for our patient's neurological conditions. We determined the location of the muscles by palpation, ultrasound, or electromyogram (EMG) needle with injection. This allowed for the noninvasive identification of muscles and surrounding structures [25], and these procedures are especially beneficial in children.

4.3. The Doses

The dose depends on the patient's weight, the severity of the spasticity, the number of muscles treated, the type of toxin, and the size of the muscle. The units were different and were not international units—there is no recognized equivalence: 1 mL for Botox, 2.5 mL for Dysport, 100 Allergan units/mL, and 100 Speywood units per 1 to 2.5 mL. The maximum recommended total dose for children was:

- For Botox, 300 units per session and 20 Allergan units/kg;
- For Dysport, 1000 units per session and 30 Speywood units/kg (professional agreement).

In adults, the recommended total dose is:

- For Botox, 500 Allergan units;

- For Dysport, 1500 Speywood units.

In the case of the first injection, lower initial doses were recommended, especially in patients with comorbidities:

- For Botox, 3 to 8 units/kg without exceeding 300 units per session;
- For Dysport, 10 units/kg in unilateral injections and 20 units/kg in bilateral injections without exceeding 1000 units per session.

We envisaged at least three months' break between two sessions [24].

4.4. Measurements

All data were collected from the Electronic Medical Records by the senior author. Medical notes were written by a multidisciplinary team that included pediatric neurologists, epidemiologists, pediatricians, orthopedic surgeons, and physiotherapists. Narrative notes were coded and filled in the database of BTX-PredictMed [9,20].

Data on diagnosis, etiology, type of spasticity, functional assessments, epilepsy, clinical history, and radiology were collected anonymously between 2005 and 2020.

CP etiology was classified as [1]:

- antenatal: cerebral malformation, genetic, prematurity, infection, vascular;
- perinatal: anoxic, infectious ischemic;
- postnatal: postnatal anoxic/ischemic injury epilepsy, cranial trauma, infectious.

Motor function was assessed using the Gross Motor Function Classification System (GMFCS) and the Manual Ability Classification System (MACS) [15,24]. Both have a 5-point classification system with higher scores indicating worse motor functioning (Figure 2).

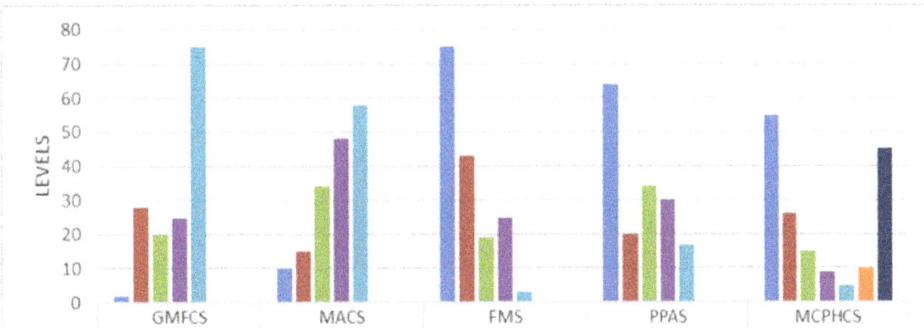

	GMFCS	MACS	FMS	PPAS	MCPHCS
I	2	10	75	64	55
II	28	15	43	20	26
III	20	34	19	34	15
IV	25	48	25	30	9
V	75	58	3	17	5
VI	0	0	0	0	10
VII	0	0	0	0	45

Figure 2. Distribution of patients according to the Manual Ability Classification System (MACS), Gross Motor Function Classification System (GMFCS), Melbourne Cerebral Palsy Hip Classification Scale (MCPHCS), Functional Mobility Scale (FMS), and Posture and Postural Ability Scale (PPAS). Nonapplicable (0).

Scoliosis was defined by a Cobb angle > 10° on the spinal radiograph and classified as severe at a Cobb angle > 40° [1,15] (Table 1).

Neurological status was assessed by the presence of hypertonia in the upper or lower extremities, the type of spastic disorder (hemiplegia, diplegia, tri/quadriplegia), the severity of epilepsy, and the presence of dystonia. The modified Ashworth Scale of Bohannon and Smith and the modified Tardieu Scale [1,15] have been used to quantify spasticity.

The severity of epilepsy has been determined by pediatric epileptologists and classified as "well-controlled" or "intractable" [36] based on the guidelines of the International League Against Epilepsy. These guidelines define intractable epilepsy as a continuous seizure despite treatment attempts with at least two antiepileptic drugs [37,38] (Table 1).

Dysplasia was estimated based on the Perkins line: 0% was assigned if the migration percentage (MP) was negative and the lateral margin of the femoral head was medial to the Perkins line. Percent migration (MP) was scored as 100% when the entire femoral head was lateral to the Perkins line.

The hips were classified as normal (MP less than 33%), subluxated (MP = 33 up to 89%) or luxated (MP $\geq$ 90%) according to the migration percentage [16,17]. Clinical measurement of the hip focused on internal rotation and hip abduction. The modified Harris score (MHHS) [16] was used to assess hip function, pain, and gait. The Melbourne Cerebral Palsy Hip Classification Scale (MCPHCS) [39,40] was used to classify hip morphology. In the case of multiple radiographs, a pediatric orthopedic surgeon evaluated the most recent one. All patients had at least one pelvic radiograph.

Trunk functional abilities were ascertained with the Functional Mobility Scale (FMS) [16], the Posture and Postural Ability Scale (PPAS), and the Lower Extremity Functional Scale (LEFS) [17]. Trunk muscle tone was assessed with the Trunk Impairment Scale (TIS) [16] (Figure 2).

The variables investigated were:

- Neurotoxins treatments (NT);
- Presence of Neuromuscular scoliosis (NS);
- Trunk muscle tone disorder (TT);
- Spasticity (SP);
- Dystonia (D);
- Epilepsy (E);
- Hip Dysplasia (HD);
- Equines foot (EF);
- Gastrostomy feeding (GA);
- Sex (SE);
- Etiology (ET);
- GMFCS;
- MACS.

ET, TT, SP, D, SE, GMFCS, MACS, and E were assessed in the first control; NT, NS, HP, GA, and EF were assessed in the last control.

4.5. Statistical Analysis

We performed Fisher's exact tests and developed contingency tables [41] for identifying the distribution frequencies and the confidence intervals of factors associated with BTX treatments. Then, we used the web-based epidemiological calculators MedCalc® statistical software 20.123 and OpenEpi software 3.01 [42,43] to calculate 95% confidence intervals, odds ratios, and *Z*-statistics (Table 2).

The glm() function of the open-source software R 4.2.2 [44–46] was used to predict each patient's probability of undergoing BTX treatment; the common thresholds for selecting relevant variables (with p-value < 0.2) [44,45] were employed as independent input variables in a bespoke multiple logistic regression model [44,46]. The binary dependent variable was the presence of BTX treatment (yes/no).

The selected Independent variables entered in BTX-PredictMed were: ET, TT, SP, D, E, NS, GMFCS, SE, MACS, and HD.

In accordance with the statistical learning theory reported by Vapnik and Chervonenkis [47], we divided the patients into a "training set" to train the LR model and a "test set" to check the performance of the model. We checked whether each subject in the "test set" was correctly predicted as a potential developer of epilepsy (or not) by calculating the sensitivity, specificity, and accuracy [43].

Table 2. Contingency table comparing subjects with and without Neurotoxin Treatments using the Fisher's exact test.

Independent Variables		Neurotoxin Treatments Yes	Neurotoxin Treatments No	Multicenter Pediatric Hospital A + Children Hospital B Odds Ratio	95% CIs	Z Statistic	*p* Value	Hospitals A *p* Value	Hospitals B *p* Value
Neuromuscular Scoliosis (NS)	Yes	39	30	2.86	1.50–5.43	3.20	0.0013	0.007	0.006
	No	30	66						
Equines Foot (EF)	Yes	45	30	4.12	2.13–7.95	4.22	<0.0001	<0.0001	<0.0001
	No	24	66						
Etiology (ET) PreNatal > Peri/PostNatal causes	Yes	31	58	0.53	0.28–0.99	1.96	0.05	0.05	0.05
	No	38	38						

To minimize the dependence on the composition of the training and test sets, cross-validation was used. Cross-validation is a technique to evaluate the generalization of the results of a statistical analysis on an independent data set. We randomly generated 20 couples of training and test sets; for each couple, we calculated the sensitivity, specificity, and accuracy of the predictions; and finally averaged all couples. Accuracy, sensitivity, and specificity were defined in terms of True Positive (TP), True Negative (TN), False Negative (FN), and False Positive (FP) [43].

Then, we compared the predictions with the patient's known status (e.g., whether he/she has BTX or not) for each patient in the "test set," calculating the sensitivity, specificity, and accuracy of the predictive logistic regression algorithm (Table 3).

Table 3. List of the logistic regression coefficients associated with the presence of Neurotoxins Treatments.

Independent Variables	Logistic Regressions: Odds Ratio Logarithm	Odds Ratio Linear	Standard Error	Z Ratio	Prob(>∣Z∣) *p* Value
Intercept	1.563	4.77	0.879	1.777	0.075
Scoliosis (NS)	0.146	0.863	0.476	−0.308	0.757
Truncal Tone Disorder (TT)	0.626	1.870	0.277	2.258	0.023
Etiology	0.077	1.080	0.316	0.246	0.805
Spasticity (SP)	0.677	1.967	0.285	2.374	0.017
Dystonia (D)	1.670	5.312	0.583	2.864	0.004
Epilepsy (E)	0.227	1.254	0.349	0.649	0.515
Gender (SE)	0.512	1.668	0.421	1.215	0.224
GMFCS score	0.299	0.741	0.312	−0.957	0.338
MACS score	1.085	2.959	0.250	−4.334	<0.001
Hip Dysplasia (HD)	1.392	4.022	0.500	2.7822	0.05

Logistic Regression: The increasing of TT, SP (Quadriplegia/triplegia >Diplegia> hemiplegia), D, MACS score, and HD are factors associated with the presence of Neurotoxin Treatments (in the "Odds Ratio-Linear" column). This means, more precisely, that for every unit increase in SP, the log odds = ln(p/1−p) increases 1.967 times (where p = probability of having Neurotoxins Treatments). The "Prob(>∣z∣)" column indicates the significant strength of the respective parameter in terms of the p-value as the presence of Neurotoxin Treatments. This means that the significance of TT, SP, D, MACS score, and HD in predicting the presence of Neurotoxin Treatments is very probable, with a p-value < 0.05. The best machine learning model score performed with an accuracy of 76%, sensitivity of 67%, specificity of 81%, and an average score of 75%.

Author Contributions: Conceptualization, C.M.B.; methodology, C.M.B., D.B. and F.S.; investigation, C.M.B., D.G., S.B., S.C.B. and M.L.; data curation, M.L. and F.S.; writing—original draft preparation, C.M.B.; writing—review and editing, C.M.B., M.L., D.B., S.B., S.C.B., D.G. and F.S.; visualization, S.B. and S.C.B.; supervision, D.G. and F.S.; project administration, C.M.B. and D.G. All authors have read and agreed to the published version of the manuscript.

Funding: This research received no external funding.

Institutional Review Board Statement: All procedures were performed in accordance with the ethical standards of the University Hos-pital and the 1964 Helsinki Declaration and its later amendments. All participants and parents consented to participate. The data were anonymized and analyzed according to the requirements of Reference Method 003, numbered "2017728 v 0". The Ethics committee approval and informed consent are registered with number "2017728 v 0- MR003 (Reference Method 003) 27 March 2017".

Informed Consent Statement: Informed consent was obtained from all subjects involved in the study.

Data Availability Statement: Not applicable.

Conflicts of Interest: The authors declare no conflict of interest.

References

1. Bertoncelli, C.M.; Solla, F.; Loughenbury, P.R.; Tsirikos, A.I.; Bertoncelli, D.; Rampal, V. Risk factors for developing scoliosis in cerebral palsy: A cross sectional descriptive study. *J. Child Neurol.* **2017**, *32*, 657–662. [CrossRef]
2. Multani, I.; Manji, J.; Hastings-Ison, T.; Khot, A.; Graham, K. Botulinum Toxin in the Management of Children with Cerebral Palsy. *Pediatr. Drugs* **2019**, *21*, 261–281. [CrossRef]
3. Lasko, T.A.; Denny, J.; Levy, M.A. Computational Phenotype Discovery Using Unsupervised Feature Learning over Noisy, Sparse, and Irregular Clinical Data. *PLoS ONE* **2013**, *8*, e66341. [CrossRef]
4. Seymour, T.; Frantsvog, D.A.; Graeber, T. Electronic Health Records (EHR). *Am. J. Health Sci.* **2012**, *3*, 201–210. [CrossRef]
5. Denny, J.C.; Spickard, A.; Johnson, K.B.; Peterson, N.B.; Peterson, J.F.; Miller, R.A. Evaluation of a Method to Identify and Categorize Section Headers in Clinical Documents. *J. Am. Med. Inform. Assoc.* **2009**, *16*, 806–815. [CrossRef]
6. Dean, B.B.; Lam, J.; Natoli, J.L.; Butler, Q.; Aguilar, D.; Nordyke, R.J. Use of electronic medical records for health outcomes research: A literature review. *Med. Care Res. Rev.* **2009**, *66*, 611–638. [CrossRef]
7. Kho, A.N.; Pacheco, J.A.; Peissig, P.L.; Rasmussen, L.; Newton, K.M.; Weston, N.; Crane, P.K.; Pathak, J.; Chute, C.G.; Bielinski, S.J.; et al. Electronic Medical Records for Genetic Research: Results of the eMERGE Consortium. *Sci. Transl. Med.* **2011**, *3*, 79re1. [CrossRef]
8. Kohane, I.S. Using electronic health records to drive discovery in disease genomics. *Nat. Rev. Genet.* **2011**, *12*, 417–428. [CrossRef]
9. Bertoncelli, C.M.; Altamura, P.; Bagui, S.; Vieira, E.R.; Costantini, S.; Monticone, M.; Solla, F.; Bertoncelli, D. Predicting osteoarthritis in adults using statistical data mining and machine learning. *Ther. Adv. Musculoskelet. Dis.* **2022**, *14*, 1759720X221104935. [CrossRef]
10. Bertoncelli, C.M.; Altamura, P.; Bertoncelli, D.; Rampal, V.; Vieira, E.R.; Solla, F. PredictMed: A Machine Learning Model for Identifying Risk Factors of Neuromuscular Hip Dysplasia: A Multicenter Descriptive Study. *Neuropediatrics* **2020**, *52*, 343–350. [CrossRef]
11. Bertoncelli, C.M.; Solla, F. Machine learning for monitoring and evaluating physical activity in cerebral palsy. *Dev. Med. Child Neurol.* **2020**, *62*, 1010. [CrossRef]
12. Hastie, T.; Tibshirani, R.; Friedman, J. Boosting and additive trees. In *The Elements of Statistical Learning*, 2nd ed.; Springer: New York, NY, USA, 2009.
13. Collins, G.S.; Reitsma, J.B.; Altman, D.G.; Moons, K.G.M. Transparent Reporting of a multivariable prediction model for Individual Prognosis or Diagnosis (TRIPOD): The TRIPOD statement. *BMJ* **2015**, *350*, g7594. [CrossRef]
14. Pereira, F.; Mitchell, T.; Botvinick, M. Machine learning classifiers and fMRI: A tutorialoverview. *Neuroimage* **2009**, *45*, S199–S209. [CrossRef]
15. Bertoncelli, C.M.; Bertoncelli, D.; Elbaum, L.; Latalski, M.; Altamura, P.; Musoff, C.; Rampal, V.; Solla, F. Validation of a Clinical Prediction Model for the Development of Neuromuscular Scoliosis: A Multinational Study. *Pediatr. Neurol.* **2018**, *79*, 14–20. [CrossRef]
16. Bertoncelli, C.M.; Altamura, P.; Vieira, V.R.; Bertoncelli, D.; Solla, F. Predicting hip dysplasia in teenagers with cerebral palsy in order to optimize prevention and rehabilitation. A longitudinal descriptive study. *Dev. Neurorehabilit.* **2020**, *24*, 166–172. [CrossRef]
17. Bertoncelli, C.M.; Altamura, P.; Vieira, E.; Bertoncelli, D.; Latalski, M.; Berthet, S.; Solla, F. Predictive Model for Gastrostomy Placement in Adolescents with Developmental Disabilities and Cerebral Palsy. *Nutr. Clin. Pr.* **2020**, *35*, 149–156. [CrossRef]

18. Bertoncelli, C.M.; Altamura, P.; Vieira, E.R.; Bertoncelli, D.; Thummler, S.; Solla, F. Identifying factors associated with intellectual disabilities in teenagers with cerebral palsy using a predictive learning model. *J. Child. Neurol.* **2019**, *34*, 221–229. [CrossRef]
19. Bertoncelli, C.M.; Altamura, P.; Vieira, E.R.; Bertoncelli, D.; Solla, F. Using Artificial Intelligence to Identify Factors Associated with Autism Spectrum Disorder in Adolescents with Cerebral Palsy. *Neuropediatrics* **2019**, *50*, 178–187. [CrossRef]
20. Bertoncelli, C.M.; Altamura, P.; Vieira, E.R.; Iyengar, S.S.; Solla, F.; Bertoncelli, D. PredictMed: A logistic regression-based model to predict health conditions in cerebral palsy. *Health Inform. J.* **2020**, *26*, 2105–2118. [CrossRef]
21. Moons, K.G.; Altman, D.G.; Reitsma, J.B.; Collins, G.S. New guideline for the reporting of studies developing, validating, or updating a multivariable clinical prediction model: The TRIPOD statement. *Adv. Anat. Pathol.* **2015**, *22*, 303–305. [CrossRef]
22. Moons, K.G.M.; Kengne, A.P.; Grobbee, D.E.; Royston, P.; Vergouwe, Y.; Altman, D.G.; Woodward, M. Risk prediction models: II. External validation, model updating, and impact assessment. *Heart* **2012**, *98*, 691–698. [CrossRef] [PubMed]
23. Altman, D.G.; Vergouwe, Y.; Royston, P.; Moons, K.G. Prognosis and prognostic research: Validating a prognostic model. *BMJ* **2009**, *338*, b605. [CrossRef] [PubMed]
24. Hareb, F.; Rampal, V.; Bertoncelli, C.M.; Rosello, O.; Solla, F. Botulinum toxin in children with cerebral palsy: An update. *Neuropediatrics* **2020**, *51*, 1–5. [CrossRef] [PubMed]
25. Mirska, A.; Kułak, W.; Okurowska-Zawada, B.; Dmitruk, E. Effectiveness of multiple botulinum toxin sessions and the duration of effects in spasticity therapy in children with cerebral palsy. *Childs Nerv Syst.* **2019**, *35*, 141–147. [CrossRef] [PubMed]
26. Barlaan Lukban, M.; Rosales, R.L.; Dressler, D. Effectiveness of botulinum toxin A for upper and lower limb spasticity in children with cerebral palsy: A summary of evidence. *J. Neural Transm.* **2009**, *116*, 319–331. [CrossRef]
27. Bohn, E.; Goren, K.; Switzer, L.; Falck-Ytter, Y.; Fehlings, D. Pharmacological and neurosurgical interventions for individuals with cerebral palsy and dystonia: A systematic review update and meta-analysis. *Dev. Med. Child Neurol.* **2021**, *63*, 1038–1050. [CrossRef] [PubMed]
28. Lin, C.Y.; Chung, C.H.; Matthews, D.J.; Chu, H.Y.; Chen, L.C.; Yang, S.S.; Chien, W.C. Long-term effect of botulinum toxin A on the hip and spine in cerebral palsy: A national retrospective cohort study in Taiwan. *PLoS ONE* **2021**, *16*, e0255143. [CrossRef]
29. Young Choi, J.; Kim, S.K.; Park, E.S. The Effect of Botulinum Toxin Injections on Gross Motor Function for Lower Limb Spasticity in Children with Cerebral Palsy. *Toxins* **2019**, *11*, 651. [CrossRef]
30. Aydil, S.; Akpinar, F.M.; Akpinar, E.; Beng, K.; Yagmurlu, M.F. Effectiveness of Multilevel Botulinum Toxin a Injection with Integrated Treatment Program on Spasticity Reduction in Non-Ambulatory Young Children with Cerebral Palsy. *Med. Princ. Pract.* **2019**, *28*, 309–314. [CrossRef]
31. Mall, V.; Heinen, F.; Linder, M.; Philipsen, A.; Korinthenberg, R. Treatment of cerebral palsy with botulinum toxin A: Functional benefit and reduction of disability. Three case reports. *Pediatr. Rehabil.* **1997**, *1*, 235–237. [CrossRef]
32. Corry, I.S.; Cosgrove, A.P.; Duffy, C.M.; McNeill, S.; Taylor, T.C.; Graham, H.K. Botulinum toxin A compared with stretching casts in the treatment of spastic equinus: A randomised prospective trial. *J. Pediatr. Orthop.* **1998**, *18*, 304–311. [CrossRef] [PubMed]
33. Chaléat-Valayer, E.; Parratte, B.; Colin, C.; Denis, A.; Oudin, S.; Bérard, C.; Bernard, J.C.; Bourg, V. A French observational study of botulinum toxin use in the management of children with cerebral palsy: BOTULOSCOPE. *Eur. J. Paediatr. Neurol.* **2011**, *15*, 439–448. [CrossRef] [PubMed]
34. Lee, Y.; Lee, S.; Jang, J.; Lim, J.; Ryu, J.S. Effect of Botulinum Toxin Injection on the Progression of Hip Dislocation in Patients with Spastic Cerebral Palsy: A Pilot Study. *Toxins* **2021**, *13*, 872. [CrossRef]
35. Biering-Soerensen, B.; Stevenson, V.; Bensmail, D.; Grabljevec, K. European expert consensus on improving patient selection for the management of disabling spasticity with intrathecal baclofen and/or botulinum toxin type A. *J. Rehabil. Med.* **2022**, *54*, jrm00241. [CrossRef] [PubMed]
36. Berg, A.T. Defining intractable epilepsy. *Adv. Neurol.* **2006**, *97*, 5–10. [PubMed]
37. Sinha, S.; Siddiqui, K.A. Definition of intractable epilepsy. *Neurosciences* **2011**, *16*, 3–9.
38. Berg, A.T. Identification of pharmacoresistant epilepsy. *Neurol. Clin.* **2009**, *27*, 1003–1013. [CrossRef]
39. Terjesen, T. The natural history of hip development in cerebral palsy. *Dev. Med. Child Neurol.* **2012**, *54*, 951–957. [CrossRef]
40. Lins, L.A.; Watkins, C.J.; Shore, B.J. Natural History of Spastic Hip Disease. *Disease J. Pediatr. Orthop.* **2019**, *39* (Suppl. 1), S33–S37. [CrossRef]
41. Solla, F.; Tran, A.; Bertoncelli, D.; Musoff, C.; Bertoncelli, C. Why a *p*-value is not enough. *Clin. Spine Surg.* **2018**, *31*, 385–388. [CrossRef]
42. Sullivan, K.; Andrew, D.; Minn Minn, S. OpenEpi: A web-based epidemiologic and statistical calculator for public health. *Public Health Rep.* **2009**, *124*, 458–471. [CrossRef] [PubMed]
43. Wen, Z.; Zeng, N.; Wang, N. Sensitivity, specificity, accuracy, associated confidence interval and ROC analysis with practical SAS® implementations. In Proceedings of the NESUG Proceedings: Health Care and Life Sciences, Baltimore, Maryland, 14–17 November 2010; pp. 1–9.
44. Robert, J.T. *An Introduction to Error Analysis: The Study of Uncertainties in Physical Measurements*; University Science Books: Sausalito, CA, USA, 1999; pp. 128–129. ISBN 0-935702-75-X.
45. Mickey, R.M.; Greenland, S. The impact of confounder selection criteria on effect estimation. *Am. J. Epidemiol.* **1993**, *129*, 125–137. [CrossRef] [PubMed]

46. Maldonado, G.; Greenland, S. Simulation study of confounder-selection strategies. *Am. J. Epidemiol.* **1993**, *138*, 923–936. [CrossRef] [PubMed]
47. Vapnik, V. *The Nature of Statistical Learning Theory*; Springer Science & Business Media: Berlin/Heidelberg, Germany, 2013.

toxins

MDPI

Article

Real-Time Documentation of the Effect of Onabotulinumtoxin A Detrusor Injection in OAB Patients—Preliminary Results

Heinrich Schulte-Baukloh [1,*], Catarina Weiß [2], Sarah Weinberger [1], Mandy Hubatsch [1], Thorsten Schlomm [1] and Bernhard Ralla [1]

1 Department of Urology, Charité Universitätsmedizin Berlin, 10117 Berlin, Germany
2 Urologic Practice, Kurfürstendamm 139, 10711 Berlin, Germany
* Correspondence: heinrich.schulte-baukloh@charite.de

Abstract: Introduction: Detrusor injection with onabotulinumtoxin A (OnabotA-DI) is an established therapy for overactive bladder (OAB). Little is known about the exact onset and course of the effect in the days after the injection therapy. By using a new type of app-controlled automated diary pod, for the first time, the precise onset of the effect of OnabotA-DI can be documented in real time. Materials and methods: Patients due for OnabotA-DI were asked to document voiding 3 days before and up to 3 weeks after therapy using the Diary Pod app. The detrusor injection was performed with onabotulinumtoxin A (Botox®), 100 units, at 20 sites of the detrusor muscle in a standardized manner. Voiding on the injection day itself was not documented. Results: A total of 17 patients (15 women, 2 men; aged 33–83 (mean 64.6; median 70) years) were included in the study. The handling of the Diary Pod app was user-friendly, and elderly patients did not encounter technical problems. The results of patients with reliably documented micturitions showed a continuous reduction in micturition frequency every day from the first day and significantly from day 5. For 24 h voiding, from 12.83 ± 5.54 in the 3 days before injection, the following mean values were found with significant ($p < 0.05$) changes after the intervention: 9.17 ± 3.19 on day 5, 8.75 ± 3.69 on day 10, 7.17 ± 2.04 on day 15, and 5.75 ± 0.5 on day 20. These changes were in similar proportions during the daytime and nighttime. Conclusions: Contrary to previous knowledge, the effect of the OnabotA-DI set in from the first postoperative days and was reflected a similar extent in day and night micturition. This study is the first to document the onset of action of OnabotA-DI in real time.

Keywords: overactive bladder; botulinum toxin; urinary incontinence; urinary bladder; injections; intramuscular

Key Contribution: Real-time documentation of the effect of onabotulinumtoxin A detrusor injection in OAB patients shows that the action starts already on the first days after injection and reaches significance on the fifth day.

Citation: Schulte-Baukloh, H.; Weiß, C.; Weinberger, S.; Hubatsch, M.; Schlomm, T.; Ralla, B. Real-Time Documentation of the Effect of Onabotulinumtoxin A Detrusor Injection in OAB Patients—Preliminary Results. *Toxins* **2023**, *15*, 30. https://doi.org/10.3390/toxins15010030

Received: 26 November 2022
Revised: 19 December 2022
Accepted: 27 December 2022
Published: 30 December 2022

1. Introduction

As defined by the International Continence Society [1], the overactive bladder (OAB) syndrome, urge syndrome, or urgency-frequency syndrome can be described by the presentation of urgency, with or without urge incontinence, usually with frequency and nocturia. These symptom combinations are suggestive of urodynamically demonstrable detrusor overactivity but can be due to other forms of urethro-vesical dysfunction. These terms can be used if there is no proven infection or other obvious pathology (e.g., bladder stones, bladder tumors, bladder outlet obstruction due to benign prostatic hyperplasia, or uterovaginal descenus). OAB is purely a patient-reported condition. These OAB symptoms can substantially limit the quality of life and cause various social, work-related, psychological, and sexual problems [2–4]. Epidemiological studies from Europe and the US have shown that the symptoms of OAB increase significantly with age and can occur in up to 16–17% of the population [5–7]. According to the current European Association of Urology (EAU) guidelines [8,9], behavioral therapy approaches such as weight loss, reducing nicotine and coffee consumption, pelvic floor therapy, and bladder training are therapeutic options. However, OAB is a domain of drug approaches such as anticholinergics or β_3-receptor agonists such as mirabegron [8,9]. If these medications are ineffective or discontinued because of side effects [10], the patient can be offered more invasive measures as second-line therapy.

In 2013, onabotulinumtoxin A detrusor injection (OnabotA-DI) was approved by the US Food and Drug Administration for the treatment of idiopathic OAB in patients whose urgency or frequency with urge incontinence did not respond adequately to anticholinergic drugs or who were intolerant to these drugs [11]. According to the EMBARK study reported by Nitti et al. [11], this therapy is carried out as part of a cystoscopy procedure and provides effective alleviation of all OAB symptoms after 12 weeks [11], to the following extent: a 47.9% reduction in urinary urge incontinence, 16.9% reduction in frequency, 31.6% reduction in urgency, and 20.2% reduction in nocturia. Clinically meaningful improvements from baseline in all I-QOL and KHQ multi-item domains indicated a positive impact on HRQOL [4]. This therapeutic approach has proven to be more effective in reducing urge incontinence than oral medication [12,13]. Numerous, generally similar data exist on the duration of the effect of onabotulinumtoxin A in the urinary bladder [14]. The effect of 100 U of onabotulinumtoxin A in a long-term study by Nitti et al. based on the time to patient re-treatment request showed a median of 7.6 months, but this varied greatly [15]: 34.2% of the patients requested re-treatment within 6 months, 37.2% after 6–12 months, and 28.5% after more than 12 months.

However, data are sparse for this approach concerning the exact onset of action in the days after the OnabotA-DI therapy, and we are not aware of any study that has explicitly dealt with this issue. The drug information and numerous studies on this topic state only that "a clinical improvement generally occurs within one to two weeks", but when exactly can the patient expect the first relief from the burdensome voiding frequency? By using a novel app-controlled automated urine collection device (Minze-Health® Diary Pod), we investigated the exact onset of the effect of OnabotA-DI in real time.

2. Results

A total of 17 patients were included in the study (15 women and 2 men). The mean/median age was 64.6/70 (33–83) years. Unfortunately, the dropout rate was high; four patients had to be excluded due to insufficient documentation ($\leq$6 documented postoperative days), two due to postoperative UTIs, one due to device failure, and one due to cancer (non-urological) diagnosed during the study period. On the other hand, nine very well-documented pre- and post-interventional datasets were evaluated.

The baseline mean micturition frequency in the 3 days before the injection therapy was 12.8 $\pm$ 5.54 micturitions/24 h. Figure 1 shows the decrease in micturition frequency over the 3 weeks. Table 1 shows the decrease in voiding frequency until a significant decrease was achieved, and then every 5 days thereafter to day 20. Due to the generally high variability, the mean micturition volume showed a delayed significant increase, from an average of 194.8 mL before the injection to 244.1 mL on the ninth day after the injection.

Table 1. Decrease in voiding frequency after injection therapy. The reduction in micturition in 24 h reached a trend on the 4th day and became significant on the 5th day. The reduction in voiding, separately during the day and at night, reached a significant level from the 7th day (not separately shown).

	Day	Frequency in 24 h (Mean ± SD)		*p*	Frequency Daytime (Mean ± SD)		*p*	Frequency Nighttime (Mean ± SD)		*p*
Average micturition BEFORE injection	−3 to −1	12.83	5.54		10.87	5.16		1.96	0.98	
Average micturition AFTER injection	1	12.29	4.39	0.767	10.14	4.02	0.636	2.14	0.90	0.689

Table 1. *Cont.*

Day	Frequency in 24 h (Mean ± SD)		*p*	Frequency Daytime (Mean ± SD)		*p*	Frequency Nighttime (Mean ± SD)		*p*
3	9.56	4.45	0.060	7.89	4.31	0.070	1.67	0.87	0.282
4	9.38	4.24	0.056	8.13	4.42	0.119	1.25	1.28	0.142
5	9.17	3.19	0.038	7.67	4.03	0.107	1.50	1.22	0.363
10	8.75	3.69	0.017	7.63	3.54	0.035	1.13	0.64	0.006
15	7.17	2.04	0.001	6.00	2.10	0.002	1.17	0.75	0.042
20	5.75	0.50	<0.001	4.25	0.96	0.001	1.50	1.29	0.049

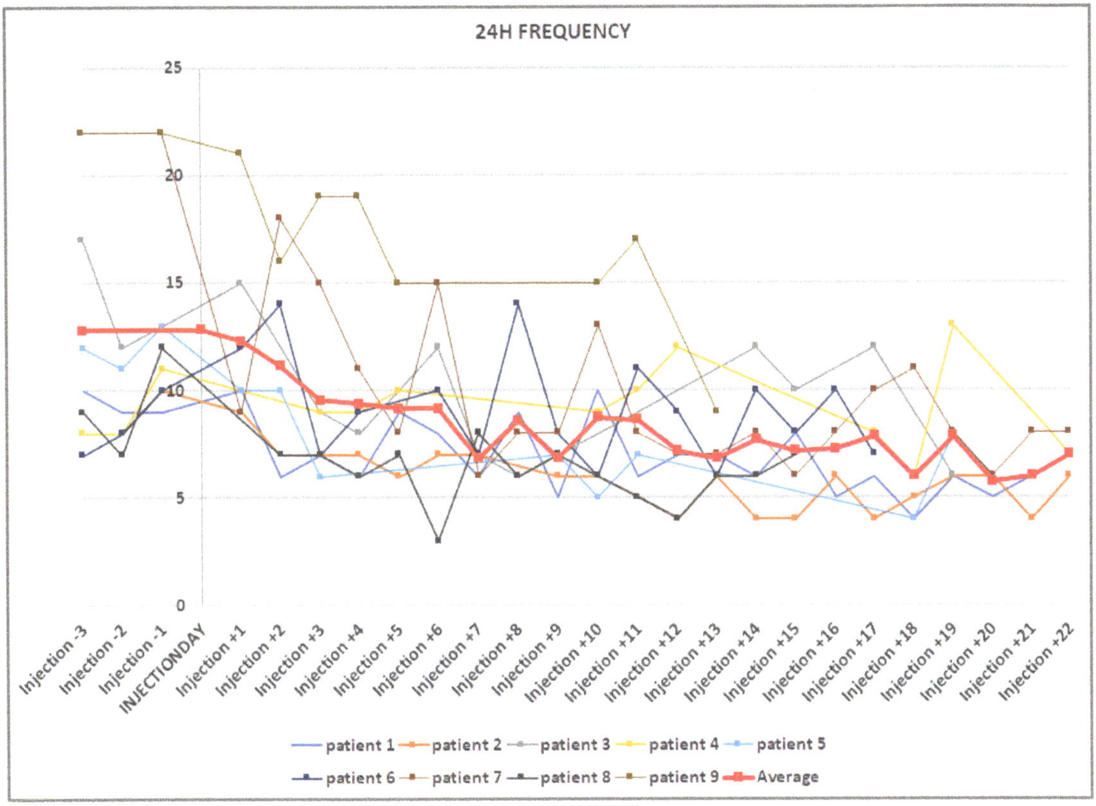

Figure 1. Graphic representation of the decrease in micturition frequencies in 24 h, before and after injection. The vertical line crosses on the day of injection. The mean value curve is in bold red.

3. Discussion

The patients evaluated in this study were representative of those typically affected with OAB symptoms and are comparable to those in larger series (e.g., the approval EMBARK study of Nitti et al. [11]). Although our patient series was small, the patients were especially homogenous and displayed typical features. Therefore, the sample was representative of average OAB patients: the increased micturition frequency of 12.8/24 h was similar to that of the large series, as was the unequal gender distribution strongly in favor of women (89.3%), and the mean age of 64.6 years [11,16]. The frequency of post-interventional UTIs in our series at 11.8% (2 of 17) fits into the variable picture described in the literature of UTI rates of 0.4–24.5% [11,16]. In our study, these patients had to be excluded because a UTI changes the micturition frequency significantly; accordingly, this would not reflect the effect of the onabotulinumtoxin A. The considerable effort for the patients of the daily permanent documentation and Diary Pod use also led to many dropping out. Unfortunately, one patient had a diagnosis of breast cancer during the study period and did not want to continue with the study, and one device did not function properly. However, despite this high dropout rate, the results properly reflect the clinical picture of OAB, and thus, they also reflect the course of action of the onabotulinumtoxinA therapy because of the high homogeneity of the representative participants. Furthermore, this study was not a proof-of-action study, but it does prove the onset of action for onabotulinumtoxinA after application in the detrusor muscle.

Effective therapeutic results of OnabotA-DI are well-documented in patients with OAB [14]. The duration of action has also been clarified: in the long-term follow-up study reported by Nitti et al. [15], the overall median duration of the effect of onabotulinumtoxinA 100 U was 7.6 months. As mentioned, the median duration was 6 months or less in 150 of the 438 patients (34.2%), between 6 and 12 months in 163 patients (37.2%), and greater than 12 months in 125 patients (28.5%). Moreover, to our knowledge, the onset of the toxin's effect has not been reported. This is the most important finding in our study: the frequency of voiding per 24 h decreased significantly from the fifth post-interventional day. Considering the absolute numbers, this also corresponds to a clinically substantial decrease in micturition frequency: from 12.8 to 9.2, i.e., −3.7 (−28.6%). In the pivotal study by Nitti [11], this value was −16.9% at week 12 post-treatment.

The good effect of onabotulinumtoxin A on urinary bladder symptoms is convincing. Its action is complex, especially with the most clinically used serotype botulinum toxin type A; the neuromuscular transmission is inhibited in not only efferent but also afferent nerves [17]. In the bladder, SV2 receptors, via which the neurotoxin is taken up into the nerve cells [18], and SNAP25, which is part of the SNARE protein complex and the primary target of botulinum toxin type A [19], are found in abundance and co-localize in parasympathetic, sympathetic, and sensory nerves. Almost all nerves of the cholinergic system (95%) present the SV2 receptor, as do 69% of the sympathetic and 58% of the sensory fibers [20,21]. The exact site of action in the urinary bladder is debated: whether exclusively or predominantly in the detrusor muscle and the suburothelial connective tissue or whether the urothelium could already be the site of attack of the botulinum toxin. While the SV2 receptor could be detected [22], SNAP25 was not detected in the human urothelium [20]. The inhibitory effect on the transmitter release of acetylcholine and, thus on, muscular overactivity is an important point of attack for botulinum toxin A. However, the release of several other transmitters after botulinum toxin injections are suspected in animal studies, e.g., ATP in healthy [22] and spinalized [23] animals. This could lead to a downregulation of the purinergic component at the suburothelial [24] and detrusor level [25]. However, the downregulation of the purinergic system is likely associated with reduced afferent activation and signaling since the decrease in suburothelial purinergic P2X3 receptors correlated well with the reduction in pathological sensations of urgency in patients [26]. Additionally, suburothelial capsaicin TRPV1 receptors, which play a role in the afferent mechano-sensation and pain pathways of the urinary bladder sensitization, were found to be downregulated after OnabotA-DI [24].

For the first time, our study documented precisely how the effect unfolds clinically—although no direct neurophysiological proof of the effect could be provided, of course (this would necessitate basic science investigations). However, regarding the onset of the effect of OnabotA-DI, all studies have only noted that it can be expected after around 1–2 weeks. Studies on the exact onset of action are rare. In striated muscle, Hamjian et al. [27] showed in the human M. extensor digitalis brevis that after injection of 10U onabotulinumtoxin A, neurophysiologically detectable activity decreased from 48 h, with a maximum decrease after 21 days. To our knowledge, no studies exist on the exact onset of action on the bladder or any other smooth muscle in humans. However, the data from our study are very helpful in the informational discussion because patients otherwise often expect an immediate onset of action.

4. Conclusions

Our study reveals for the first time the exact onset of the action of OnabotA-DI, using the example of voiding frequency because this was the most reliable documented parameter in the Diary Pod app. A trend of reduction can be seen from the 4th day, and a significant and clinically meaningful change and alleviation can be expected from the 5th day after OnabotA-DI. This is important information for the expectations of patients who choose this minimally invasive therapy.

5. Materials and Methods

Adult patients with OAB symptoms, as described, were included in this study. All patients had received oral anticholinergic or β_3-receptor agonist premedication, but this was discontinued because of insufficient effect or significant side effects, or contraindications existed. OnabotA-DI was recommended. In order to make a per se good OnabotA-DI effect likely, the EAU guideline recommendation [8,9] for the diagnosis of OAB and the exclusion of other potential causes of the symptoms were carefully observed. This was performed meticulously to minimize the rate of potential treatment failures due to OnabotA-DI misindication, which would lead to misinterpretations of the onset of action. Another inclusion criterion was the presence and confident use of a smartphone (subjective assessment by the patients) to be able to load the Minze-Health app. The patients were instructed in this app and also received instructions for use in their native language. The procedure was scheduled for 2–4 weeks after this briefing.

The patients began using the Diary Pod app 3 days before the agreed injection date to collect the pre-therapeutic basic data. The Minze Diary Pod (Figure 2) is an automated bladder diary solution that includes a measuring device connected to a mobile app. The Diary Pod is a device for all ages and genders that measures volume and time, and the app assists the patient in keeping a bladder diary in a compliant way. The collected data are visualized in a dashboard on the Minze Clinician Portal (Figure 3 left and right graph), where the results can be analyzed, compared over time, or exported as raw data for further research. The first morning void must be marked to stop the counting of nocturnal voids (nocturia). Voiding related to defecation could not be documented and had to be ignored. In principle, patients can also document drinking quantities and episodes of urinary incontinence in the app, but this was not the primary aim of the study since we had to assume that the manual registration of incontinence events in the app documentation could be incomplete.

On the day of the injection (day 0), no micturition was documented. A urine test was performed before the operation to rule out an acute UTI. The injection itself was always carried out using a rigid cystoscope (17 or 21 French) in a specified manner [28]: the onabotulinumtoxin A was dissolved in 10 mL of sodium chloride. The application was carried out either under local anesthesia with 50 mL of lidocaine and a 20 min exposure time or under general anesthesia (by an anesthetist), according to the patient's preferences. Twenty injections were made on the side walls of the bladder, the posterior wall, the base of the bladder, and in the trigonum vesicae, sparing the ostia areas. The intervention was always performed under perioperative antibiotic prophylaxis.

The automated Diary Pod was used again from the first postoperative day, ideally continued for 21 days. A control visit was made around 1–2 weeks after the injection day to rule out a UTI or significant residual urine formation.

Patients who showed a UTI in the control visit 1–2 weeks after the injection day were withdrawn from the study because this has a very significant and negative impact on the frequency of micturition. Likewise, patients who had not documented at least 2 preoperative and 7 postoperative days via the app could not be evaluated. Patients who had several gap days in a row in the visualized diary on the Minze Clinician Portal documentation were also excluded.

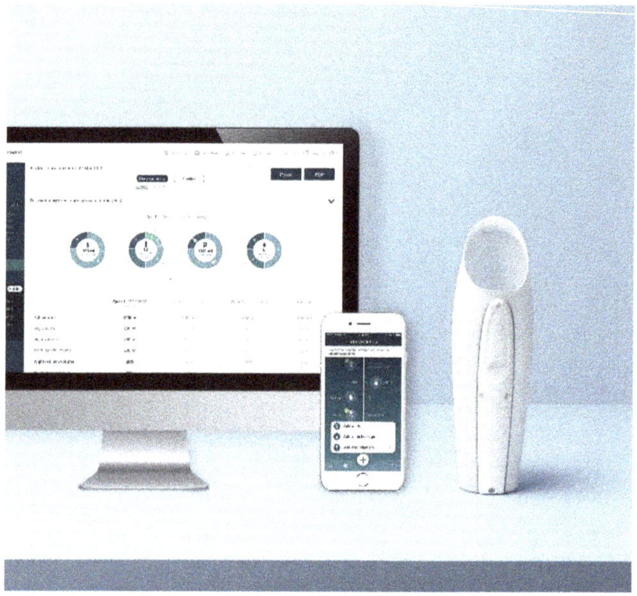

Figure 2. The Diary Pod, an automated bladder diary solution that includes a measuring device connected to a mobile app. The collected data are visualized in a dashboard on the Minze Clinician Portal.

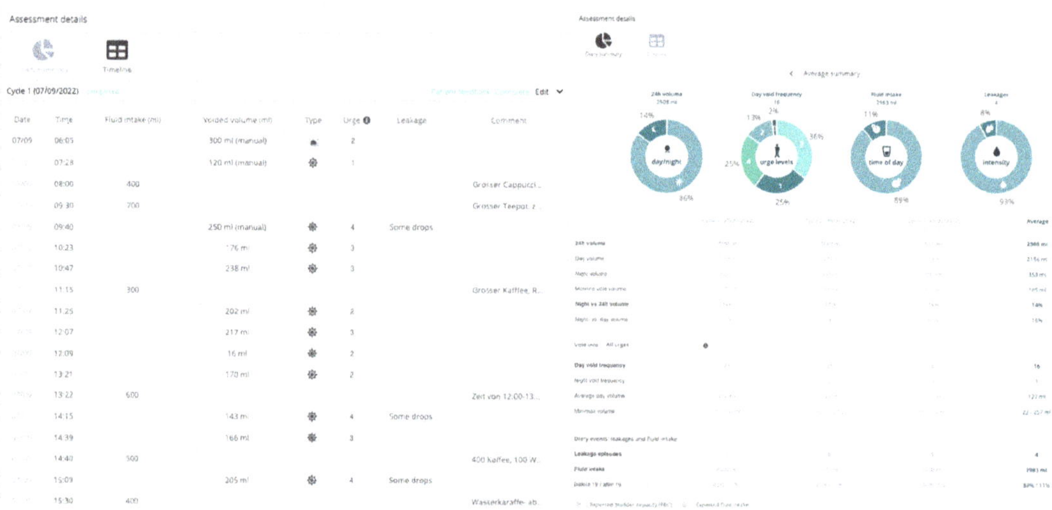

Figure 3. (Left side) Example of documentation of the kind and amount of drinking, micturitions, urgencies, and leakages and (right side) the summaries of these values.

The ethics committee of the Charité Universitätsmedizin Berlin was consulted for the study.

Statistics: The mean voiding frequency value was calculated from the 3 days before the injection therapy to obtain a picture of the pre-therapeutic state that was as representative as possible. This mean represented the reference value for the post-therapeutic voiding frequencies. The values of the patients were compared with the baseline value (=average of the first 3 days) per single day (after injection) using a one-sample *t*-test. All statistical analyzes were performed using the program R-4.2.2. for Windows. Because of the several potential disruptive factors in interpreting the onset of action (e.g., UTI, missing documentation) and because of the relatively high effort for the patients (the motivation for app documentation was expected to decrease once the therapy was completed and the desired relief occurred), in the calculation of the patients to be included, we considered a high dropout rate. This study was not supposed to prove the effect of OnabotulinumtoxinA but to document the onset of the effect, and the number of completely evaluable patients was calculated accordingly.

Author Contributions: Conceptualization, H.S.-B. and C.W.; methodology, H.S.-B.; software, S.W.; validation, H.S.-B., C.W. and M.H.; formal analysis, H.S.-B. and C.W.; investigation, H.S.-B., C.W. and T.S.; resources, H.S.-B. and T.S.; H.S.-B. and S.W.; writing—original draft preparation, H.S.-B. and B.R.; writing—review and editing, H.S.-B. and B.R.; H.S.-B. and M.H.; supervision, H.S.-B. and B.R.; project administration, H.S.-B.; funding acquisition, T.S. All authors have read and agreed to the published version of the manuscript.

Funding: This research received no external funding.

Institutional Review Board Statement: The study was approved by the Ethics Committee of Charité University Hospital (protocol code EA4/203/22 and date of approval on 16 February 2022).

Informed Consent Statement: Informed consent was obtained from all subjects involved in the study.

Data Availability Statement: Not appliable.

Conflicts of Interest: The authors declare no conflict of interest.

References

1. Abrams, P.; Cardozo, L.; Fall, M.; Griffiths, D.; Rosier, P.; Ulmsten, U.; van Kerrebroeck, P.; Victor, A.; Wein, A. Standardisation Sub-committee of the International Continence Society. The standardisation of terminology of lower urinary tract function: Report from the Standardisation Sub-committee of the International Continence Society. *Neurourol. Urodyn.* **2002**, *21*, 167–178. [CrossRef] [PubMed]
2. Fowler, C.J.; Auerbach, S.; Ginsberg, D.; Hale, D.; Radziszewski, P.; Rechberger, T.; Patel, V.D.; Zhou, J.; Thompson, C.; Kowalski, J.W. OnabotulinumtoxinA improves health-related quality of life in patients with urinary incontinence due to idiopathic overactive bladder: A 36-week, double-blind, placebo-controlled, randomized, dose-ranging trial. *Eur. Urol.* **2012**, *62*, 148–157. [CrossRef] [PubMed]
3. Bientinesi, R.; Gavi, F.; Coluzzi, S.; Nociti, V.; Marturano, M.; Sacco, E. Neurologic Urinary Incontinence, Lower Urinary Tract Symptoms and Sexual Dysfunctions in Multiple Sclerosis: Expert Opinions Based on the Review of Current Evidences. *J. Clin. Med.* **2022**, *11*, 6572. [CrossRef] [PubMed]
4. Chapple, C.; Sievert, K.D.; MacDiarmid, S.; Khullar, V.; Radziszewski, P.; Nardo, C.; Thompson, C.; Zhou, J.; Haag-Molkenteller, C. OnabotulinumtoxinA 100 U significantly improves all idiopathic overactive bladder symptoms and quality of life in patients with overactive bladder and urinary incontinence: A randomised, double-blind, placebo-controlled trial. *Eur. Urol.* **2013**, *64*, 249–256. [CrossRef] [PubMed]
5. Milsom, I.; Abrams, P.; Cardozo, L.; Roberts, R.G.; Thüroff, J.; Wein, A.J. How widespread are the symptoms of an overactive bladder and how are they managed? A population-based prevalence study. *BJU Int.* **2001**, *87*, 760–766. [CrossRef]
6. Stewart, W.F.; Van Rooyen, J.B.; Cundiff, G.W. Prevalence and burden of overactive bladder in the United States. *World J. Urol.* **2003**, *20*, 327–336. [CrossRef]
7. Irwin, D.E.; Milsom, I.; Hunskaar, S. Population-based survey of urinary incontinence, overactive bladder, and other lower urinary tract symptoms in five countries: Results of EPIC study. *Eur. Urol.* **2006**, *50*, 1306–1314. [CrossRef]
8. Nambiar, A.K.; Arlandis, S.; Bø, K.; Cobussen-Boekhorst, H.; Costantini, E.; de Heide, M.; Farag, F.; Groen, J.; Karavitakis, M.; Lapitan, M.C.; et al. European Association of Urology Guidelines on the Diagnosis and Management of Female Non-neurogenic Lower Urinary Tract Symptoms. Part 1: Diagnostics, Overactive Bladder, Stress Urinary Incontinence, and Mixed Urinary Incontinence. *Eur. Urol.* **2022**, *82*, 49–59. [CrossRef]
9. Gacci, M.; Sakalis, V.I.; Karavitakis, M.; Cornu, J.N.; Gratzke, C.; Herrmann, T.R.W.; Kyriazis, I.; Malde, S.; Mamoulakis, C.; Rieken, M.; et al. European Association of Urology Guidelines on Male Urinary Incontinence. *Eur. Urol.* **2022**, *82*, 387–398. [CrossRef]

10. Wagg, A.; Compion, G.; Fahey, A.; Siddiqui, E. Persistence with prescribed antimuscarinic therapy for overactive bladder: A UK experience. *BJU Int.* **2012**, *110*, 1767–1774. [CrossRef]
11. Nitti, V.W.; Dmochowski, R.; Herschorn, S.; Sand, P.; Thompson, C.; Nardo, C.; Yan, X.; Haag-Molkenteller, C.; EMBARK Study Group. OnabotulinumtoxinA for the treatment of patients with overactive bladder and urinary incontinence: Results of a phase 3, randomized, placebo controlled trial. *J. Urol.* **2013**, *189*, 2186–2193. [CrossRef] [PubMed]
12. Visco, A.G.; Brubaker, L.; Richter, H.E. Anticholinergic therapy vs. onabotulinumtoxinA for urgency urinary incontinence. *N. Engl. J. Med.* **2012**, *367*, 1803–1813. [CrossRef] [PubMed]
13. Drake, M.J.; Nitti, V.W.; Ginsberg, D.A. Comparative assessment of the efficacy of onabotulinumtoxinA and oral therapies (anticholinergics and mirabegron) for overactive bladder: A systematic review and network meta-analysis. *BJU Int.* **2017**, *120*, 611–622. [CrossRef] [PubMed]
14. Eldred-Evans, D.; Sahai, A. Medium- to long-term outcomes of botulinum toxin A for idiopathic overactive bladder. *Ther. Adv. Urol.* **2017**, *9*, 3–10. [CrossRef] [PubMed]
15. Nitti, V.W.; Ginsberg, D.; Sievert, K.D.; Sussman, D.; Radomski, S.; Sand, P.; De Ridder, D.; Jenkins, B.; Magyar, A.; Chapple, C.; et al. Durable Efficacy and Safety of Long-Term OnabotulinumtoxinA Treatment in Patients with Overactive Bladder Syndrome: Final Results of a 3.5-Year Study. *J. Urol.* **2016**, *196*, 791–800. [CrossRef]
16. Hamid, R.; Lorenzo-Gomez, M.F.; Schulte-Baukloh, H.; Boroujerdi, A.; Patel, A.; Farrelly, E. OnabotulinumtoxinA is a well tolerated and effective treatment for refractory overactive bladder in real-world practice. *Int. Urogynecol. J.* **2021**, *32*, 65–74. [CrossRef]
17. Apostolidis, A.; Dasgupta, P.; Fowler, C.J. Proposed mechanism for the efficacy of injected botulinum toxin in the treatment of human detrusor overactivity. *Eur. Urol.* **2006**, *49*, 644–650. [CrossRef]
18. Dong, M.; Yeh, F.; Tepp, W.H.; Dean, C.; Johnson, E.A.; Janz, R.; Chapman, E.R. SV2 is the protein receptor for botulinum neurotoxin A. *Science* **2006**, *312*, 592–596. [CrossRef]
19. Dolly, O. Synaptic transmission: Inhibition of neurotransmitter release by botulinum toxins. *Headache* **2003**, *43* (Suppl. S1), S16–S24. [CrossRef]
20. Coelho, A.; Dinis, P.; Pinto, R.; Gorgal, T.; Silva, C.; Silva, A.; Silva, J.; Cruz, C.D.; Cruz, F.; Avelino, A. Distribution of the high-affinity binding site and intracellular target of botulinum toxin type A in the human bladder. *Eur. Urol.* **2010**, *57*, 884–890. [CrossRef]
21. Coelho, A.; Cruz, F.; Cruz, C.D.; Avelino, A. Spread of onabotulinumtoxinA after bladder injection. Experimental study using the distribution of cleaved SNAP-25 as the marker of the toxin action. *Eur. Urol.* **2012**, *61*, 1178–1184. [CrossRef] [PubMed]
22. Hanna-Mitchell, A.T.; Wolf-Johnston, A.S.; Barrick, S.R.; Kanai, A.J.; Chancellor, M.B.; de Groat, W.C.; Birder, L.A. Effect of botulinum toxin A on urothelial-release of ATP and expression of SNARE targets within the urothelium. *Neurourol. Urodyn.* **2015**, *34*, 79–84. [CrossRef] [PubMed]
23. Khera, M.; Somogyi, G.T.; Kiss, S.; Boone, T.B.; Smith, C.P. Botulinum toxin A inhibits ATP release from bladder urothelium after chronic spinal cord injury. *Neurochem. Int.* **2004**, *45*, 987–993. [CrossRef] [PubMed]
24. Apostolidis, A.; Popat, R.; Yiangou, Y.; Cockayne, D.; Ford, A.P.; Davis, J.B.; Dasgupta, P.; Fowler, C.J.; Anand, P. Decreased sensory receptors P2X3 and TRPV1 in suburothelial nerve fibers following intradetrusor injections of botulinum toxin for human detrusor overactivity. *J. Urol.* **2005**, *174*, 977–982. [CrossRef] [PubMed]
25. Schulte-Baukloh, H.; Priefert, J.; Knispel, H.H.; Lawrence, G.W.; Miller, K.; Neuhaus, J. Botulinum toxin A detrusor injections reduce postsynaptic muscular M2, M3, P2X2 and P2X3 receptors in children and adolescents who have neurogenic detrusor overactivity: A single-blind study. *Urology* **2013**, *81*, 1052–1057. [CrossRef] [PubMed]
26. Apostolidis, A.; Rahnama'i, M.S.; Fry, C.; Dmochowski, R.; Sahai, A. Do we understand how botulinum toxin works and have we optimized the way it is administered to the bladder? ICI-RS 2014. *Neurourol. Urodyn.* **2016**, *35*, 293–298. [CrossRef]
27. Hamjian, J.A.; Walker, F.O. Serial neurophysiological studies of intramuscular botulinum-A toxin in humans. *Muscle Nerve* **1994**, *17*, 1385–1392. [CrossRef]
28. Available online: https://www.accessdata.fda.gov/drugsatfda_docs/label/2017/103000s5302lbl.pdf (accessed on 26 November 2022).

toxins MDPI

Article

Xeomin®, a Commercial Formulation of Botulinum Neurotoxin Type A, Promotes Regeneration in a Preclinical Model of Spinal Cord Injury

Valentina Mastrorilli, Federica De Angelis, Valentina Vacca, Flaminia Pavone, Siro Luvisetto * and Sara Marinelli *

Institute of Biochemistry and Cell Biology, National Council of Research of Italy, Via Ercole Ramarini 32, 00015 Monterotondo, Italy

* Correspondence: siro.luvisetto@cnr.it (S.L.); sara.marinelli@cnr.it (S.M.)

Abstract: Xeomin® is a commercial formulation of botulinum neurotoxin type A (BoNT/A) clinically authorized for treating neurological disorders, such as blepharospasm, cervical dystonia, limb spasticity, and sialorrhea. We have previously demonstrated that spinal injection of laboratory purified 150 kDa BoNT/A in paraplegic mice, after undergoing traumatic spinal cord injury (SCI), was able to reduce excitotoxic phenomena, glial scar, inflammation, and the development of neuropathic pain and facilitate regeneration and motor recovery. In the present study, as proof of concept in view of a possible clinical application, we studied the efficacy of Xeomin® in the same preclinical SCI model in which we highlighted the positive effects of lab-purified BoNT/A. Data comparison shows that Xeomin® induces similar pharmacological and therapeutic effects, albeit with less efficacy, to lab-purified BoNT/A. This difference, which can be improved by adjusting the dose, can be attributable to the different formulation and pharmacodynamics. Although the mechanism by which Xeomin® and laboratory purified BoNT/A induce functional improvement in paraplegic mice is still far from being understood, these results open a possible new scenario in treatment of SCI and are a stimulus for further research.

Keywords: botulinum neurotoxin; spinal cord injury; regeneration; motor recovery; sciatic static index; neuropathic pain; glial cells; mice

Key Contribution: Spinal injection of Xeomin® in paraplegic mice, subjected to traumatic injury at the thoracic level of the spinal cord, induced spinal cord regeneration and motor recovery. These beneficial effects were achieved by reducing excitotoxic phenomena, glial scar and inflammation.

Citation: Mastrorilli, V.; De Angelis, F.; Vacca, V.; Pavone, F.; Luvisetto, S.; Marinelli, S. Xeomin®, a Commercial Formulation of Botulinum Neurotoxin Type A, Promotes Regeneration in a Preclinical Model of Spinal Cord Injury. *Toxins* **2023**, *15*, 248. https://doi.org/10.3390/toxins15040248

Received: 1 March 2023
Revised: 17 March 2023
Accepted: 27 March 2023
Published: 28 March 2023

1. Introduction

Botulinum neurotoxins (BoNTs) are produced by *Clostridium botulinum* in eight different serotypes, named by letters from A to G, and X [1,2]. A number of different subtypes, together with chimeric molecules, complete the large family of these toxins [3,4]. All serotypes consist of a 100 kDa di-chain molecule, called heavy chain (HC), which binds to nerve membrane receptors, and a 50 kDa molecule, called light chain (LC), which enters the cytosol where it cleaves the soluble NSF (N-ethylmaleimide-sensitive factor) attachment receptor (SNARE) proteins, the key components whose integrity is required for the formation of the protein complex responsible for the fusion of synaptic vesicles with the cell membrane [1,5–7]. All BoNTs act by specifically cleaving a different peptide bond on one of the SNARE proteins: BoNT/A and /E cleave SNAP-25; BoNT/B, /D, /G and /F cleave VAMP/synaptobrevin; whereas BoNT/C cleaves both syntaxin and SNAP-25 [8].

Several studies evidenced the use of BoNTs, mainly BoNT/A and /B serotypes, for therapy in a variety of human diseases [9,10]. Today, BoNT/A and /B are licensed for treatment of several autonomic nervous system and movement disorders, such as dystonias, muscle spasms, spasticity, excessive sweating, overactive urinary bladder, along with many off-label uses in other neurological pathologies [11,12], besides the well-known treatment for aesthetic purposes [13]. A potential role of BoNT/A as a novel agent in pain relief has also been demonstrated [14–20], and the use of BoNT/A for the prophylactic treatment of migraine has recently been approved [21–23].

In a previous study [24], we demonstrated the beneficial effect of spinal injection of lab-purified 150 kDa BoNT/A protein in a mouse model of spinal cord injury (SCI) at two different degrees of severity [25,26]. In severe SCI, traumatic spinal cord injury resulted in complete hindlimb paralysis, while in moderate SCI, the damage was partial and the mice retained some movement and sensation in the hindlimbs. Within one hour from SCI, i.e., during the acute phase of the injury, we injected a single dose of BoNT/A (15 pg dissolved in 5 μL of saline) and observed: (i) an extraordinary motor recovery from paralysis with reconstruction of the damaged spinal cord in mice with severe SCI; and, (ii) in addition to motor recovery, a prevention of the development of neuropathic pain, a comorbidity often associated with SCI, in mice with moderate SCI.

The purpose of the present study was to investigate whether the beneficial effects of lab-purified BoNT/A in counteracting SCI [24] could also be obtained using a commercial formulation of BoNT/A. Among the various formulations of BoNT/A on the market, we chose Xeomin® (IncobotulinumtoxinA; Merz Pharmaceuticals GmbH, Frankfurt am Main, Germany) because this drug does not contain complexing proteins, which keeps the molecule similar to the one purified in the lab.

2. Results

To evaluate the effects of Xeomin® on motor paralysis, we tested mice with both severe and moderate SCI; to evaluate the effect of Xeomin® on sensory deficits and neuropathic pain symptoms, we tested only mice with moderate SCI. Furthermore, to avoid the entry of toxin into the blood stream, as a consequence of the destruction of the blood-spinal barrier by SCI at the thoracic level (T9–T11), we injected Xeomin® (single dose of 2.5 U dissolved in 5 μL of saline) not directly to the lesion site but at the lumbar level (L4–L5). The dose of toxin was chosen on the basis of the conversion ratio suggested by the manufacturer: 2.5U of Xeomin® corresponds to approximately 15 pg of lab-purified 150 kDa BoNT/A, a single dose used in our previous study [24]. As the control group, some SCI mice were injected with saline (0.9% NaCl).

Figure 1A shows the BMS score, expressed as incremental recovery (ΔBMS) starting from day 3 after SCI (D3), obtained in severe SCI mice treated with saline or Xeomin®. As inclusion criteria, all mice that at D3 performed a BMS score in the range 0–3 were considered in this group. At D3, ΔBMS was calculated with respect to baseline value of BMS = 9 (normal movement of the hindlimbs) and the days from SCI, which in our experimental protocol is equal to 3, by using the Equation (1):

$$\Delta\text{BMS at D3} = ((\text{BMS value at D3} - 9)/3) \quad (1)$$

while at Dx (x = 7, 10, 14, 21, 28, 35), ΔBMS was calculated with respect to BMS value at D3 and the days from D3, by using the Equation (2):

$$\Delta\text{BMS at Dx} = ((\text{BMS value at Dx} - \text{BMS value at D3})/\text{days from D3}) \quad (2)$$

Although an increasing trend was observed, the mean ΔBMS of saline- or Xeomin®-treated SCI mice was not different from D3 to D21. Starting from D28, Xeomin®-treated mice showed a significant amelioration of motor deficit when compared with saline-treated mice (ANOVA for repeated measures: interaction time x treatment $F_{6,54} = 7.557$, $p < 0.0001$) (see also Supplementary Video-S1 and Supplementary Video-S2).

Figure 1B shows the ΔBMS score, as a function of days after SCI, obtained in moderate SCI mice treated with saline or Xeomin®. All mice that at D3 performed a BMS score in the range of 3 to 6 were included in this group. Both saline- and Xeomin®-treated mice were not significantly different.

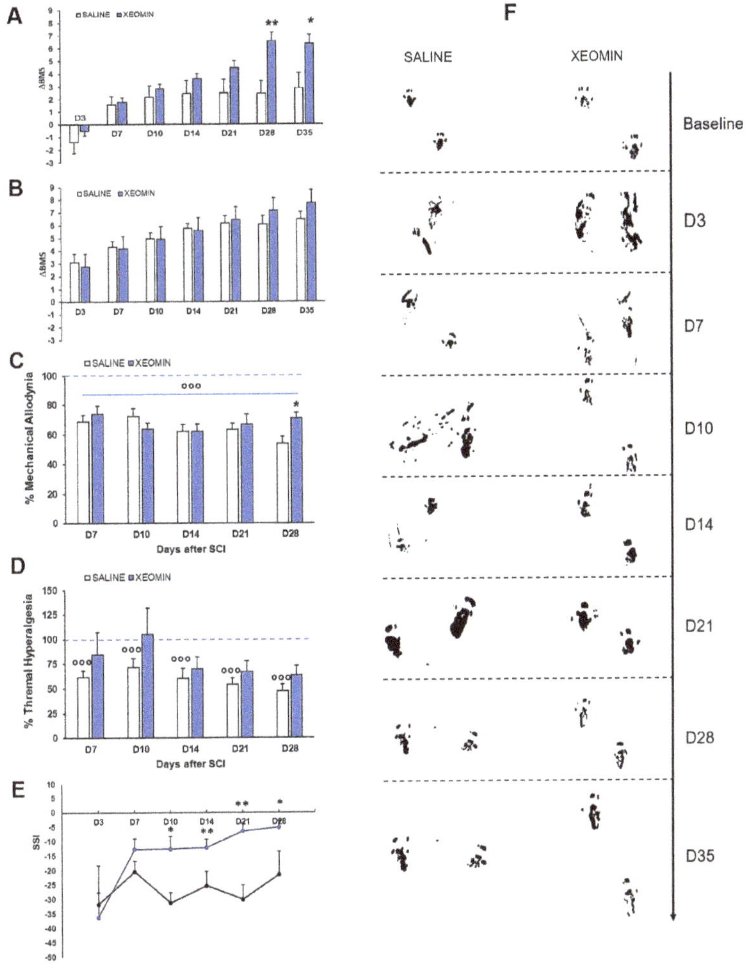

Figure 1. (**A**) Incremental ratio (ΔBMS) of motor recovery, calculated with respect to baseline for D3 (first day of the BMS measurement) or D3 for the subsequent days, in saline- or Xeomin®-treated severe SCI mice (N = 5–6/group). Statistics: Tukey-Kramer post-hoc test: (*) $p < 0.05$, (**) $p < 0.001$ vs. saline. (**B**) ΔBMS of motor recovery in saline- or Xeomin®-treated moderate SCI mice (N = 5/group). (**C**) Percentage of mechanical allodynia threshold in saline- or Xeomin®-treated moderate SCI mice (N = 5/group) with respect to the threshold before SCI. Statistics: Tukey-Kramer post-hoc test: (°°°) $p < 0.0001$ vs. baseline (blue dashed line at 100%); unpaired t-test at D28: (*) $p < 0.05$ vs. saline; (**D**) Percentage of thermal hyperalgesia threshold in saline- or Xeomin®-treated moderate SCI mice (N = 5/group) with respect to the threshold before SCI. Statistics: unpaired *t*-test: (°°°) $p < 0.0001$ vs. baseline (blue dashed line at 100%). (**E**) Sciatic Static Index (SSI) calculated from hindpaws' footprints in saline- (black line) or Xeomin®-treated (blue line) moderate SCI mice (N = 5/group). Values of SSI = 0 represent normal walking, while negative values are an index of walking deficits. Statistics: Tukey-Kramer post-hoc test: (*) $p < 0.05$, (**) $p < 0.001$ vs. saline. (**F**) Representative examples of footprint walking track in saline- or Xeomin®-treated moderate SCI mice.

Because the hindlimbs of moderate SCI mice were not completely paralyzed, these mice retained nociceptive sensitivity and thus could respond to nociceptive stimuli. For this reason, on moderate SCI mice, we were able to perform a behavioral analysis to ascertain whether Xeomin® was able to counteract the onset of neuropathic pain. Figure 1C,D shows the percentages of mechanical allodynia threshold (panel C) and thermal hyperalgesia threshold (panel D) in moderate SCI mice, at different days post SCI, calculated for each mouse with respect to the baseline threshold before SCI (BL) using Equation (3):

$$\% \text{ of threshold} = (\text{threshold at Dx/BL}) \times 100 \quad (3)$$

Both saline- and Xeomin®-treated mice developed and maintained mechanical allodynia for all time-points considered. ANOVA one-way comparing treatments with their baseline shows a significant effect for both the saline ($F_{5,50} = 13.645$, $p < 0.0001$) and Xeomin® ($F_{5,46} = 8.685$, $p < 0.0001$) group. Post-hoc comparison between groups evidenced that Xeomin® was significantly different from saline at D28.

More evident is the effect of Xeomin® on thermal hyperalgesia. In detail, the threshold-to-thermal stimuli in saline-treated SCI mice was significantly reduced, indicating greater sensitivity to thermal pain, compared with baseline values at each testing day. In contrast, the thermal pain response in Xeomin®-treated SCI mice was not significantly different from baseline values. One-way ANOVA comparing treatments with their baseline shows a significant effect for saline ($F_{5,49} = 13.336$, $p < 0.0001$) but not for the Xeomin® group. This finding indicates that Xeomin® was able to prevent the worsening of thermal sensitivity due to the development of neuropathy induced by SCI.

Figure 1E reports the Sciatic Static Index (SSI) calculated from the analysis of footprint parameters in saline- or Xeomin®-treated moderate SCI mice. Figure 1F shows two representative examples of footprint during walking. The SSI allows to evaluate motor function and ambulation recovery because it analyzes footprint parameters not easily detectable with BMS analysis. The SSI analysis shows that saline- and Xeomin®-treated moderate SCI mice had an approximately 35% walking deficit at D3 after SCI, and only Xeomin® treatment was able to induce rapid and significant recovery. Two-way ANOVA for repeated measures shows a significant main effect for treatment ($F_{1,18} = 7.18$, $p = 0.015$) and time ($F_{3,5} = 3.69$, $p = 0.0044$). Moreover, note that the footprint of Xeomin®-treated mice at D28 and D35 showed a regular gait compared with saline-treated mice.

Figure 2 shows examples of immunofluorescence (IF) images of glial fibrillary acid protein (GFAP, astrocyte marker) expression in spinal cord sections taken from the impact zone in saline- or Xeomin®-treated severe SCI mice, at D35 after SCI. As previously demonstrated [24], large areas of the spinal cord of saline-treated SCI mice were damaged with glial scarring and astrocyte hyperactivation, and the spinal cord size was markedly reduced with the spinal horns completely enveloped by the glial scar. In the Xeomin® treated SCI mouse, although astrocyte activation and small glial scarring areas were still present, especially in the impact zone, the spinal cord morphology was preserved, with clearly recognizable dorsal and ventral horns.

To test whether Xeomin® could exert a neuroprotective action against excitotoxic phenomena, as previously observed for lab-purified BoNT/A [24], and considering that glutamate induces excitotoxic cell death [27], we analyzed the expression of vesicular glutamate transporter type 1 (vGLUT1) in colocalization with GFAP expression in severe SCI mice, at D35 after SCI. Figure 3A,B shows representative examples of IF images of the colocalization of vGLUT1 with GFAP expression in spinal cord sections taken from the impact area. In the spinal cord of saline-treated SCI mice (panels A), a striking expression of vGLUT1 was detected together with a strong colocalization with GFAP. In contrast, in spinal cord of Xeomin®-treated mice (panels B), both vGLUT1 expression and its colocalization with GFAP were strongly reduced.

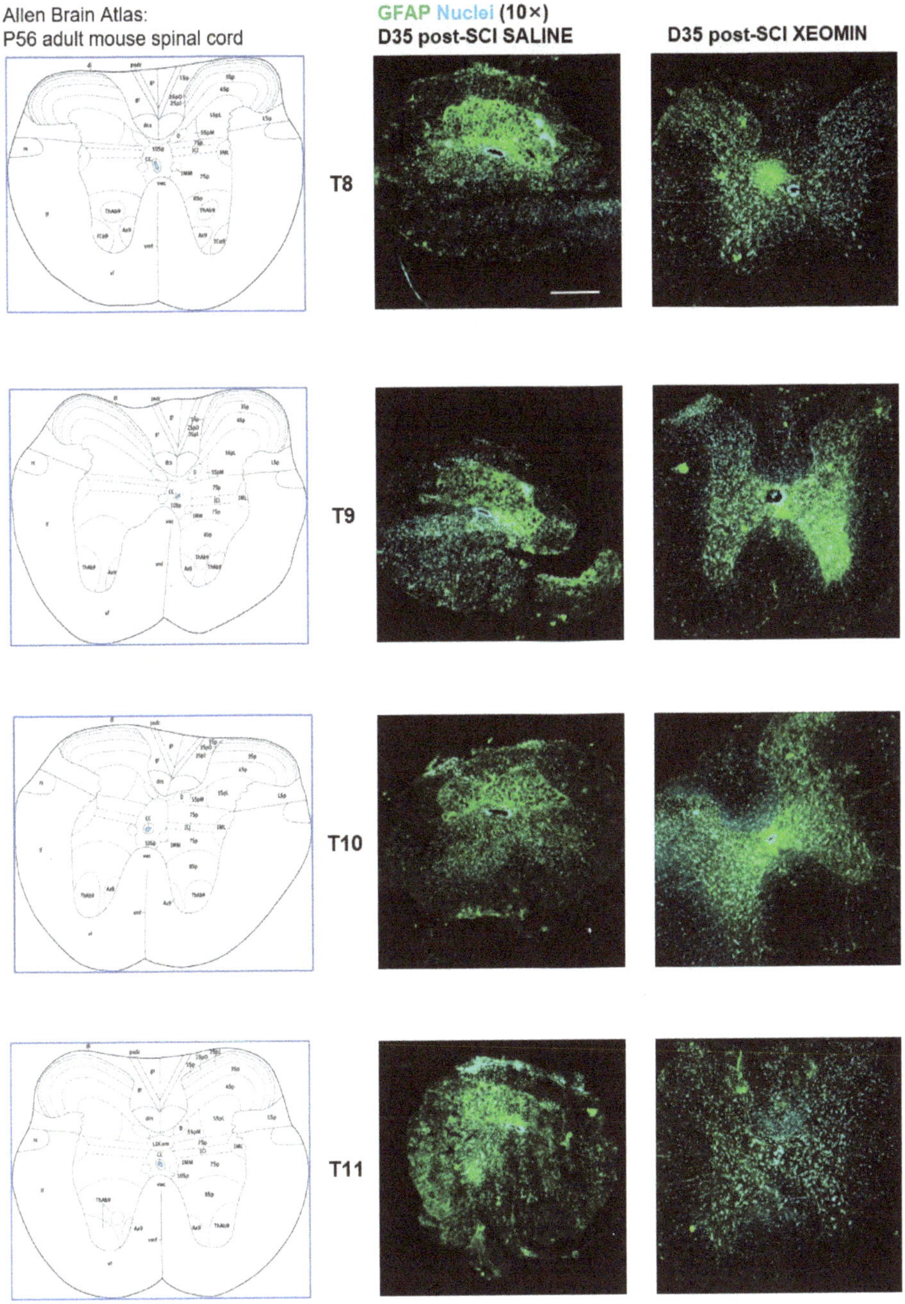

Figure 2. Neuroprotective effects of Xeomin® on injured spinal cord. Representative examples of confocal images (10×) deriving from thoracic (T8/T11) spinal cord of saline- or Xeomin®-treated mice, collected at D35 after severe SCI. GFAP (green) marker evidences astroglia reactivity and/or glial scar. With respect to the impact: EPI indicates the epicenter zone of injury, i.e., the core lesion at T9, while PERI indicates the peri-lesioned zone of injury. Bar 300 µm.

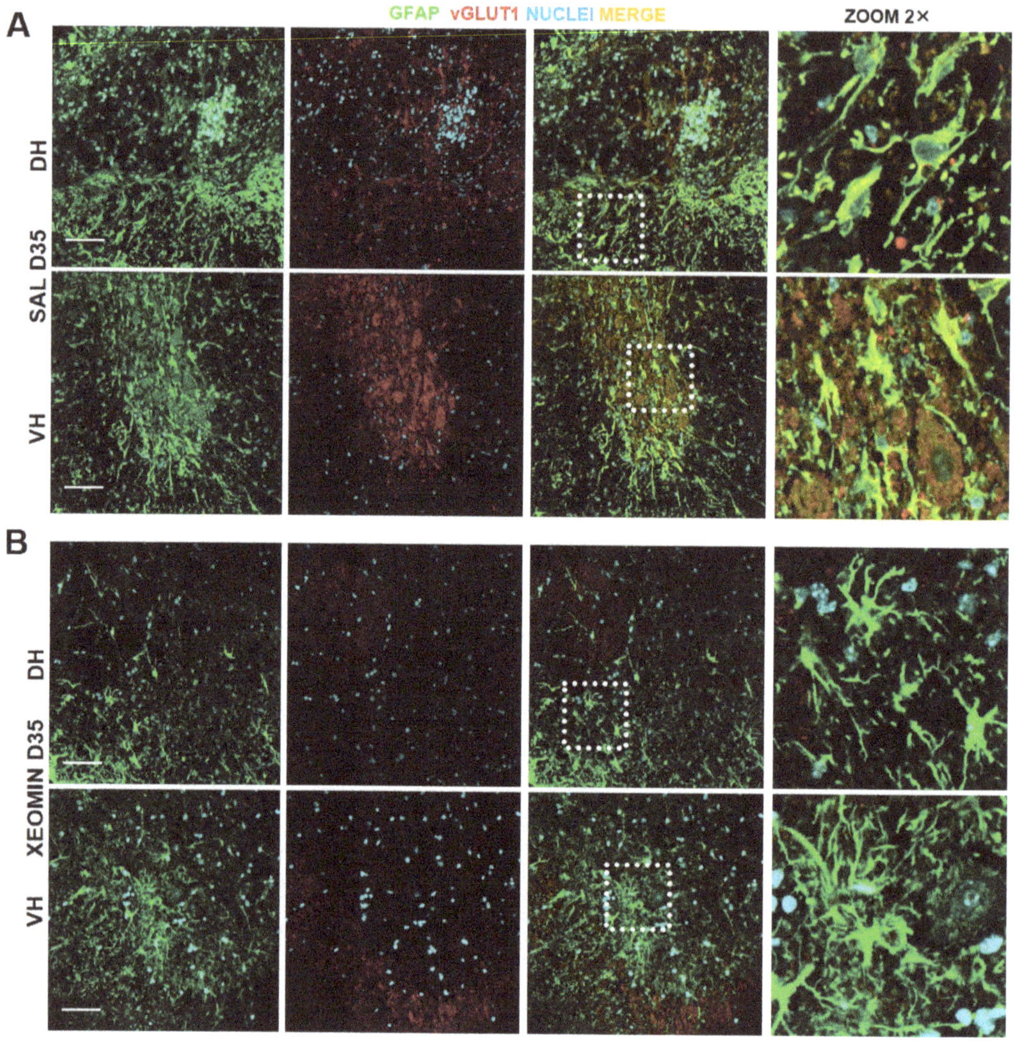

Figure 3. Xeomin® modulates astrocytes reactivity and glial scar formation. Representative examples of confocal images (40× and 40×-zoom2×) deriving from thoracic (T9) spinal cord of: (**A**) saline- or (**B**) Xeomin®-treated mice, collected at D35 after severe SCI. Immunostaining in dorsal (DH) and ventral horn (VH): GFAP (green) marker stains for astrocytes gliosis and glial scar, vGLUT1 (red) marker stains for the glutamate transporter as indirect index of glutamate release, while merge (yellow) indicates colocalization of the two markers. Dotted white boxes indicate areas where the zoom was taken. Bar 60 μm.

3. Discussion and Conclusions

In a previous article [24], we demonstrated that a spinal injection of lab-purified 150 kDa BoNT/A was able to: (i) induce spinal nerve regeneration and motor recovery in a severe SCI mice model; and (ii) counteract the development of neuropathic pain in a moderate SCI mice model. These effects correlated with reduced activation of spinal glia and reduced formation of glial scar, events that positively contrast the cascade of adverse phenomena occurring during SCI-induced secondary injury [28–32]. Thus, by modulating the glial response, BoNT/A allows a gradual progression toward a facilitated recovery of motor and sensory function.

In the present study, we tested Xeomin® to evaluate its possible therapeutic potential in counteracting SCI with a view to its possible clinical application. We chose Xeomin® because, of all the BoNTs available on the market, it is the commercial formulation that most closely matches the lab-purified BoNT/A. Although our findings unequivocally demonstrate a therapeutic effect of Xeomin on SCI, we also highlighted some substantial differences in comparison with the effects of lab-purified BoNT/A. The most evident regards the pharmacokinetic. In fact, in comparison with the lab-purified BoNT/A, we observed a delay in the regenerative and neuroprotective effects on SCI with Xeomin®, as well as modest effectiveness. In detail, the time of onset of therapeutic effects was the first day after administration in the case of lab-purified BoNT/A [24], while in the case of Xeomin®, it was three weeks after administration.

Another discrepancy concerns motor recovery, estimated by the BMS parameter. In fact, motor recovery was almost complete in the case of lab-purified BoNT/A, with BMS values between 8 and 9 [24], while motor recovery was not complete in the case of Xeomin®, with BMS values between 6 and 8. This is particularly evident in the moderate SCI model in which it was not possible to highlight better motor recovery of Xeomin® compared to saline. This is probably caused by the ability of the murine axons to regenerate. When the neurological insult is moderate and does not completely affect the spinal nervous system, the natural functional recovery can mask the limited therapeutic effect of Xeomin®. It is also evident that, although Xeomin® appears to be ineffective in moderate SCI, analysis of footprints during walking demonstrates a strong ability of Xeomin® to induce early functional recovery.

Comparing the results obtained from lab-purified BoNT/A [24] with results obtained from Xeomin®, another difference immediately emerges in the effects on neuropathic pain. In fact, although Xeomin® was able to mitigate the symptoms of neuropathic pain, it did not completely prevent the onset of neuropathic pain or restore the physiological threshold, especially in mechanical allodynia. On the other hand, from an overview of spinal morphology, astrocytosis and excitotoxicity, Xeomin® was effective in inducing neuroprotection and reduction of glial scar.

All these discrepancies can be reasonably attributed again to pharmacodynamics. It has been already demonstrated that different BoNT/A formulations influence the effectiveness requiring a dose adjustment [33–35]. In particular, Byun et al. [35] evidenced that BoNT/A with high efficacy and duration is that one lacking human-derived component (coretox®), such as human albumin, which are substituted by polysorbate 20 and L-methionine. Also the higher efficacy of our lab-purified BoNT/A, with respect to Xeomin®, could be explained by the absence of these animal components. This is an important point to be considered because pharmacodynamics and pharmacocinetics of the toxin are influenced by the immunesystem response that can be activated by the exogenous antigens, such as the effect exerted by human albumin in mice.

Moreover, the different onset of therapeutic effects of Xeomin®, the BoNT/A formulation used in this study, compared to lab-purified 150 kDa BoNT/A protein used in [24] may originate from the difficulty obtaining the exact correspondence between doses of toxin, expressed by units of Xeomin® and pico-grams of BoNT/A in [24]. Dose correspondence between picograms and units of Xeomin® was calculated considering that vials of Xeomin® contained approximately 600 pg of toxin per 100 units, thus 15 pg of lab-purified 10 kDa BoNT/A corresponded to 2.5 units of Xeomin®. Another discrepancy between the two BoNT/A formulations resides in the solvent and/or excipients used in preparation of lab-purified BoNT/A and Xeomin®. Both BoNT/A are prepared by dilution of stock solution in saline (0.9%), but Xeomin® also contains human albumin and sucrose.

Overall, our results demonstrate that Xeomin® may be a possible candidate in clinical application as a therapy against SCI, although a dose-response study would be desirable. In conclusion, the present study represents proof of concept for the clinical application of BoNT/A in the therapy of traumatic SCI, as we validated and confirmed the pro-regenerative and neuroprotective action of Xeomin®, one of the most used commercial compounds in neurological disorders.

4. Materials and Methods

4.1. Animals

Four-month-old CD1 female mice (EMMA Infrafrontier, Monterotondo, Italy) were used. Mice were housed in groups of 4 in standard cages under a 12/12 h light/dark cycle (7:00 a.m.–7:00 p.m.), with food and water available ad libitum. Thirty minutes before surgery, mice were moved to a surgical room and were randomly assigned to different experimental groups. The groups' size for in vivo experiments was calculated by implementing a power analysis (Gpower 3.1), and the number of mice used is reported in the figure legends. Testing was done by blind investigators as for treatment groups. Care and handling of mice were in accordance with the guidelines of the Committee for Research and Ethical Issues of IASP [36]. In vivo procedures were approved by the Italian Ministry of Health (PR122/2019, 10 February 2019) on the use of animals for research.

4.2. Surgery

Different groups of mice were subjected to SCI with severe or moderate traumatic injury. Detailed surgical procedures and postoperative care were as described in [24,25]. Spinal cord contusion was done at the thoracic level (T9-T11) using a cortical PinPoint precision impactor device (Hatteras Instruments Inc., Cary, NC, USA). Impactor parameters were as follows: middle, round, and flat tip (#4); depth: 5 mm; velocity: 3 m/s (severe SCI) or 1 m/s (moderate SCI); dwell time: 800 ms (severe SCI) or 75 ms (moderate SCI).

4.3. Drugs

Within 1 h from contusion, a single dose (2.5U in 5 μL saline) of Xeomin® (Merz Pharmaceuticals GmbH, Frankfurt am Main, Germany), or saline (5 μL), was spinally injected through intervertebral space at the lumbar level (L4–L5), using an automatic injector (KDS 310 Plus; KD Scientific; Holliston, MA, USA) equipped with a 10 μL syringe (Hamilton #701; Biosigma; Cona, Italy) with a needle of 30 μm internal diameter (Eppendorf, Hamburg, Germany). Injections were made at 2 μL/min and, to avoid liquid reflux, the needle was maintained in situ for 3 min after the end of the injection, as already described in [24,25].

4.4. Behavioral Test: Basso Mouse Scale (BMS)

For all mice groups, the hindlimb functions were assessed by estimating the BMS score in an open field arena as previously described [24]. The BMS score is a parameter, ranging from 0 to 9, where 0 indicates complete paralysis while 9 indicates normal hindlimbs movement. Each mouse was tested at D3, D7, D10, D14, D21, D28, and D35 post-SCI, and BMS scores of left and right hindlimbs were averaged together. Mice with a BMS score between 0 and 3 at D3 were included in the severe SCI group, while mice with a BMS score between >3 and 6 at D3 were included in the moderate SCI group. Mild-lesioned mice with BMS > 6 were excluded. To better appreciate amelioration in BMS values, we calculated the incremental BMS performance for the first day of BMS measurement (D3).

4.5. Nociceptive Tests

Mechanical allodynia was tested using a Dynamic Plantar Aesthesiometer (Model 37,400, Ugo Basile Srl, Comerio, Italy) as described in [37]. Thermal hyperalgesia was tested using an automatic plantar test instrument (Plantar Test, Ugo Basile, Comerio, Italy) as described in [38]. For the thermal hyperalgesia test, a cut-off time of 15 s was imposed to avoid damage of hindpaw skin tissue. Both the mechanical and thermal threshold were measured before injury (baseline value) and at D3, D7, D10, D14, and D28 after SCI, on the same moderate SCI mice, with an interval between the two tests of one hour. For each mouse, two values of mechanical and thermal threshold after SCI were obtained because the two hindpaws, the right and the left, can develop different degrees of neuropathic pain. At each testing day, threshold values were averaged from 3 consecutive measurements per hindpaw and reported as a percentage with corresponding the baseline value.

4.6. Sciatic Static Index

As well as the BMS score, recovery of the hindpaw functionality was also analysed by the the recording of the walking footprints in moderate SCI mice, as described in [39]. Footprints were collected by painting with ink on the plantar surface of the hindpaws. The toe spread (TS), from the 1st to 5th toe, and the paw length (PL), from the tip of the 3rd toe to the most posterior aspect of the paw, were considered to calculate the Sciatic Static Index (SSI). These parameters were measured from at least five footprints, taken from three different walking tracks. At each different walking track, the plantar surface of the hindpaws were repainted with ink, to avoid that footprint becoming weaker after a few runs. As suggested by [40], SSI was calculated by the equation:

$$SSI = +101.3 \times (ITS - NTS)/NTS - 54.03 \times (IPL - NPL)/NPL - 9.5$$

where ITS and NTS are the injured and normal toe spread, respectively, while IPL and NPL are the injured and normal paw length, respectively. SSI values range between 0 (normal fuction) to 100 (complete loss of hindpaw functionality).

4.7. Immunohistochemistry and Confocal Images

Spinal cords from saline- or Xeomin-treated animals were harvested at D35 for IF analysis as described in [24]. Briefly, sacrify mice were immediately perfused with saline and 4% paraformaldehyde (PFA) in phosphate buffer solution (PBS; pH 7.4). The spinal cord was removed, maintained 48 h in 4% PFA in PBS and, after cryoprotection with a solution of 30% (*w*/*v*) sucrose in PBS, conserved at −80 °C. A selection of T9–T11 tissue slices (40 microns) were collected in PBS up to the IF procedure. For double IF staining, slices were incubated 48 h with the following primary antibodies diluted in PBS with 0.3% Triton (Sigma-Aldrich St. Louis, Missouri, USA): (i) anti-GFAP (astrocytes marker; mouse monoclonal 1:100; Sigma-Aldrich); (ii) anti-vGLUT1 (vesicular glutamate transporter 1; guinea pig 1:200; Millipore AB5905). After three washes in PBS, the slices were incubated 2 h at room temperature with ALEXA Fluor 488 donkey anti-mouse (1:100; Jackson ImmunoResearch Europe Ltd, Cambridge, UK) and Rhodamine goat anti-guinea pig (1:100; Jackson ImmunoResearch Europe Ltd, Cambridge, UK). After 2 washings in PBS, slices were incubated 10 min with bisBenzimide, DNA-fluorochrome (Hoechst, 1:1000, Sigma-Aldrich Italia, Milano, Italy) in PBS.

Low (10×) and high (40×) magnification IF images were recorded by a confocal microscopy using a TCS SP5 microscope (Leica Microsystems Srl, Milan, Italy), in sequential laser scanning mode, to rule out cross-bleeding between channels. Images were analysed by I.A.S. software (Leica Microsystems Srl, Milan, Italy).

4.8. Data Analysis

Experimental data are expressed as mean ± sem. Group comparisons were conducted by one-way or two-way ANOVA for repeated measures or by Student's *t*-test. Post hoc comparisons were made with the Tukey–Kramer (statistical significance at $p < 0.05$). Data analysis was performed by StatView SAS (version 5.0, Cary, NC, USA).

Supplementary Materials: The following supporting material can be downloaded from the websites: https://www.mdpi.com/article/10.3390/toxins15040248/s1, Video-S1: Video of a saline-treated severe SCI mouse recorded at increasing days post-SCI; Video-S2: Video of a Xeomin®-treated severe SCI mouse recorded at increasing days post SCI.

Author Contributions: Conceptualization, S.M. and F.P.; data curation and formal analysis, V.M., F.D.A., V.V. and S.M.; funding acquisition, S.M.; investigation, V.M., V.V., F.D.A. and S.M.; methodology, V.M. and S.M.; supervision, F.P., S.L. and S.M.; writing-original draft, F.P., S.L. and S.M.; writing-review and editing, F.P., S.L. and S.M. All authors have read and agreed to the published version of the manuscript.

Funding: This research was funded by the grant AMICO Program-PoC, Ministero dello Sviluppo Economico (SCIBTX-A) 2021–2022.

Institutional Review Board Statement: The animal study protocol was approved by the Institutional Review Board OpBA (Organizazione per il Benessere Animale) of Institute of Biochemistry and Cell Biology and by Istituto Superiore di Sanità (Ministero della Salute; protocol code 122/2019PR, date of approval 12 February 2019).

Data Availability Statement: Data presented in this study are available on request from the corresponding authors.

Acknowledgments: Special thanks to Alessia Naso (Struttura di Particolare Rilievo "Valorizzazione della Ricerca" (SPRVR)—CNR) to support in patent development.

Conflicts of Interest: The funders had no role in the design of the study; collection, analyses, or interpretation of data; writing of the manuscript; or decision to publish the results. F.P., S.L., S.M., and V.V. are the inventors of the patent Nr. WO2016170501A1 "A new therapeutic use of the botulinum neurotoxin serotype A".

References

1. Montecucco, C.; Schiavo, G. Structure and function of tetanus and botulinum neurotoxins. *Q. Rev. Biophys.* **1995**, *28*, 423–472. [CrossRef]
2. Masuyer, G.; Zhang, S.; Barkho, S.; Shen, Y.; Henriksson, L.; Kosenina, S.; Dong, M.; Stenmar, P. Structural characterisation of the catalytic domain of botulinum neurotoxin X—High activity and unique substrate specificity. *Sci. Rep.* **2018**, *8*, 4518. [CrossRef]
3. Rossetto, O.; Pirazzini, M.; Montecucco, C. Botulinum neurotoxins: Genetic, structural and mechanistic insights. *Nat. Rev. Microbiol.* **2014**, *12*, 535–549. [CrossRef] [PubMed]
4. Steward, L.; Brin, M.F.; Brideau-Andersen, A. Novel Native and Engineered Botulinum Neurotoxins. *Handb. Exp. Pharmacol.* **2021**, *263*, 63–89. [PubMed]
5. Lacy, D.B.; Tepp, W.; Cohen, A.C.; DasGupta, B.R.; Stevens, R.C. Crystal structure of botulinum neurotoxin type A and implications for toxicity. *Nat. Struct. Biol.* **1998**, *5*, 898–902. [CrossRef]
6. Schiavo, G.; Poulain, B.; Rossetto, O.; Benfenati, F.; Tauc, L.; Montecucco, C. Tetanus toxin is a zinc protein and its inhibition of neurotransmitter release and protease activity depend on zinc. *EMBO J.* **1992**, *11*, 3577–3583. [CrossRef] [PubMed]
7. Söllner, T.; Whiteheart, S.W.; Brunner, M.; Erdjument-Bromage, H.; Geromanos, S.; Tempst, P.; Rothman, J.E. SNAP receptors implicated in vesicle targeting and fusion. *Nature* **1993**, *362*, 318–324. [CrossRef]
8. Pirazzini, M.; Rossetto, O.; Eleopra, R.; Montecucco, C. Botulinum neurotoxins: Biology, pharmacology, and toxicology. *Pharmacol. Rev.* **2017**, *69*, 200–235. [CrossRef] [PubMed]
9. Jankovic, J. Botulinum toxin: State of the art. *Mov. Disord.* **2017**, *32*, 1131–1138. [CrossRef]
10. Choudhury, S.; Baker, M.R.; Chatterjee, S.; Kumar, H. Botulinum Toxin: An Update on Pharmacology and Newer Products in Development. *Toxins* **2021**, *13*, 58. [CrossRef] [PubMed]
11. Spiegel, L.L.; Ostrem, J.L.; Bledsoe, I.O. FDA Approvals and Consensus Guidelines for Botulinum Toxins in the Treatment of Dystonia. *Toxins* **2020**, *12*, 332. [CrossRef]
12. Lewandowski, M.; Świerczewska, Z.; Barańska-Rybak, W. Off-Label Use of Botulinum Toxin in Dermatology-Current State of the Art. *Molecules* **2022**, *27*, 3143. [CrossRef]
13. Yi, K.H.; Lee, J.H.; Hu, H.W.; Kim, H.J. Anatomical Proposal for Botulinum Neurotoxin Injection for Glabellar Frown Lines. *Toxins* **2022**, *14*, 268. [CrossRef]
14. Jabbari, B. Botulinum neurotoxins in the treatment of refractory pain. *Nat. Clin. Pract. Neurol.* **2008**, *4*, 676–685. [CrossRef]
15. Pavone, F.; Luvisetto, S. Botulinum neurotoxin for pain management: Insights from animal models. *Toxins* **2010**, *2*, 2890–2913. [CrossRef] [PubMed]
16. Matak, I.; Lacković, Z. Botulinum toxin A, brain and pain. *Prog. Neurobiol.* **2014**, *119–120*, 39–59. [CrossRef]
17. Matak, I.; Bölcskei, K.; Bach-Rojecky, L.; Helyes, Z. Mechanisms of Botulinum Toxin Type A Action on Pain. *Toxins* **2019**, *11*, 459. [CrossRef] [PubMed]
18. Luvisetto, S.; Gazerani, P.; Cianchetti, C.; Pavone, F. Botulinum Toxin Type A as a Therapeutic Agent against Headache and Related Disorders. *Toxins* **2015**, *7*, 3818–3844. [CrossRef]
19. Egeo, G.; Fofi, L.; Barbanti, P. Botulinum Neurotoxin for the Treatment of Neuropathic Pain. *Front. Neurol.* **2020**, *11*, 716. [CrossRef]

20. Reyes-Long, S.; Alfaro-Rodríguez, A.; Cortes-Altamirano, J.L.; Lara-Padilla, E.; Herrera-Maria, E.; Romero-Morelos, P.; Salcedo, M.; Bandala, C. The Mechanisms of Action of Botulinum Toxin Type A in Nociceptive and Neuropathic Pathways in Cancer Pain. *Curr. Med. Chem.* **2021**, *28*, 2996–3009. [CrossRef] [PubMed]
21. Frampton, J.E.; Silberstein, S. OnabotulinumtoxinA: A Review in the Prevention of Chronic Migraine. *Drugs* **2018**, *78*, 589–600. [CrossRef] [PubMed]
22. Burstein, R.; Blumenfeld, A.M.; Silberstein, S.D.; Manack Adams, A.; Brin, M.F. Mechanism of Action of OnabotulinumtoxinA in Chronic Migraine: A Narrative Review. *Headache* **2020**, *60*, 1259–1272. [CrossRef]
23. Raciti, L.; Raciti, G.; Militi, D.; Casella, C.; Calabrò, R.S. Chronic Migraine: A Narrative Review on the Use of Botulinum Toxin with Clinical Indications and Future Directions. *J. Integr. Neurosci.* **2022**, *21*, 141. [CrossRef]
24. Vacca, V.; Madaro, L.; De Angelis, F.; Proietti, D.; Cobianchi, S.; Orsini, T.; Puri, P.L.; Luvisetto, S.; Pavone, F.; Marinelli, S. Revealing the Therapeutic Potential of Botulinum Neurotoxin Type A in Counteracting Paralysis and Neuropathic Pain in Spinally Injured Mice. *Toxins* **2020**, *12*, 491. [CrossRef]
25. Marinelli, S.; Vacca, V.; De Angelis, F.; Pieroni, L.; Orsini, T.; Parisi, C.; Soligo, M.; Protto, V.; Manni, L.; Guerrieri, R.; et al. Innovative mouse model mimicking human-like features of spinal cord injury: Efficacy of Docosahexaenoic acid on acute and chronic phases. *Sci. Rep.* **2019**, *9*, 8883. [CrossRef] [PubMed]
26. Kundi, S.; Bicknell, R.; Ahmed, Z. Spinal cord injury: Current mammalian models. *Am. J. Neurosci.* **2013**, *4*, 1–12. [CrossRef]
27. Mills, C.D.; Fullwood, S.D.; Hulsebosch, C.E. Changes in Metabotropic Glutamate receptor expression following spinal cord injury. *Exp. Neurol.* **2001**, *170*, 244–257. [CrossRef] [PubMed]
28. Yiu, G.; He, Z. Glial inhibition of CNS axon regeneration. *Nat. Rev. Neurosci.* **2006**, *7*, 617–627. [CrossRef]
29. Karimi-Abdolrezaee, S.; Billakanti, R. Reactive Astrogliosis after spinal cord injury—Beneficial and detrimental effects. *Mol. Neurobiol.* **2012**, *46*, 251–264. [CrossRef]
30. Hausmann, O.N. Post-traumatic inflammation following spinal cord injury. *Spinal Cord* **2003**, *41*, 369–378. [CrossRef] [PubMed]
31. Ju, G.; Wang, J.; Wang, Y.; Zhao, X. Spinal cord contusion. *Neural Regen. Res.* **2014**, *9*, 789–794. [CrossRef] [PubMed]
32. Walters, E.T. Neuroinflammatory contributions to pain after SCI: Roles for central glial mechanisms and nociceptor-mediated host defense. *Exp. Neurol.* **2014**, *258*, 48–61. [CrossRef]
33. Rupp, D.; Nicholson, G.; Canty, D.; Wang, J.; Rhéaume, C.; Le, L.; Steward, L.E.; Washburn, M.; Jacky, B.P.; Broide, R.S.; et al. OnabotulinumtoxinA Displays Greater Biological Activity Compared to IncobotulinumtoxinA, Demonstrating Non-Interchangeability in Both In Vitro and In Vivo Assays. *Toxins* **2020**, *12*, 393. [CrossRef] [PubMed]
34. Car, H.; Bogucki, A.; Bonikowski, M.; Dec-Ćwiek, M.; Drużdż, A.; Koziorowski, D.; Rudzińska-Bar, M.; Sarzyńska-Długosz, I.; Sławek, J. Botulinum toxin type-A preparations are not the same medications—Basic science (Part 1). *Neurol. Neurochir. Pol.* **2021**, *55*, 133–140. [CrossRef] [PubMed]
35. Byun, J.; Kwak, S.; Kwon, J.H.; Shin, M.; Lee, D.K.; Rhee, C.H.; Kang, W.H.; Oh, J.W.; Cruz, D.J.M. Comparative Pharmacodynamics of Three Different Botulinum Toxin Type A Preparations following Repeated Intramuscular Administration in Mice. *Toxins* **2022**, *14*, 365. [CrossRef] [PubMed]
36. Zimmermann, M. Ethical guidelines for investigations of experimental pain in conscious animals. *Pain* **1983**, *16*, 109–110. [CrossRef] [PubMed]
37. Luvisetto, S.; Marinelli, S.; Cobianchi, S.; Pavone, F. Anti-allodynic efficacy of botulinum neurotoxin A in a model of neuropathic pain. *Neuroscience* **2007**, *145*, 1–4. [CrossRef] [PubMed]
38. Vacca, V.; Luvisetto, S.; Marinelli, S.; Pavone, F. Botulinum toxin A increases analgesic effects of morphine, counters development of morphine tolerance and modulates glia activation and μ opioid receptor expression in neuropathic mice. *Brain Behav. Immun.* **2013**, *32*, 40–50. [CrossRef] [PubMed]
39. Cobianchi, S.; Marinelli, S.; Florenzano, F.; Pavone, F.; Luvisetto, S. Short but not long-lasting treadmill running reduces allodynia and improves functional recovery after peripheral nerve injury. *Neuroscience* **2010**, *168*, 273–287. [CrossRef]
40. Baptista, A.F.; de Souza Gomes, J.R.; Oliveira, J.T.; Santos, S.M.; Vannier-Santos, M.A.; Martinez, A.M. A new approach to assess function after sciatic nerve lesion in the mouse adaptation of the sciatic static index. *J. Neurosci. Methods* **2007**, *161*, 259–264. [CrossRef]

toxins

MDPI

Article

Efficacy of Intra-Articular Injection of Botulinum Toxin Type A (IncobotulinumtoxinA) in Temporomandibular Joint Osteoarthritis: A Three-Arm Controlled Trial in Rats

Marie Béret [1], Florent Barry [2], Maria-Jose Garcia-Fernandez [1], Henry Chijcheapaza-Flores [1], Nicolas Blanchemain [1], Feng Chai [1] and Romain Nicot [2,*]

[1] Univ. Lille, INSERM, U1008-Advanced Drug Delivery Systems, F59000 Lille, France; marie.beret@chu-lille.fr (M.B.); maria-jose.garcia-fernandez@univ-lille.fr (M.-J.G.-F.); henry.chijcheapaza@univ-lille.fr (H.C.-F.); nicolas.blanchemain@univ-lille.fr (N.B.); feng.hildebrand@univ-lille.fr (F.C.)

[2] Univ. Lille, CHU Lille, INSERM, Department of Oral and Maxillofacial Surgery, U1008-Advanced Drug Delivery Systems, F59000 Lille, France; fl.barry38@gmail.com

* Correspondence: romain.nicot@gmail.com; Tel.: +33-320-446360; Fax: +33-320-445860

Abstract: Temporomandibular disorders (TMD) are complex pathologies responsible for chronic orofacial pain. Intramuscular injection of botulinum toxin A (BoNT/A) has shown effectiveness in knee and shoulder osteoarthritis, as well as in some TMDs such as masticatory myofascial pain, but its use remains controversial. This study aimed to evaluate the effect of intra-articular BoNT/A injection in an animal model of temporomandibular joint osteoarthritis. A rat model of temporomandibular osteoarthritis was used to compare the effects of intra-articular injection of BoNT/A, placebo (saline), and hyaluronic acid (HA). Efficacy was compared by pain assessment (head withdrawal test), histological analysis, and imaging performed in each group at different time points until day 30. Compared with the rats receiving placebo, those receiving intra-articular BoNT/A and HA had a significant decrease in pain at day 14. The analgesic effect of BoNT/A was evident as early as day 7, and lasted until day 21. Histological and radiographic analyses showed decrease in joint inflammation in the BoNT/A and HA groups. The osteoarthritis histological score at day 30 was significantly lower in the BoNT/A group than in the other two groups ($p = 0.016$). Intra-articular injection of BoNT/A appeared to reduce pain and inflammation in experimentally induced temporomandibular osteoarthritis in rats.

Keywords: osteoarthritis; temporomandibular joint disorders; injection intra-articular; botulinum toxins; type A

Key Contribution: Intra-articular injection of BoNT/A appears to reduce pain and inflammation in a rat model of temporomandibular joint osteoarthritis. This treatment could have a place in the management of temporomandibular joint dysfunction in humans.

Citation: Béret, M.; Barry, F.; Garcia-Fernandez, M.-J.; Chijcheapaza-Flores, H.; Blanchemain, N.; Chai, F.; Nicot, R. Efficacy of Intra-Articular Injection of Botulinum Toxin Type A (IncobotulinumtoxinA) in Temporomandibular Joint Osteoarthritis: A Three-Arm Controlled Trial in Rats. *Toxins* **2023**, *15*, 261. https://doi.org/10.3390/toxins15040261

Received: 2 March 2023
Revised: 27 March 2023
Accepted: 28 March 2023
Published: 1 April 2023

1. Introduction

Temporomandibular disorders (TMD) are a heterogeneous group of conditions that involve the temporomandibular joint (TMJ) and associated tissues, causing chronic pain, joint noises, limitation of mandibular movement, and impaired quality of life. About 5–12% of the population in industrialized countries are estimated to suffer from TMD [1], with the most common problems being TMJ-related myalgia, arthralgia, and headache, as well as intra-articular pathologies such as disc displacement, degenerative joint disease (osteoarthrosis and osteoarthritis), and subluxation. The management of intra-articular TMD is complex and multidisciplinary. Treatment should be non-invasive in the first instance, with painkillers, physiotherapy, and stress management [2]. Oxygen–ozone therapy is the subject of some studies in the treatment of TMD and shows interesting results [3,4]. For patients not responding to noninvasive measures, pain relief may be obtained with intra-articular injection of hyaluronic acid (HA), platelet-rich plasma (PRP), or corticosteroids, but the long-term effectiveness is limited, and repeated injections increase the risk of adverse effects [5].

Chronic pain results in peripheral and central sensitization due to excess pain fiber activity, with resultant lowering of the pain threshold. Botulinum toxin type A (BoNT/A) reduces peripheral and central sensitization by decreasing the release of neuropeptides and neurotransmitters from cells or nerve endings [6,7]. Intramuscular injection of BoNT/A has been used for more than 20 years for the treatment of chronic pain and has proven efficacy in TMD with a predominantly muscular component or mixed origin [8,9].

In patients with knee or shoulder osteoarthritis resistant to systemic treatment and intra-articular HA or corticosteroid injections, intra-articular BoNT/A has shown very promising results, reducing pain and improving quality of life without causing major adverse effects [10–18]. In some studies, pain relief after articular BoNT/A lasted up to 8 weeks after the injection [11,13]. Furthermore, in a study of patients with non-neuropathic nociceptive knee pain, Arendt et al. [19] demonstrated a correlation between pain severity and response to BoNT/A. However, some authors are more reserved regarding the efficacy of intra-articular BoNT/A. In a study of 105 patients with knee osteoarthritis, Mendes et al. [20] showed that the pain reduction achieved with intra-articular BoNT/A was not significantly different from that achieved with intra-articular corticosteroid or saline. In a multicenter, double-blind, randomized, placebo-controlled study of 158 patients with knee osteoarthritis, McAlindon et al. [21] found no significant difference in pain reduction between patients treated with intra-articular BoNT/A versus saline. High-powered randomized controlled studies are needed to evaluate the effectiveness of intra-articular BoNT/A in the shoulder, knee, and ankle. The indications must be clearly specified, particularly the osteoarthritis stage, as BoNT/A seems to be more effective in patients with advanced disease and severe pain [22].

There is limited literature on the use of intra-articular BoNT/A in TMJ osteoarthritis (TMJOA) [8]. Two previous studies that evaluated the effect of intra-articular BoNT/A in albumin-induced TMJOA in rats reported a decrease in inflammatory mediators after the injection [23,24]. One of these studies assessed neuropeptide release and pain response (using a behavioral scale) and showed that the peripheral inhibition of release of glutamate, substance P, and calcitonin gene-related peptide (CGRP)—all neuropeptides involved in the pain signal—was responsible for the decrease in arthritis and persistent hypernociception [24]. The first clinical evaluation of intra-articular BoNT/A in humans was conducted by Batifol et al. in 2018 [25]. In this non-controlled study about patients suffering from TMD with severe chronic pain resistant to systemic treatment and intra-articular injections of HA and other drugs, intra-articular BoNT/A (30 IU administered unilaterally or bilaterally) brought about a significant reduction in temporomandibular joint pain up to 3 months after injection, but had no effect on mouth opening and joint noises. In 2022, Sari et al. [26] showed an improvement in pain relief and mouth opening with intra-articular injection of BoNT/A following arthrocentesis in patients with anterior disc displacement.

Given the poor literature but promising data on the effectiveness of BoNT/A in TMJOA, in this three-arm controlled trial we aimed to compare the effect of intra-articular BoNT/A with that of saline (placebo) and HA (as a reference molecule with proven efficacy in the treatment of articular forms of TMD) in a rat model of TMJOA induced by monosodium iodoacetate (MIA) injection.

2. Results

2.1. Clinical Observation (Body Weight Change)

All of the rats in the placebo and HA groups gained body weight steadily over time; in contrast, the body weight of the rats in the BoNT/A group slightly decreased initially (between day 2 and day 7) and then gradually increased until day 30 (Figure 1).

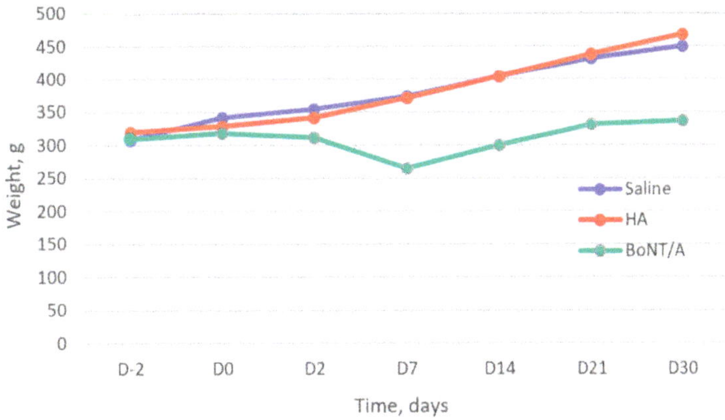

Figure 1. Evolution of mean weight of rats over time: before (D-2 and D0) and after (D2 to D30) intra-articular injection of placebo (saline), hyaluronic acid 1% (HA, Ostenil Mini®; TRB Chemedica SAS, Archamps, France), and botulinum toxin A (BoNT/A, incobotulinumtoxinA/Xeomin®; Merz Pharma, Frankfurt am Main, Germany).

2.2. Nociception Assessment

Table 1 and Figure 2 show the left TMJ HWT values in the three groups. In each group, the HWT values were significantly lower at day 0 (two days after induction of TMJOA) than at day −2, highlighting that the TMJOA model was well induced. In addition, at day −2 and at day 0 (two days after induction of TMJOA), the HWT values were comparable between the placebo, HA, and BoNT/A groups. At day 2, there were also no significant differences in HWT values between the three groups. At day 7, the mean HWT was significantly lower in the placebo group than in the BoNT/A group (17.56 ± 9.50 vs. 58.06 ± 18.42; $p = 0.05$). At day 14, the mean HWT was significantly lower in the placebo group than in the HA group (16.75 ± 10.29 vs. 65.88 ± 11.62; $p = 0.028$) and the BoNT/A group (16.75 ± 10.29 vs. 66.06 ± 22.53, $p = 0.019$); the difference between the HA group and the BoNT/A group was not statistically significant ($p = 0.422$). At day 21, the mean HWT was significantly lower in the placebo group than in the BoNT/A group (30.91 ± 13.64 vs. 86.01 ± 20.42; $p = 0.048$), but the differences between the placebo group and HA group or between the HA group and BoNT/A group were not statistically significant. At day 30, the mean HWT was comparable between the three groups.

Table 1. Left temporomandibular joint head-withdrawal test values (in grams) of rats with temporomandibular joint osteoarthritis at different time points: before (D-2 and D0) and after (D2 to D30) intra-articular injection of placebo (saline), hyaluronic acid 1% (HA, Ostenil Mini®; TRB Chemedica SAS, Archamps, France), and botulinum toxin A (BoNT/A, incobotulinumtoxinA/Xeomin®; Merz Pharma, Frankfurt am Main, Germany). * significant *p* value.

Left HWT, Grams Mean ± SD	D-2	D0	D2	D7	D14	D21	D30
Saline	91.99 ± 15.03	29.81 ± 15.48	26.71 ± 9.96	17.56 ± 9.50	16.75 ± 10.29	30.91 ± 13.64	47.54 ± 14.19
HA	88.02 ± 9.59	44.85 ± 16.68	62.96 ± 21.69	67.75 ± 15.04	65.88 ± 11.62	63.03 ± 7.31	61.47 ± 6.15
p	0.99	0.41	0.20	0.17	0.03 *	0.16	0.52
Saline	91.99 ± 15.03	29.81 ± 15.48	26.71 ± 9.96	17.56 ± 9.50	16.75 ± 10.29	30.91 ± 13.64	47.54 ± 14.19
BoNT/A	103.87 ± 11.63	45.92 ± 15.39	43.73 ± 17.73	58.06 ± 18.42	66.06 ± 22.53	86.01 ± 20.42	87.48 ± 11.92
p	0.90	0.47	0.45	0.05 *	0.02 *	0.05 *	0.22
HA	88.02 ± 9.59	44.85 ± 16.68	62.96 ± 21.69	67.75 ± 15.04	65.88 ± 11.62	63.03 ± 7.31	61.47 ± 6.15
BoNT/A	103.87 ± 11.63	45.92 ± 15.39	43.73 ± 17.73	58.06 ± 18.42	66.06 ± 22.53	86.01 ± 20.42	87.48 ± 11.92
p	0.91	0.83	0.99	0.70	0.42	0.28	0.46

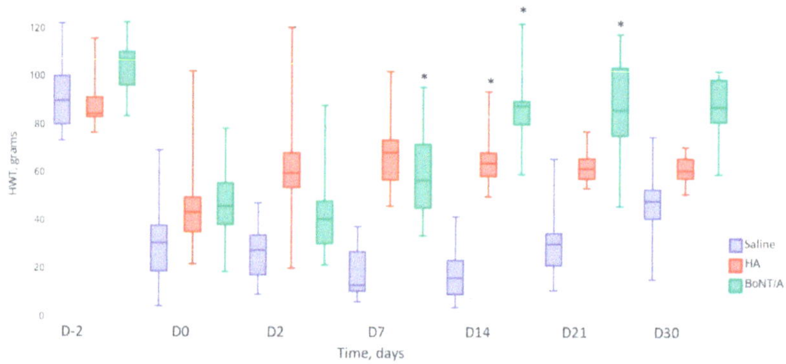

Figure 2. Box plot of the left temporomandibular joint head-withdrawal test values (HWT; in grams) of rats with temporomandibular joint osteoarthritis before (D-2 and D0) and after (D2 to D30) intra-articular injection of placebo (saline), hyaluronic acid 1% (HA, Ostenil Mini®; TRB Chemedica SAS, Archamps, France), and botulinum toxin A (BoNT/A, incobotulinumtoxinA/Xeomin®; Merz Pharma, Frankfurt am Main, Germany). * significant *p* value.

2.3. Histological Analysis

Figure 3 shows the results of the histological assessment (modified Mankin histological scores) of the left TMJ (with MIA induced osteoarthritis). Changes in the profile of the histological scores over time were different in the three groups. In the placebo group, the Mankin scores were largely similar at different time points: 4.50 ± 0.87 (n = 3) at day 2, 3.83 ± 1.15 (n = 3) at day 14, and 4.15 ± 1.71 (n = 14) at day 30. In the HA group, the Mankin score increased progressively: 2.83 ± 0.76 (n = 3) at day 2, 3.50 ± 0.87 (n = 3) at day 14, and 4.00 ± 2.11 (n = 14) at day 30. Similarly, in the BoNT/A group, the Mankin score increased over time: 1.00 ± 0.87 (n = 3) at day 2, 2.00 ± 0.00 (n = 3) at day 14, and 2.31 ± 1.09 at day 30 (n = 14). The mean Mankin score at day 30 was significantly lower in the BoNT/A group than in the other two groups ($p = 0.016$).

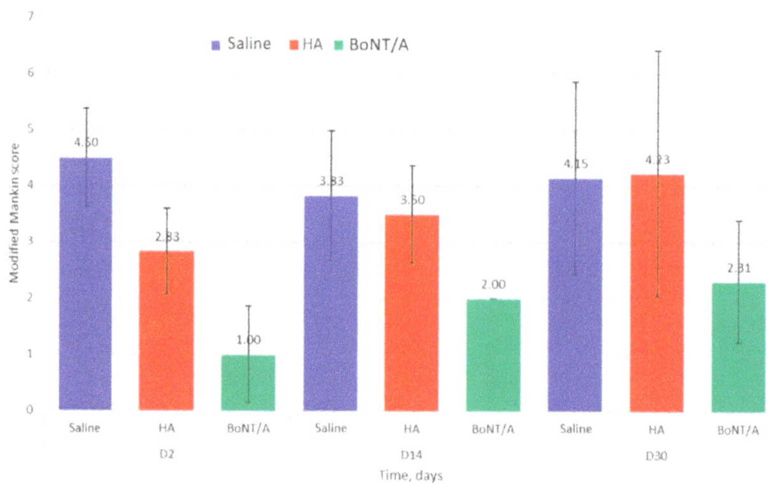

Figure 3. Modified Mankin histological scores of rats with temporomandibular joint osteoarthritis at day 2, day 14, and day 30 after intra-articular injection of placebo (saline), hyaluronic acid 1% (HA, Ostenil Mini®; TRB Chemedica SAS, Archamps, France), and botulinum toxin A (BoNT/A, incobotulinumtoxinA/Xeomin®; Merz Pharma, Frankfurt am Main, Germany).

2.4. ^{18}FDG PET Imaging

Table 2 and Figure 4 show the standard uptake value (SUV) on ^{18}FDG PET at day 30 in each group. SUV expresses the rate of ^{18}FDG consumption in the area of interest, standardized by the weight of the animal and the dose injected, showing the degree of inflammation and thus osteoarthritis. The median SUV in the left TMJ was comparable between the placebo and HA groups (1.09 [0.95, 1.13] vs. 1.01 [0.86, 1.06], respectively); the median SUV in the BoNT/A group was slightly lower (0.89 [0.88, 0.91]) than the two other groups.

Table 2. Standard uptake value measured on ^{18}FDG positron emission tomography scans performed at day 30 after intra-articular injection of placebo (saline), hyaluronic acid 1% (HA, Ostenil Mini®; TRB Chemedica SAS, Archamps, France), and botulinum toxin A (BoNT/A, incobotulinumtoxinA/Xeomin®; Merz Pharma, Frankfurt am Main, Germany).

	SUV–Mdn (Q1;Q3)	
	Left	**Right**
Saline (n = 5)	1.09 (0.95;1.13)	0.9 (0.79;0,90)
HA (n = 5)	1.01 (0.86;1.06)	1.01 (0.89;1.08)
BoNT/A (n = 5)	0.89 (0.88;0.91)	0.92 (0.83;0.93)

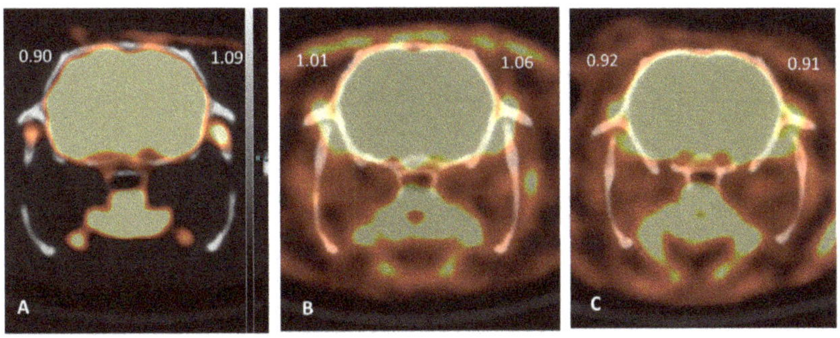

Figure 4. Representative ^{18}FDG PET scans in a rat with temporomandibular joint osteoarthritis at day 30 after intra-articular injection of placebo (saline) (**A**), hyaluronic acid 1% (HA, Ostenil Mini®; TRB Chemedica SAS, Archamps, France) (**B**), and botulinum toxin A (BoNT/A, incobotulinumtoxinA/Xeomin®; Merz Pharma, Frankfurt am Main, Germany) (**C**).

3. Discussion

This study evaluates the effect of intra-articular injection of BoNT/A (incobotulinumtoxinA/Xeomin®) versus intra-articular injection of saline or HA in a rat model of TMJOA. The use of the MIA-induced osteoarthritis model, based on the work of Barry et al. [27], allowed for assessment of the effect of these three substances over time. Reduction in TMJOA-related pain after intra-articular injection was similar in the BoNT/A and HA groups at day 14, with both groups having significantly better pain relief than the placebo group. Moreover, the BoNT/A group also had significantly better pain relief than the placebo group at day 7 and day 21.

3.1. Generalization

The results of this study are consistent with most previous studies on the subject, finding an effectiveness of BoNT/A in reducing osteoarthritis-related pain [10–18]. Indeed, we showed prolonged improvement in osteoarthritis-related pain from day 7 to day 21 with intra-articular injection of BoNT/A compared with intra-articular injection of the placebo (saline). Some studies also showed prolonged pain relief after intra-articular injection of BoNT/A, lasting up to 8 weeks after the injection [11,13]. Nevertheless, McAlindon et al. [21] showed contradictory results in human knee osteoarthritis, concluding to no significant difference between the intra-articular injection of BoNT/A and placebo (saline) in reducing the daily average numeric rating scale pain score over 7 days at 8 weeks. Their results are consistent with Mendes et al. [20], who found, in their randomized controlled trial, a higher short-term effectiveness of intra-articular injection of triamcinolone hexacetonide than the intra-articular injection of BoNT/A in reducing pain. It should be noted that the study involved only one dose of botulinum toxin (100 IU in a human knee), which may constitute a bias by underestimating the effectiveness of intra-articular injections of botulinum toxin. Focusing on TMJ, two previous studies on animal models of TMJOA have shown a decrease in pain mediators after intra-articular BoNT/A injection [23,24] and, in addition, Lora et al. [24] demonstrated decrease in pain in behavioral tests. Our results are in line with these previous findings in TMJOA.

Rezasoltani et al. [28] showed that intra-articular BoNT/A was more effective than HA for controlling pain and recovering function in patients in knee osteoarthritis. Conversely, Anil et al. [29] found that intra-articular stromal vascular factor, PRP, and HA were superior to BoNT/A for pain control (assessed by visual analog scale score) and functional outcomes (WOMAC score) in knee osteoarthritis. Our results showed no significant difference in pain improvement in TMJOA treated with intra-articular BoNT/A and HA; however, while pain in the BoNT/A group was significantly lower than in the placebo group at day 7, day 14, and day 21, pain in the HA group was significantly lower than in the placebo group only at day 14. Thus, our results suggest that both BoNT/A and HA can relieve TMJOA-related pain, but the effect of BoNT/A acted earlier and was more prolonged.

3.2. *Interpretation*

The observed effect of intra-articular BoNT/A on TMJOA-related pain relief is consistent with its known pharmacologic properties. Intra-articular BoNT/A inhibits the release of nociceptive neurotransmitters such as glutamate, substance P, and CGRP, leading to a reduction in pain and inflammation [24]. Glutamate is the main excitatory neurotransmitter in the nervous system of adult mammals and is involved in both pain neurotransmission and central sensitization. Glutamate release has been shown to result in inflammation, pain, and edema [7]. Meanwhile, animal models of adjuvant arthritis and of chronic inflammatory pain have shown marked upregulation of CGRP and mRNA in dorsal root ganglia neurons, as well as elevation of CGRP levels in primary afferent terminals of the spinal dorsal horn [30]. Furthermore, blocking of CGRP has been shown to block behavioral and electrophysiologic signs of enhanced pain in animals with inflammation [31]. In addition, Shi et al. [32] recently reported that the anti-inflammatory effect of BoNT/A in chronic arthritis may also be due to the inhibition of microglial cell activation and the release of microglia-derived tumor necrosis factor α (TNF-α). It is known that microglial cells are activated in chronic pain and release proinflammatory cytokines such as interleukin 6, TNFα, and interleukin 1ß, and thereby cause neuroinflammation. Moreover, P2 $\times$ 4 receptors (P2 $\times$ 4R) expressed in microglia are involved in neuropathic and inflammatory pain. All of these mechanisms may explain the pain reduction achieved by the intra-articular injection of BoNT/A.

The histological findings in this study offer further evidence in support of the efficacy of BoNT/A in treatment of TMJOA, with the mean modified Mankin score in the BoNT/A group being significantly lower than in the other two groups. The pattern of improvement of osteoarthritis over time was similar in the BoNT/A group and the HA group, but the mean Mankin scores were lower in the BoNT/A group at each time point that those in the HA group. Our findings also suggest that BoNT/A may have an early direct action on the histology, as the modified Mankin score at day 2 was 1.00 ± 0.87 in the BoNT/A group versus 4.50 ± 0.87 in the placebo group. This may be via an effect on the early inflammatory phase of osteoarthritis, with a decrease in the release of inflammatory neuropeptides and the expression of inflammatory cytokines limiting joint degradation [31–33]. Our results are consistent with the literature, but the mechanism of action of BoNT/A in the TMJ needs to be clarified in future studies.

The PET scan performed at day 30 in each group provided additional supportive information. SUV values were similar in the placebo and HA groups; however, they were slightly lower in the BoNT/A group. Increased ^{18}FDG tracer uptake was not specific to inflammation, but it could be seen in any area with a significant glycolytic activity, for example, areas with active repair processes. This made interpretation difficult, especially in the HA group. Nevertheless, the lower bilateral SUV values in the BoNT/A group were in favor of a decrease in TMJOA at day 30 and corroborated the clinical and histological findings.

3.3. Limitations of the Study

This study has several limitations. The first was the choice of HA as the reference intra-articular treatment for TMJOA. We chose HA because it is currently the most widely used intra-articular treatment for TMJOA because of its viscosupplementation properties [5]. Other injectable substances such as corticosteroids and PRP are also used. Intra-articular corticosteroids, alone or in combination with other drugs, have not shown better results than intra-articular HA and, moreover, are associated with a risk of condylar resorption [5]. Several studies have shown good results with intra-articular PRP in TMJOA, but the manufacturing protocol is not standardized, and time and special equipment are required to obtain PRP [5,29,34–36]. Second, the weight of the rats in the BoNT/A group initially decreased due to feeding difficulties, probably due to muscle weakness caused by the diffusion of BoNT/A into the masticatory muscles. Change from a hard to a soft diet allowed the rats to eat normally and regain weight. This change in the weight and diet of the rats may have induced stress and behavioral changes, which may have resulted in an underestimation of the pain improvement in the BoNT/A group. In addition, the volume and concentration of the injected BoNT/A was based on the articles by Lora et al. [23,24], and recent studies in humans on the use of high doses of toxin showed a rare occurrence of adverse effects [37,38]. Third, the study sample size of the study was calculated for the statistical analysis of nociception; this sample may have been too small for the statistical analysis of histological and PET imaging data.

4. Conclusions

Intra-articular injection of BoNT/A (incobotulinumtoxinA/Xeomin®) appears to be effective for reducing pain in experimentally induced TMJOA in rats. Histological and PET imaging findings support these results. The mean Mankin osteoarthritis histological score at day 30 was significantly lower in the BoNT/A group than in the other two groups.

More high-powered preclinical and clinical studies are needed to determine the place of intra-articular injection of BoNT/A for the treatment of temporomandibular joint osteoarthritis and to draw a firm conclusion.

5. Material and Methods

5.1. Animals

Sixty male Wistar rats (6-weeks-old; weight of 250–300 g) were included in this study. The rats were housed in individual cages in a temperature-controlled room (22 °C ± 1 °C) with a 12-h dark–light cycle and allowed for free access to food and water. Manipulations started after ten days of quarantine.

All of the procedures in this study were approved by Ministère de l'enseignement supérieur, de la recherche et de l'innovation (APAFIS#25897, 29/10/2020).

5.2. Induction of Temporomandibular Joint Osteoarthritis and Injection Protocol

The animals were anesthetized by the inhalation of isoflurane mixed with pure oxygen (flow rate: 1.5 L/min) for 2–3 min in a Plexiglas® chamber. TMJOA was induced in the left TMJ of all rats by intra-articular injection of monosodium iodoacetate (MIA) into the upper compartment in normal saline (0.5 mg/50 µL of saline; Sigma, Saint Louis, MI, USA) [27,39]. The injection protocol was based on the work of Fuentes et al. [40]. Two days after MIA injection, the rats were anesthetized by the same technique and then randomly divided into three groups: 20 rats (the BoNT/A group) received intra-articular injection of 2.5 UI/50 µL BoNT/A (incobotulinumtoxinA; Xeomin®; Merz Pharma, Frankfurt am Main, Germany) in the left and right joint of each rat; 20 rats (the HA group) received intra-articular injection of 50 µL of 1% HA (Ostenil Mini®; TRB Chemedica SAS, Archamps, France) in the left and right joint of each rat; and 20 rats (the placebo group or saline group) received intra-articular injection of 50 µL of 0.9% saline in the left and right joint of each rat. Both Xeomin® and Ostenil® were selected because they are used in clinical practice. In addition, Ostenil® has European certification for use in small joints, including TMJ. Neither Merz Pharma or TRB Chemedica were sponsors of the study.

5.3. Nociception Assessment

The head-withdrawal test (HWT) was used to assess pain. According to the systematic review by Nicot et al. [39], long-term pain related to TMJOA has mostly been assessed by measuring the threshold pressure value (in grams) that triggers a pain response. In this study, a Von Frey aesthesiometer (Figure 5) was applied with gradually increasing pressure to the rat TMJ till head withdrawal, vocalization, or both occurred, indicating nociception; the threshold value was defined as the lowest pressure that induced the response. After each measurement, the rats were weighed (in grams) to monitor their general wellbeing before being returned to their cages.

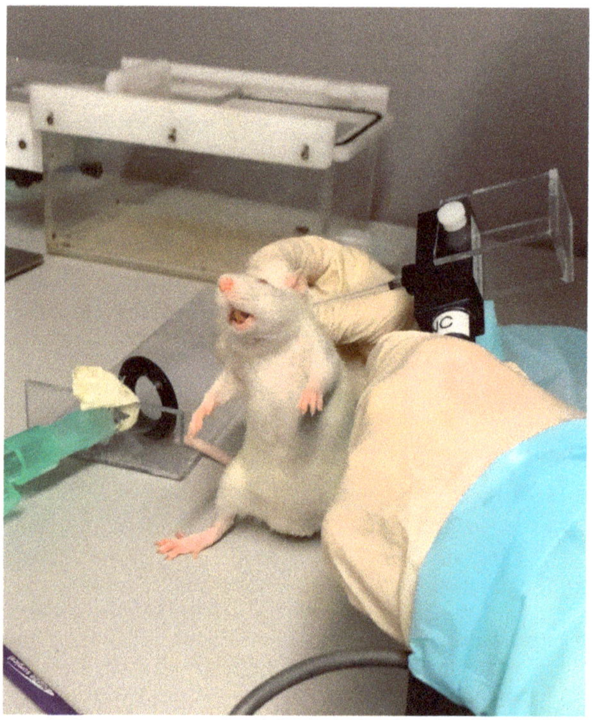

Figure 5. Head withdrawal test method: gradually increasing pressure was applied using a Von Frey esthesiometer to the temporomandibular joint area until the animal withdrew its head, vocalized, or both, and the lowest value of pressure (in grams) that induced a response was recorded. The gesture was performed on each temporomandibular joint of each rat.

5.4. Histological Analysis

In each group, randomly selected animals were humanely killed at day 2 (n = 3), day 14 (n = 3), and day 30 (n = 14) by intracardiac injection of 0.2 mL of T61 under isoflurane anesthesia and then immediately stored at −20 °C for at least 48 h. Then, clean-cut samples of approximately 5 mm thickness were obtained from the TMJ area. The samples were first placed in cassettes and fixed in 4% formaldehyde solution for 24 h, and then decalcified by immersion in 15% ethylenediaminetetraacetic acid (EDTA; pH 7.2) solution for 5 days. The prepared samples were stored in 70% ethanol solution at 4 °C until histological processing (paraffin embedding, cutting, and staining) and analysis.

Briefly, the sections were first stained with hematoxylin and eosin (HE) staining to select the slides of interest. The selected slides were thus stained with toluidine blue stain (TB) and examined under the microscope for determining the histological osteoarthritis score using the modified Mankin scale (Table 3). The higher the final score on the Mankin scale, the more advanced the TMJOA stage [41,42].

Table 3. Modified Mankin scale for histological scoring of temporomandibular joint osteoarthritis.

Structure	
Normal	0
Surface irregularities	1
Pannus	2
Cleft to transitional zone	3
Cleft to radial zone	4
Cleft to calcified zone	5
Complete disorganization	6
Tidemark integrity	
Intact	0
Crossed by blood vessels	1
Proteoglycan staining	
Normal	0
Slight reduction	1
Moderate reduction	2
Severe reduction	3
No dye noted	4
Cellularity	
Normal	0
Diffuse hypercellularity	1
Cloning	2
Hypocellularity	3

5.5. Positron Emission Tomography (PET) Imaging

PET with 2-deoxy-2-[18F] fluoro-D-glucose (^{18}FDG) was carried out for monitoring the stage of inflammation. ^{18}FDG radiotracer was used to visualize the areas of high glucose consumption, caused in this case by synovitis and TMJOA bone lesions [43,44]. Imaging was performed at day 30 on five randomly selected rats in each group. The rats were fasted the day before the examination. Intravenous administration of the ^{18}FDG radiotracer (30–35 MBq) and image acquisition were carried out under general anesthesia. Manipulations were performed in compliance with the rules of human radioprotection [45]. The animals were isolated the first 24 h after radiotracer injection.

5.6. Full Protocol Frame of the Study

Figure 6 summarizes the basic frame structure of the full protocol of analysis described below: from TMJOA induction (day −2) to day 30 after therapeutic joint injection.

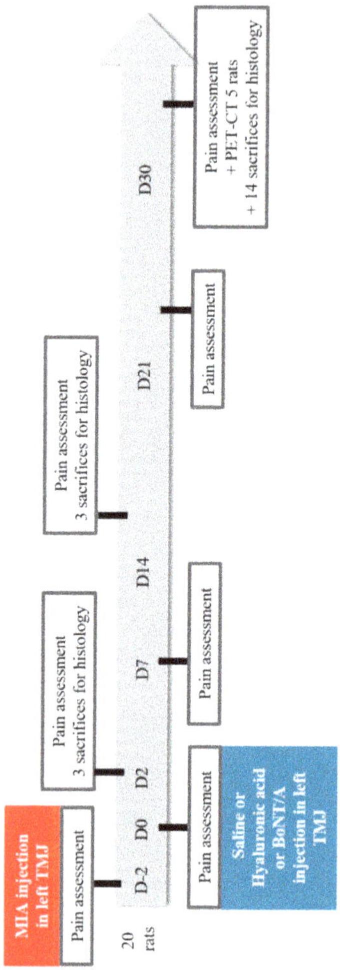

Figure 6. Full experimental protocol from temporomandibular joint osteoarthritis induction (day −2) to day 30 after therapeutic joint injection.

5.7. Statistical Analysis

The sample size was calculated for three-group one-way analysis of variance (ANOVA) using G*Power 3.1 (Heinrich-Heine-Universität Düsseldorf, Düsseldorf, Germany), assuming $\alpha = 0.05$, $\beta = 0.2$, standard deviation (SD) = 10, and effect size = 0.42. The calculated sample size was 19 per group. The final sample size was set at 20 rats per group, keeping in mind potential animal losses and the 3R's rule for experimental procedures in animals [46]. Quantitative variables were expressed as means (±standard deviation) or medians (interquartile range; Q1, Q3), depending on the normality of the distribution. The normality of distributions was assessed using histograms and the Shapiro–Wilk test. Categorical variables were expressed as numbers (percentage). The mean weights and HWT values on day −2 and day 0 were first compared to check the comparability of the three groups. One-way

ANOVA was used to compare the HWT values in the three groups. Levene's test was used to test the homogeneity assumption required by ANOVA. Multiple comparisons within the experimental groups were performed using Tukey's test. One-way ANOVA followed by Dunnett's test was used to compare the placebo group with the experimental groups. Kruskal–Wallis test was used to compare the left TMJ Mankin score at day 30 because the assumptions of one-way ANOVA were not met. All of the statistical analyses were performed using XLSTAT 2022.5.1 (Addinsoft, New York, NY, USA). Statistical significance was at $p < 0.05$.

Author Contributions: Conceptualization, R.N.; Methodology, F.C. and R.N.; Formal analysis, M.B. and R.N.; Investigation, M.B., F.B. and R.N.; Writing—original draft, M.B. and R.N.; Writing—review & editing, F.B., M.-J.G.-F., H.C.-F., N.B., F.C. and R.N.; Supervision, N.B., F.C. and R.N.; Funding acquisition, M.-J.G.-F., N.B. and R.N. All authors have read and agreed to the published version of the manuscript.

Funding: This research was funded by Conseil Régional des Hauts-de-France: grant number DRESS-2019-015910.

Institutional Review Board Statement: The animal study protocol was approved by the Ethics Committee of Ministère de l'enseignement supérieur, de la recherche et de l'innovation (APAFIS#25897, 29/10/2020).

Informed Consent Statement: Not applicable.

Data Availability Statement: Data available on request due to restrictions eg privacy or ethical.

Acknowledgments: The authors would like to thank the Platform Experimental Resources, D.H.U.R.E, University of Lille, for their support on animal studies. We also thank Mrs A-S. Drucbert for her technical help in histological analysis.

Conflicts of Interest: Author Romain Nicot has been involved as expert consultant in Company Merz Pharma from May 2020 to December 2021 and conducted an educational training program on botulinum toxin injection techniques in validated indications. No conflict of interest with Merz Pharma has to be declared for the present study. The other authors declare no conflict of interest.

References

1. Schiffman, E.; Ohrbach, R.; Truelove, E.; Look, J.; Anderson, G.; Goulet, J.-P.; List, T.; Svensson, P.; Gonzalez, Y.; Lobbezoo, F.; et al. Diagnostic Criteria for Temporomandibular Disorders (DC/TMD) for Clinical and Research Applications: Recommendations of the International RDC/TMD Consortium Network and Orofacial Pain Special Interest Group. *J. Oral Facial Pain Headache* **2014**, *28*, 6–27. [CrossRef]
2. Garrigós-Pedrón, M.; Elizagaray-García, I.; Domínguez-Gordillo, A.A.; Del-Castillo-Pardo-De-Vera, J.L.; Gil-Martínez, A. Temporomandibular disorders: Improving outcomes using a multidisciplinary approach. *J. Multidiscip. Health* **2019**, *12*, 733–747. [CrossRef]
3. de Sire, A.; Marotta, N.; Ferrillo, M.; Agostini, F.; Sconza, C.; Lippi, L.; Respizzi, S.; Giudice, A.; Invernizzi, M.; Ammendolia, A. Oxygen-Ozone Therapy for Reducing Pro-Inflammatory Cytokines Serum Levels in Musculoskeletal and Temporomandibular Disorders: A Comprehensive Review. *Int. J. Mol. Sci.* **2022**, *23*, 2528. [CrossRef] [PubMed]
4. Oshaghi, S.; Haghighat, S. Effectiveness of ozone injection therapy in temporomandibular disorders. *Adv. Biomed. Res.* **2020**, *9*, 73. [CrossRef] [PubMed]
5. Liu, Y.; Wu, J.-S.; Tang, Y.-L.; Fei, W.; Liang, X.-H. Multiple Treatment Meta-Analysis of Intra-Articular Injection for Temporomandibular Osteoarthritis. *J. Oral Maxillofac. Surg.* **2020**, *78*, 373.e1–373.e18. [CrossRef]
6. Aoki, K.R.; Francis, J. Updates on the antinociceptive mechanism hypothesis of botulinum toxin A. *Park. Relat. Disord.* **2011**, *17*, S28–S33. [CrossRef] [PubMed]
7. Matak, I.; Bölcskei, K.; Bach-Rojecky, L.; Helyes, Z. Mechanisms of Botulinum Toxin Type A Action on Pain. *Toxins* **2019**, *11*, 459. [CrossRef]
8. Delcanho, R.; Val, M.; Nardini, L.G.; Manfredini, D. Botulinum Toxin for Treating Temporomandibular Disorders: What is the Evidence? *J. Oral Facial Pain Headache* **2022**, *36*, 6–20. [CrossRef]
9. Villa, S.; Raoul, G.; Machuron, F.; Ferri, J.; Nicot, R. Improvement in quality of life after botulinum toxin injection for temporomandibular disorder. *J. Stomatol. Oral Maxillofac. Surg.* **2019**, *120*, 2–6. [CrossRef]
10. Najafi, S.; Sanati, E.; Khademi, M.; Abdorrazaghi, F.; Mofrad, R.K.; Rezasoltani, Z. Intra-articular botulinum toxin type A for treatment of knee osteoarthritis: Clinical trial. *Toxicon* **2019**, *165*, 69–77. [CrossRef]

11. Zhai, S.; Huang, B.; Yu, K. The efficacy and safety of Botulinum Toxin Type A in painful knee osteoarthritis: A systematic review and meta-analysis. *J. Int. Med. Res.* **2020**, *48*, 0300060519895868. [CrossRef] [PubMed]
12. Mahowald, M.L.; Singh, J.A.; Dykstra, D. Long term effects of intra-articular botulinum toxin a for refractory joint pain. *Neurotox. Res.* **2006**, *9*, 179–188. [CrossRef] [PubMed]
13. Wu, T.; Song, H.-X.; Dong, Y.; Ye, Y.; Li, J.-H. Intra-articular injections of botulinum toxin a for refractory joint pain: A systematic review and meta-analysis. *Clin. Rehabil.* **2017**, *31*, 435–443. [CrossRef]
14. Courseau, M.; Salle, P.V.; Ranoux, D.; Lachatre, A.D.P. Efficacy of Intra-Articular Botulinum Toxin in Osteoarticular Joint Pain: A meta-analysis of randomized controlled trials. *Clin. J. Pain* **2018**, *34*, 383–389. [CrossRef]
15. Cinone, N.; Letizia, S.; Santoro, L.; Gravina, M.; Amoruso, L.; Molteni, F.; Ranieri, M.; Santamato, A. Intra-articular injection of botulinum toxin type A for shoulder pain in glenohumeral osteoarthritis: A case series summary and review of the literature. *J. Pain Res.* **2018**, *11*, 1239–1245. [CrossRef] [PubMed]
16. Boon, A.J.; Smith, J.; Dahm, D.L.; Sorenson, E.J.; Larson, D.R.; Fitz-Gibbon, P.D.; Dykstra, D.D.; Singh, J.A. Efficacy of Intra-Articular Botulinum Toxin Type A in Painful Knee Osteoarthritis: A Pilot Study. *PM&R* **2010**, *2*, 268–276. [CrossRef]
17. Chou, C.-L.; Lee, S.-H.; Lu, S.-Y.; Tsai, K.-L.; Ho, C.-Y.; Lai, H.-C. Therapeutic Effects of Intra-articular Botulinum Neurotoxin in Advanced Knee Osteoarthritis. *J. Chin. Med. Assoc.* **2010**, *73*, 573–580. [CrossRef]
18. Shukla, D.; Sreedhar, S.K.; Rastogi, V. A comparative study of botulinum toxin: A with triamcinolone compared to triamcinolone alone in the treatment of osteoarthritis of knee. *Anesth. Essays Res.* **2018**, *12*, 47–49. [CrossRef]
19. Arendt-Nielsen, L.; Jiang, G.-L.; DeGryse, R.; Turkel, C. Intra-articular onabotulinumtoxinA in osteoarthritis knee pain: Effect on human mechanistic pain biomarkers and clinical pain. *Scand. J. Rheumatol.* **2017**, *46*, 303–316. [CrossRef]
20. Mendes, J.G.; Natour, J.; Nunes-Tamashiro, J.C.; Toffolo, S.R.; Rosenfeld, A.; Furtado, R.N.V. Comparison between intra-articular Botulinum toxin type A, corticosteroid, and saline in knee osteoarthritis: A randomized controlled trial. *Clin. Rehabil.* **2019**, *33*, 1015–1026. [CrossRef]
21. McAlindon, T.E.; Schmidt, U.; Bugarin, D.; Abrams, S.; Geib, T.; DeGryse, R.E.; Kim, K.; Schnitzer, T. Efficacy and safety of single-dose onabotulinumtoxinA in the treatment of symptoms of osteoarthritis of the knee: Results of a placebo-controlled, double-blind study. *Osteoarthr. Cartil.* **2018**, *26*, 1291–1299. [CrossRef] [PubMed]
22. Khenioui, H.; Houvenagel, E.; Catanzariti, J.F.; Guyot, M.A.; Agnani, O.; Donze, C. Usefulness of intra-articular botulinum toxin injections. A systematic review. *Jt. Bone Spine* **2016**, *83*, 149–154. [CrossRef] [PubMed]
23. Muñoz-Lora, V.R.M.; Abdalla, H.B.; Cury, A.A.D.B.; Clemente-Napimoga, J.T. Modulatory effect of botulinum toxin type A on the microglial P2X7/CatS/FKN activated-pathway in antigen-induced arthritis of the temporomandibular joint of rats. *Toxicon* **2020**, *187*, 116–121. [CrossRef]
24. Lora, V.R.M.M.; Clemente-Napimoga, J.T.; Abdalla, H.B.; Macedo, C.G.; Canales, G.D.L.T.; Barbosa, C.M.R. Botulinum toxin type A reduces inflammatory hypernociception induced by arthritis in the temporomadibular joint of rats. *Toxicon* **2017**, *129*, 52–57. [CrossRef]
25. Batifol, D.; Huart, A.; Finiels, P.J.; Nagot, N.; Jammet, P. Effect of intra-articular Botulinum toxin injections on temporo-mandibular joint pain. *J. Stomatol. Oral Maxillofac. Surg.* **2018**, *119*, 319–324. [CrossRef]
26. Sari, B.C.; Develi, T. The effect of intraarticular botulinum toxin-A injection on symptoms of temporomandibular joint disorder. *J. Stomatol. Oral Maxillofac. Surg.* **2022**, *123*, e316–e320. [CrossRef]
27. Barry, F.; Chai, F.; Chijcheapaza-Flores, H.; Garcia-Fernandez, M.J.; Blanchemain, N.; Nicot, R. Comparison of chemical-induced temporomandibular osteoarthritis rat models (monosodium iodoacetate versus collagenase type II) for the study of prolonged drug delivery systems. *PLoS ONE* **2023**, *18*, e0281135. [CrossRef] [PubMed]
28. Rezasoltani, Z.; Azizi, S.; Najafi, S.; Sanati, E.; Dadarkhah, A.; Abdorrazaghi, F. Physical therapy, intra-articular dextrose prolotherapy, botulinum neurotoxin, and hyaluronic acid for knee osteoarthritis: Randomized clinical trial. *Int. J. Rehabil. Res.* **2020**, *43*, 219–227. [CrossRef]
29. Anil, U.; Markus, D.H.; Hurley, E.T.; Manjunath, A.K.; Alaia, M.J.; Campbell, K.A.; Jazrawi, L.M.; Strauss, E.J. The efficacy of intra-articular injections in the treatment of knee osteoarthritis: A network meta-analysis of randomized controlled trials. *Knee* **2021**, *32*, 173–182. [CrossRef]
30. Calzà, L.; Pozza, M.; Arletti, R.; Manzini, E.; Hökfelt, T. Long-Lasting Regulation of Galanin, Opioid, and Other Peptides in Dorsal Root Ganglia and Spinal Cord during Experimental Polyarthritis. *Exp. Neurol.* **2000**, *164*, 333–343. [CrossRef]
31. Iyengar, S.; Ossipov, M.H.; Johnson, K.W. The role of calcitonin gene–related peptide in peripheral and central pain mechanisms including migraine. *Pain* **2017**, *158*, 543–559. [CrossRef] [PubMed]
32. Shi, X.; Gao, C.; Wang, L.; Chu, X.; Shi, Q.; Yang, H.; Li, T. Botulinum toxin type A ameliorates adjuvant-arthritis pain by inhibiting microglial activation-mediated neuroinflammation and intracellular molecular signaling. *Toxicon* **2019**, *178*, 33–40. [CrossRef] [PubMed]
33. Wang, L.; Wang, K.; Chu, X.; Li, T.; Shen, N.; Fan, C.; Niu, Z.; Zhang, X.; Hu, L. Intra-articular injection of Botulinum toxin A reduces neurogenic inflammation in CFA-induced arthritic rat model. *Toxicon* **2017**, *126*, 70–78. [CrossRef] [PubMed]
34. Bousnaki, M.; Bakopoulou, A.; Koidis, P. Platelet-rich plasma for the therapeutic management of temporomandibular joint disorders: A systematic review. *Int. J. Oral Maxillofac. Surg.* **2018**, *47*, 188–198. [CrossRef] [PubMed]

35. Dasukil, S.; Arora, G.; Boyina, K.K.; Jena, A.K.; Jose, A.; Das, S. Intra-articular injection of hyaluronic acid versus platelet-rich plasma following single puncture arthrocentesis for the management of internal derangement of TMJ: A double-blinded randomised controlled trial. *J. Cranio-Maxillofac. Surg.* **2022**, *50*, 825–830. [CrossRef] [PubMed]
36. Rajput, A.; Bansal, V.; Dubey, P.; Kapoor, A. A Comparative Analysis of Intra-articular Injection of Platelet-Rich Plasma and Arthrocentesis in Temporomandibular Joint Disorders. *J. Maxillofac. Oral Surg.* **2022**, *21*, 168–175. [CrossRef] [PubMed]
37. Baricich, A.; Picelli, A.; Santamato, A.; Carda, S.; de Sire, A.; Smania, N.; Cisari, C.; Invernizzi, M. Safety Profile of High-Dose Botulinum Toxin Type A in Post-Stroke Spasticity Treatment. *Clin. Drug Investig.* **2018**, *38*, 991–1000. [CrossRef]
38. Mueller, J.; Langbein, T.; Mishra, A.; Baum, R.P. Safety of High-Dose Botulinum Toxin Injections for Parotid and Submandibular Gland Radioprotection. *Toxins* **2022**, *14*, 64. [CrossRef] [PubMed]
39. Nicot, R.; Barry, F.; Chijcheapaza-Flores, H.; Garcia-Fernandez, M.J.; Raoul, G.; Blanchemain, N.; Chai, F. A Systematic Review of Rat Models With Temporomandibular Osteoarthritis Suitable for the Study of Emerging Prolonged Intra-Articular Drug Delivery Systems. *J. Oral Maxillofac. Surg.* **2021**, *79*, 1650–1671. [CrossRef] [PubMed]
40. Fuentes, R.; Veuthey, C.; Arias, A.; Saravia, D.; Ottone, N.E. Injection in temporomandibular joint of rats. Description of technical protocol. *Pol. J. Veter. Sci.* **2017**, *20*, 207–211. [CrossRef]
41. Mankin, H.J.; Dorfman, H.; Lippiello, L.; Zarins, A. Biochemical and metabolic abnormalities in articular cartilage from osteo-arthritic human hips. II. Correlation of morphology with biochemical and metabolic data. *J. Bone Jt. Surg. Am.* **1971**, *53*, 523–537. [CrossRef]
42. Zhang, S.; Teo, K.Y.W.; Chuah, S.J.; Lai, R.C.; Lim, S.K.; Toh, W.S. MSC exosomes alleviate temporomandibular joint osteoarthritis by attenuating inflammation and restoring matrix homeostasis. *Biomaterials* **2019**, *200*, 35–47. [CrossRef] [PubMed]
43. Hayashi, D.; Roemer, F.W.; Guermazi, A. Imaging for osteoarthritis. *Ann. Phys. Rehabil. Med.* **2016**, *59*, 161–169. [CrossRef]
44. Suh, M.S.; Park, S.H.; Kim, Y.-K.; Yun, P.-Y.; Lee, W.W. 18 F-NaF PET/CT for the evaluation of temporomandibular joint disorder. *Clin. Radiol.* **2018**, *73*, 414.e7–414.e13. [CrossRef] [PubMed]
45. Marengo, M.; Martin, C.J.; Rubow, S.; Sera, T.; Amador, Z.; Torres, L. Radiation Safety and Accidental Radiation Exposures in Nuclear Medicine. *Semin. Nucl. Med.* **2022**, *52*, 94–113. [CrossRef]
46. Cozigou, G.; Crozier, J.; Hendriksen, C.; Manou, I.; Ramirez-Hernandez, T.; Weissenhorn, R. The European Partnership for Alternative Approaches to Animal Testing (EPAA): Promoting alternative methods in Europe and beyond. *J. Am. Assoc. Lab. Anim. Sci.* **2015**, *54*, 209–213. [PubMed]

MDPI
St. Alban-Anlage 66
4052 Basel
Switzerland
www.mdpi.com

Toxins Editorial Office
E-mail: toxins@mdpi.com
www.mdpi.com/journal/toxins

www.ingramcontent.com/pod-product-compliance
Lightning Source LLC
Chambersburg PA
CBHW040716270926
51850CB00023B/2060
9783036588285